D0126198

Tree Maintenance

Originally published in 1941 as Maintenance of Shade and Ornamental Trees, and revised in 1948.

P. P. Pirone

Plant Pathologist, The New York Botanical Garden, and formerly Associate Professor of Plant Pathology at Cornell and Rutgers Universities.

New York OXFORD UNIVERSITY PRESS 1959

© 1959 by Oxford University Press, Inc.

Originally published in 1941 as *Maintenance of Shade and Ornamental Trees,* and revised in 1948; copyright 1941, 1948 by Oxford University Press, Inc.

Library of Congress Catalogue Card Number: 59-6260

Printed in the United States of America

To My Mother and Father

Foreword *

The care of trees is a never ending obligation, more insistent each year because appreciation of them is increasing. The owner of the tree carries the responsibility of his own account and he then naturally extends his concern to trees that stand on public property and need the interest of citizens.

One's responsibility is met only when one recognizes the cause of injury and disease and is informed on the remedies and proper procedures. Even if one has been taught these subjects in school he nevertheless needs to be informed on the newest difficulties and the latest approved practices. The difficulties increase. Thus it comes that new and authoritative books are always needed.

Therefore I am glad to see this book on Maintenance of Shade and Ornamental Trees by Dr. P. P. Pirone, whose work I have known for many years. He is investigator and teacher. He has an eye to the immediately practical, as evidenced by the simplicity of the writing, clearness of the directions, and the significant pictures. It is an important presentation of the subject.

L. H. Bailey

Ithaca, N.Y.
March 1, 1941.

* As presented in the original edition.

Preface

Nearly twenty years have passed since I wrote the first edition of this book on tree maintenance. In that period great changes have occurred in the field of arboriculture. Not only have more effective pesticides been developed but also more efficient machines to apply them. And new machinery for feeding, transporting, and pruning trees has since been introduced. Our whole concept of feeding trees has also changed, as is evidenced by the present widespread interest in foliage feeding.

Perhaps the most unusual development in the last 20 years has been the change in our views of what constitutes an ideal streetside or lawn tree. This is emphasized in Chapter 8, where I have relegated such old standbys as oaks, elms, and planes to secondary parts and have cast the so-called small- or low-growing trees in the leading role. There will be some who will disagree with me, but I believe in time they will see the light and join the distinguished company of such nationally known arboricultural personalities as Edward Scanlon of Olmsted Falls, Ohio, editor and publisher of *Trees Magazine*; Dr. Henry Skinner, Director of the National Arboretum, Washington, D.C.; Professor Clarence Lewis of Michigan State College; Jake Gerling of the Rochester, New York Department of Parks; and Dr. Donald Wyman, Horticulturist of the Arnold Arboretum—all of whom have done much to foster the use of smaller-growing trees.

The extensive home-building program following World War II has led to greater interest in trees and their care. Perhaps the lack of trees in new housing developments has helped to make young, newly married couples more tree-conscious. The do-it-yourself fad may also have contributed to home owners' increased interest in trees.

As a result, tree maintenance companies have sprung up like mushrooms. Even the older, nationally known tree preservation companies have changed their approach to the business of caring for trees. No longer does the major part of their work come from large estates. An important part is now done on properties of middle-income families.

In preparing this book I have tried to incorporate my experiences with tree maintenance problems of the last 20 years—experiences involving personal examination of literally hundreds of trees.

In presenting the information, I have attempted to follow a course midway between the technical and the popular. That is, I have tried to be not too technical for the beginner in tree maintenance work, and not too elementary for the professional arborist, nurseryman, landscape architect, and the college student enrolled in a course on tree maintenance. The student, in particular, should like the extensive bibliography at the end of the book.

I trust that I have satisfied most of my readers in both groups.

Many persons have lent a hand in the preparation of this book. My dear wife heads the list, because without her constant help, encouragement, and care I could not have prepared the book and carried on my regular duties at The New York Botanical Garden, as well, during these many months. Next is my highly efficient secretary for the last 11 years, Mary Louise Sutton, who gave unstintingly of her time to the manuscript. Then comes a good friend, Mrs. Herminie B. Kitchen of Royalton, Vermont, Specialist Emeritus in Agricultural Information at Rutgers University, who helped edit the present volume and who made many valuable suggestions.

I am also greatly indebted to Edward H. Scanlon for many excellent suggestions in the preparation of Chapter 8.

My daughter Mary Grace, my son Joseph, and my assistant Helen Karpik also had a hand in checking and proofreading the manuscript and in preparing the index.

My sincerest thanks are also extended to my friend Gordon S. King, Assistant Professor in Arboriculture at the University of Massachusetts, Amherst, for his critical reading of the manuscript and for making many valuable suggestions. Professor King conducts one of the best courses for training arborists in the United States.

I also want to express my gratitude to the many professional arborists and agricultural scientists whose publications are listed in the bibliography. Without their contributions, a book of this type could not have been written.

Finally I wish to thank the following persons and organizations for the illustrations listed:

Bartlett Tree Research Laboratories, Stamford, Conn.: Figs. 16, 74B, 79, 81, 84, 88, 91, 102, 117, and 137.

Dr. J. C. Carter, Plant Pathologist, Urbana, Ill.: Figs. 62, 73, 75, and 85.

Colonial Gardens, Delaware, Ohio: Fig. 12.

Ross Daniels, Inc., Des Moines, Ia.: Fig. 17.

Douglas Eastman, Belmar, N.J.: Fig. 106.

T. H. Everett, Horticulturist, New York Botanical Garden: Figs. 6, 7, 25, 26, 27, 30, 31, 53, 54, 56, 57, 60, 90, and 147.

Dr. C. C. Hamilton, Entomologist, Rutgers University, New Brunswick, N.J.: Figs. 93 and 141.

Lucille Kopp, Huntington, Long Island: Figs. 9 and 18.

E. N. Mitchell, Staff Photographer, New York Botanical Garden: Figs. 15, 65, 66, 68, 69, 70, 78, 80, 82, 97, 100, 107, 109, 112, 127, 146, and 148.

New Jersey Federation of Shade Tree Commissions: Fig. 13.

New York Botanical Garden: Figs. 86, 89, 95, 132, and 142.

Edward H. Scanlon, publisher, *Trees Magazine*, Olmsted Falls, Ohio: Figs. 52, 55, and 59.

Siebenthaler Co., Dayton, Ohio: Fig. 58.

Dr. G. E. Thompson, Athens, Ga.: Figs. 98 and 99.

U. S. Department of Agriculture: Figs. 77A, 77B, 114, 121, 124, and 129.

P. P. Pirone

New York, New York
November 1958

Contents

FOREWORD *by L. H. Bailey* vii

Part I. GENERAL MAINTENANCE PRACTICES

1. *The Normal Tree* 3

Value of individual trees. What is a normal tree? Parts of a tree. Structure of the stem. Function of the stem. Structure of leaves. Function of leaves. Structure of roots. Function of roots.

2. *The Soil and Its Relation to Trees* 12

Soil components. The soil air. Soil water. The soil minerals. Soil organic matter. Soil reaction. Soil temperatures. The soil as an anchor. Soil improvement for trees. Mycorrhiza. Soil reaction adaptations of some trees.

3. *Transplanting Trees* 27

Importance of good transplanting. Selection of the tree. Season for transplanting. Preparing trees for transplanting. Digging deciduous trees and evergreens. Container-grown trees. Transporting. Planting. Setting the tree. Supporting. Pruning. Watering. Mulching. Wrapping. Fertilizing. Anti-transpirants.

4. *Fertilizers and Their Use* 46

How fertilizers act. The essential elements for food manufacture. The influence of elements on growth. Fertilizing materials and formulas. Rates of application. Fertilizing narrow-leaved evergreens. Fertilizing broad-leaved evergreens. Methods of applying fertilizers to trees. When and how frequently trees should be fertilized. Cost of fertilizers. Feeding above-ground parts of the tree. Tree injection.

5. *Pruning Trees and Treating Wounds* 65

Need for pruning. Pruning equipment. How to prune. Shaping the final cuts. Removing large branches. Removal of V-crotches. Dehorning. Pruning suckers or water-sprouts. Pruning storm-damaged trees. Pruning for disease control. When to prune. Pruning evergreen trees. Pruning for line clearance. Disposing of prunings. Safety hints to pruners. Treatment of wounds. Wound dressings.

6. *Cavity Treatments* 85

History of cavity work. How cavities start. Objectives of cavity treatments. Method of treating cavities. Open cavities. Covered cavities. Filled cavities. Factors determining whether to fill a cavity. Filling the cavity. Specifications for a good filler. Filling materials. Advice to tree owners.

7. *Bracing and Cabling* 104

When is artificial support justified? Types of artificial supports. Rigid bracing. Flexible bracing. Materials and installation procedure.

8. *Trees Suitable for Various Locations* 112

Low-growing trees for streetsides and lawns: Ashes-Cherries-Crabapples-Dogwoods-Magnolias-Maples. Miscellaneous trees: Crape Myrtle-Dove Tree-Franklin Tree-Fringe Tree-Goldenchain Tree-Goldenrain Tree-Hawthorns-Hophornbeam-Katsura Tree-Redbud-Russian Olive-Silk Tree-Silverbell-Smoke Tree-Snowbell-Sorrel Tree-Stewartia. Less common species: Cucumber Tree-Dawn Redwood-Kentucky Coffee Tree-Japanese Pagoda Tree. Trees with upright growth habit. Standard trees for streetside and lawn planting: Ash-Cork trees-Elm-Hackberry-Horsechestnut-Linden-Locust-Maidenhair Tree-Maple-Oak-Plane-Poplar-Sweetgum-Tulip Tree-Yellowwood. Trees for seashore planting. Trees as windbreaks. Planting streetside trees. Selection of species for city streets. Trees for different areas.

Part II. SPECIFIC ABNORMALITIES OF TREES

9. *Diagnosing Tree Troubles* 151

Qualifications of a tree diagnostician. Correct diagnosis important. Procedure in diagnosing. Diagnostic tools.

10. *Non-Parasitic Injuries* 159

Damage caused by grade changes around trees. Slime flux. Injuries caused by low temperatures. Lightning injury. Other electrical injuries. Gas injury. Injury by girdling roots. Smoke and soot injury. Lichen injury. Chemical injuries. Hail injury. Mechanical injuries. Leaf scorch. Chlorosis. Fire damage. Rodent damage.

11. *Insect Control* 200

Life history of insects. Structure of insects. Control methods. Insecticides and miticides. Spreading and sticking agents. Borers. Termites and ants in trees. Biological control of insects.

12. *Spraying Equipment and Practices* 215

Equipment. Spraying practices. Safety precautions.

13. *Tree Diseases and Their Control* 222

Symptoms. Nature of parasitic agents. Dissemination of bacteria and fungi. Overwintering of bacteria and fungi. Principles of tree-disease control. Manipulation of the environment. Spraying for disease control. Fungicides and bactericides. Spray compatibilities. Spray formulas.

14. *General Parasitic Diseases* 238

Fire blight. Crown gall. Powdery mildews. Shoestring root rot. Verticillium wilt. Other general diseases.

15. *Diseases and Insect Pests of Low-Growing Trees* 253

ASH: Rust. Cankers. Sawfly. Lilac borer. Carpenter worm. Fall webworm. Oyster shell scale. Scurfy scale. Lilac leaf miner. Flower gall.

BIRCH: Leaf spot. Leaf blister. Leaf rust. Cankers. Wood decay. Die-back. Case bearer. Leaf skeletonizer. Bronze birch borer.

CHERRY: Leaf spot. Yellowing. Asiatic garden beetle. Peach borer. West Indian peach scale. San Jose scale. Pear-slug sawfly. Japanese beetle.

CRABAPPLE: Rusts and scab. Insect control.

CRAPE MYRTLE: Powdery mildew. Black spot. Aphid. Florida wax scale.

DOGWOOD: Crown canker. Flower and twig blight. Leaf spot. Spot anthracnose. Borers. White fly. Scales. Club gall. Twig girdler.

EMPRESS TREE: Leaf spot. Wood decay.

FRANKLIN TREE: Leaf spot. Black mildew.

FRINGE TREE: Leaf spot. Powdery mildew.

GOLDENCHAIN TREE: Twig blight. Leaf spot. Aphids.

GOLDENRAIN TREE: Leaf spot. Canker. Wilt.

HAWTHORNS: Leaf blight. Fire blight. Cedar-hawthorn rust. Leaf spot. Powdery mildew. Leaf skeletonizer. Western tent caterpillar. Other caterpillars. Pear-slug sawfly.

HORNBEAM: Twig blight. Canker. Leaf spot. Scale.

JAPANESE PAGODA TREE: Canker. Twig blight.

KENTUCKY COFFEE TREE: Leaf spot.

MAGNOLIA: Leaf blight. Die-back. Nectria canker. Wood decay. Scale.

MAPLE: Wilt. Basal canker. Bleeding canker. Leaf spot. Other leaf spots. Leaf blisters. Cankers. Die-back. Ganoderma rot. Norway maple aphid. Leafhoppers. Cottony maple scale. Terrapin scale. Woolly maple scale. Bladder gall mite. Leopard moth. Sugar maple borer. Maple and oak twig pruner. Forest tent caterpillar. Boxelder bug. Flat-headed borer. Alder blight aphid. Bagworm. Carpenter worm. Japanese beetle. Spring and fall cankerworms. Tussock moth. White fly.

MOUNTAIN ASH: Leaf rust. Cankers. Crown gall. Pear leaf blister mite. Round-headed borer. Scales. Sawfly. Woolly apple aphid.

REDBUD: Leaf spot. Botryosphaeria canker. Two-marked tree hopper. Caterpillars. Scales.

RUSSIAN OLIVE: Leaf spot. Cankers. Oleander-thistle aphid. Scales.

SASSAFRAS: Canker. Leaf spot. Yellows. Promethia moth. Weevil.

SERVICEBERRY: Witches'-broom. Leaf blight. Rust. Mildews. Borers. Pear leaf blister mite. Pear-slug sawfly. Willow scurfy scale.

SILK TREE: Wilt. Canker. Webworm. Scale.

SILVERBELL: Leaf spot.

SORRELL TREE: Twig blight. Leaf spot.

SMOKE TREE: Rusts. Leaf spot. Oblique-banded leaf roller.

TREE-OF-HEAVEN: Wilt. Shoestring root rot. Leaf spot. Twig blight. Cynthia moth. Ailanthus webworm.

16. *Diseases and Insect Pests of Tall-Growing Trees* 303

ALDER: Leaf curl. Powdery mildew. Canker. Alder blight aphid. Alder flea beetle. Alder lace bug. Alder psyllid.

BEECH: Leaf mottle. Bleeding canker. Beech bark disease. Beech blight aphid. Woolly beech aphid. Brown wood borer. Beech scale. Two-lined chestnut borer. Scales. Apple tree tent caterpillar. Datana caterpillar. Gypsy moth.

CATALPA: Leaf spot. Powdery mildew. Wood decay. Mealy bug. Midge. Sphinx.

CHESTNUT: Blight. Twig canker of Asiatic chestnuts. Brown wood borer. Datana caterpillar. Japanese beetle. Lecanium scale. Maple and oak twig pruner. Spring cankerworm. Tussock moth. Two-lined chestnut borer.

ELM: Dutch elm disease. Phloem necrosis. Mosaic. Dothiorella wilt. Cankers. Leaf spot. Other leaf diseases. Wetwood. Wood decays. Leaf beetle. Tussock moth. Spiny elm caterpillar. Spring cankerworm. Fall cankerworm. Elm case bearer. Cockscomb gall. Woolly apple aphid. European elm scale. Gypsy moth. European bark beetle. Elm borer. Twig girdler. Leaf miner. Lace bug. Mites. Carpenter worm. Fall webworm. Leopard moth. Scurfy scale.

HACKBERRY: Witches'-broom. Leaf spot. Ganoderma rot. Nipple gall. Scales. Painted hickory borer. Spiny elm caterpillar.

HICKORY: Leaf spot. Cankers. Witches'-broom. Crown gall. Hickory gall aphid. Bark beetle. Cigar case bearer. Painted hickory borer. Twig girdler. June bugs. Mite.

HORSECHESTNUT: Leaf blotch. Powdery mildew. Sapwood rot. Comstock mealy bug. Scales. Bagworm. Flat-headed borer. Japanese beetle. Tussock moth. Walnut scale.

LARCH: Needle rusts. Leaf cast. European larch canker. Wood decay. Larch case bearer. Sawfly. Woolly larch aphid.

LINDEN: Leaf blotch. Leaf spot. Powdery mildew. Nectria canker. Japanese beetle. Aphid. Leaf beetle. Cecropia moth. Linden borer. Scales. Spring and fall cankerworms. Gypsy moth. Tussock moth. Twig girdler.

LOCUST AND HONEY LOCUST: Wood decays. Leaf spot. Witches'-broom. Borer. Locust leaf beetle. Locust gall maker. Mite. Webworm. Pod gall. Leafhopper. Honey locust borer. Scales. Bagworm. Carpenter worm. Cottony maple scale. Spring and fall cankerworms. San Jose scale. Walnut datana.

MAIDENHAIR TREE: Leaf spot. Insects.

OAK: Wilt. Twig blight. Anthracnose. Canker. Basal canker. Leaf blister. Leaf spot. Wood decay. Shoestring root rot. Datana caterpillar. Saddleback caterpillar. White blotch oak leafminer. Obscure scale. Lecanium scale. Kermes scale. Golden oak scale. Oak galls. Two-lined chestnut borer. Oak mite. Lace bug. Pin oak sawfly. Asiatic oak weevil. Cecropia moth. Carpenter worm. June bugs. Leopard moth. Twig pruner.

PLANE: Blight. Cankerstain. Botryosphaeria canker. Other diseases. Lace bug. Sycamore plant bug. Scales. Bagworm. White fly.

POPLAR: Dothichiza canker. Cytospora canker. Hypoxylon canker. Septoria canker. Branch gall. Leaf rusts. Leaf blister. Leaf spot. Die-back. Poplar inkspot. Canker. Red-humped caterpillar. Tentmaker. Poplar borer. Willow leaf beetle. Bronze birch borer. Fall webworm.

SWEETGUM: Bleeding necrosis. Other diseases. Leaf tier. Caterpillars. Bagworm. Fall webworm. Forest tent caterpillars.

TULIP TREE: Leaf yellowing. Leaf spot. Powdery mildew. Die-back. Nectria canker. Tulip scale. Tulip tree aphid.

TUPELO: Leaf spot. Rust. Leaf miner.

WALNUT: Canker. Bacterial blight. Die-back. Leaf spot. Trunk decay. Witches'-broom. Walnut datana. Walnut scale.

WILLOW: Leaf blight. Cytospora canker. Bacterial twig blight. Leaf spot. Gray scab. Leaf rusts. Apple tree tent caterpillar. Cone gall. Willow shoot sawfly. European willow gall midge. Mottled willow borer. Willow scurfy scale. Willow flea beetle. Willow lace bug. Bagworm. Carpenter worm. Willow leaf beetle. Red-humped caterpillar.

YELLOWWOOD: Wood decay. Botryosphaeria canker. Verticillium. Powdery mildew.

17. *Diseases and Insect Pests of Evergreen Trees* 382

ARBORVITAE: Leaf blight. Tip blight. Juniper blight. Browning and shedding of leaves. Bagworm. Scale. Leaf miner. Cedar tree borer. Spider mite.

BALD CYPRESS: Twig blight. Wood decay. Cypress moth.

BOXWOOD: Canker. Blight. Leaf spot. Winter injury and sun scald. Leaf miner. Mealy bugs. Scales. Boxwood psyllid. Giant hornet. Boxwood mite. Nematode.

CEDARS: Tip blight. Root rot. Black scale. Deodar weevil.

CRYPTOMERIA: Leaf blight. Leaf spot.

CYPRESS: Canker. Cytospora canker. Crown gall. Blight. Twig blight. Cypress aphid. Cypress meay bug. Caterpillars. Bark scale.

FIRS: Needle and twig blight. Leaf cast. Needle rusts. Cankers. Sooty mold. Witches'-broom. Shoestring root rot. Wood decays. Balsam twig aphid. Bark beetle. Caterpillars. Cedar tree borer. Cooley gall aphid. Spider mite.

HEMLOCK: Leaf blight. Needle rusts. Cankers. Sapwood rot. Hemlock looper. Spotted hemlock borer. Fiorinia hemlock scale. Grape scale. Spider mites. Bagworm. Cypress moth.

HOLLY: Tar spot. Other leaf spots. Canker. Die-back. Spot anthracnose. Twig die-back. Spine spot. Leaf scorch. Leaf miner. Bud moth. Holly scale. Pit-making scale. Southern red mite. White fly.

JUNIPER: Twig blight. Cedar rusts. Wood decays. Juniper scale. Juniper midge. Juniper webworm. Red cedar aphid. Spider mite. Bagworm. Taxus mealy bug.

PINE: Diplodia tip blight. White pine blister rust. Needle rusts. Other rusts. Leaf casts. Cenangium twig blight. Cankers. Needle blight. Blight of white pine. Wood decay. Pine bark aphid. Pine leaf aphid. Sawflies. Webworm. European pine shoot moth. Pine leaf scale. Pine needle miner. Bark beetles. Spittle bug. White pine weevil. Red pine scale. Pales weevil. Bagworm.

REDWOOD: Canker. Needle blights. Shoestring root rot. Cedar tree borer. Sequoia pitch moth. Mealy bugs. Scales.

SPRUCE: Cytospora canker. Needle cast. Needle rusts. Wood decay. Spruce gall aphid. Cooley gall aphid. Spider mite. Spruce bud scale. Budworm. Spruce Epizeuxis. Spruce needle miners. Bagworm.

YEW: Die-back. Twig blight. Twig browning. Needle blight. Crown gall. Black vine weevil. Strawberry root weevil. Taxus mealy bug. Grape mealy bug. Cottony taxus scale. Other scales. Ants and termites.

APPENDIX 429

SELECTED BIBLIOGRAPHY 447

INDEX 467

Part I

GENERAL MAINTENANCE PRACTICES

1

THE NORMAL TREE

Almost everyone likes trees for one reason or another. Some like them for the cool shade they provide during hot summer days. Others like them because they add to the beauty or to the value of a property. Still others like them for sentimental reasons—perhaps because grandpa planted the trees when he and grandma moved into the neighborhood many years ago. No matter what the reason, whether aesthetic, financial, or sentimental, a tree is a sound investment. A house surrounded by large trees is worth more money than a house without them. A house with no trees near it looks hot in summer, appears unbalanced, and suggests, however unjustly, a lack of interest on the part of those who live in it.

Old residential areas are usually more inviting and more restful than most brand-new housing developments for one important reason: they are well supplied with large trees.

The character of a city is changed by an abundance or a dearth of trees. Cities that spend liberal amounts of money to maintain their old trees and to plant new ones are generally considered nicer places in which to live.

Trees planted along streets and parkways of towns and cities represent a considerable financial investment. For example, the tree evaluation of Greenwich, Connecticut, has been estimated at $25 million, and each of the 60,000 municipally owned trees is valued at $200, or a total of $12 million. No one has estimated the value of trees in large cities like Cleveland, Los Angeles, and New York, but it certainly runs into hundreds of millions of dollars!

New York City, usually considered a place of steel and concrete, has upwards of two million trees within its borders. Of these, more than 250,000 grow along city streets and in parks. Under the jurisdiction of its Department of Parks and Public Property, Cleveland has 130,000 trees, most of which are planted along the streets. The Department of Parks and Recreation in Los Angeles takes care of some 380,000 trees.

The average per capita expenditure for tree planting and maintenance in cities and towns has been estimated at 53 cents a year. A few communities, such as Escanaba and Oak Park, Michigan, spend more than a dollar, whereas cities like Denver, New York, and Columbus, Ohio, spend less than 20 cents.

Because shade and ornamental trees represent such a great investment, it behooves all Americans to do what they can to keep the trees in good health. There are many ways, as this book shows, to increase the beauty, usefulness, longevity, and value of trees.

Value of Individual Trees

Occasionally it becomes necessary to appraise the value of a tree. This may be needed to establish a casualty loss to determine the damages one must pay for injuring or killing someone else's tree. Or an insurance company may want to know the value of an insured tree that died. Or the homeowner may want to know how much to deduct from his income tax for a tree damaged or destroyed by hurricane, ice-storm, lightning, or fire.

A committee headed by Mr. Norman Armstrong of Chapel Hill, North Carolina, for many years one of America's best commercial arborists, was asked by two organizations interested in tree care—the National Shade Tree Conference and the National Arborist Association—to devise a method for arriving at the values of shade trees. After ten years of study the committee published their results in July 1957. The highlights of this publication are presented in the Appendix of this book.

The basic factors of this particular evaluation system are the kind of tree, its size, and its condition. The diameter at breast height (DBH), which is about 4½ feet above ground level, determines the tree's size. This is translated by means of a conversion table into a measurement of the cross section of the tree. A basic valuation of $5 per square inch of trunk cross section has been established as a fair price. This figure is then taken at 100, 80, 60, 40, or 20 per cent, depending on the kind of tree involved. An oak tree or a dogwood obviously is worth much more than a silver maple or an ailanthus. Finally, this adjusted figure is taken at 100, 80, 60, 40, or 20 per cent, depending on the condition of the tree as determined by a competent arborist or forester. A tree with an extensive trunk cavity is certainly far less valuable than a sound, uninfected one.

The Shade Tree Evaluation guide contains extensive lists which divide trees into five classes for determining what percentage of the basic evaluation they should have. Separate lists are given for trees grown in six regions of the United States and Canada.

A little common sense is also needed in adjusting a local situation to this guide. Obviously a 12-inch DBH white oak growing on the front lawn of a $50,000 house is worth more than the same kind of tree growing in a wooded, isolated plot of ground.

What Is a Normal Tree?

Before anyone can successfully maintain old trees or plant new ones, he must understand the anatomy and functions of a normal, healthy tree. Without such knowledge it is difficult to diagnose accurately the many abnormalities that commonly occur in trees or to appreciate various practices undertaken to maintain or restore their vigor.

Normality in any living organism is so vague and relative it almost defies definition. Perhaps a scientifically accurate definition might be derived by applying to the tree something of the psychologists' concept of the normal human being. By their rule, the normal tree would fall within certain limits of health and vigor that include the majority of trees in a specified range.

For our purposes, however, we cannot go far astray if we are guided in our evaluation of a healthy tree by the desirable external appearances with which we are all familiar. Since a tree is most admired when it has a symmetrical branch system and abundant, attractive foliage, we may include these two characteristics among our standards of perfection. Healthy roots must also be included, for without them proper branch and foliage growth cannot continue.

To gain a working knowledge of the proper care of trees, we must know something of the structure of each of the different parts that constitute a normal shade tree, the functions these parts perform in the growth process, and the relation between the parts.

Parts of a Tree

The tree is divided into three major parts or organs: the stem, the leaves, and the roots. Each of these has its important and specific functions to perform, and each by its peculiar structure is remarkably well adapted to the performance of these functions.

Structure of the Stem. Tree trunks and their branches are constituted to endure the strains placed upon them by the varying conditions of their environment. The largest part of the stem or trunk is composed of a wood cylinder made up of tissues called *xylem.* The xylem is composed chiefly of long, thick-walled tubes or ducts, through which water containing nutrient salts is conducted from the roots to the leaves, and wood fibers, which lend tensile strength to the tree. Xylem makes up the greater part of the stem and in trees is commonly known as wood. Wood, then, is that part of a tree that gives it strength and stability. The inner central portion, called *heartwood,* is entirely composed of dead cells and does not conduct water or nutrients. The outer portion of the wood cylinder, the *sapwood,* has many living cells and is the region in which upward conduction of water and nutrient takes place.

In the center of every tree is a region known as *pith,* small in most trees, relatively large in the alders. The pith is composed of the remnants of the soft, cellular tissues of the growing tip. It has no specific function and is frequently so crowded by growth pressure that in older stems it is entirely lost.

Inspection of the cross section of any tree trunk (Fig. 1) reveals numerous concentric lines, the *annual rings,* which are plainly visible because of the alternating large and small cells. These rings represent the annual increase in the diameter of the tree. The large cells are formed by rapid growth in the spring; the smaller, by slower growth in the summer. As the season progresses, the new cells become smaller and smaller, until growth stops altogether in the fall. This compact region of small cells, known as *summer wood,* makes the annual ring. It is in sharp contrast to the large cells, or *spring wood,* formed by rapid growth during the following spring.

Across the annual rings many radial lines are visible, running from the outer edge of the wood cylinder toward the center. These lines, called *medullary rays,* are groups of cells extending through the many cylinders of wood that provide for radial transfer of water and food.

Immediately outside the xylem is a narrow band of cells known as the *cambium.* During the growing season, the cells of the cambium are stimulated, perhaps by certain natural hormones, to active and continuous division, giving rise to xylem tissue on the interior, and *phloem* tissue on the exterior. The phloem tissue is mainly conductive in function, but, unlike the xylem, which conducts water and nutrients upward, it chiefly conducts the elaborated food material from the leaves to the growing region in a downward direction. Outside of the phloem lies another type of cambium, the *cork cambium*, which gives rise to cork cells, the *outer bark*. The bark serves the very important function of preventing loss of moisture from the inner tissues. It also provides some protection from mechanical injuries, from the claws and teeth of animals, and from insects. Rapid growth of the inner, woody cylinder frequently causes the bark to crack. In some trees the outer layers of corky

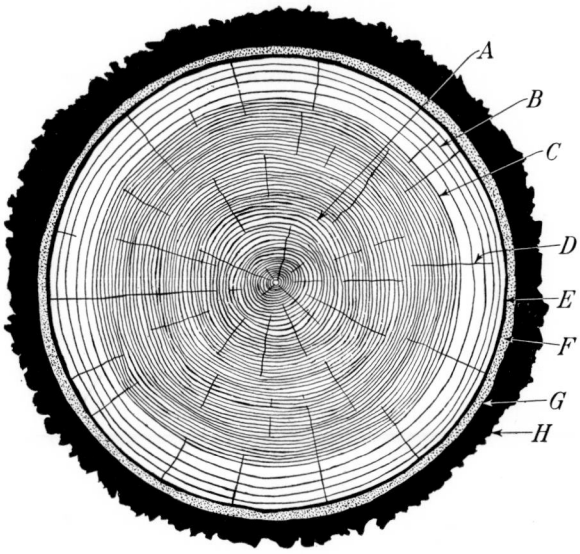

FIG. 1. Cross section of a tree trunk, showing *A*, the heartwood, composed of dead cells; *B*, the sapwood, which has many living cells that conduct water and nutrients upward; *C*, the annual rings composed of summer and spring wood; *D*, the medullary rays, which transfer water and food radially; *E*, the cambium, from which are formed xylem cells on the inside and phloem tissue on the outside; *F*, the phloem cells, which conduct elaborated food downward; *G*, the cork cambium, which produces cork cells to form the outer bark, H.

tissues soon become detached from the stem, leaving the surface of the bark smooth and relatively thin, as in buttonwood. Trees like oak and elm, on the other hand, may retain the cork layers for many years, forming a thick furrowed bark. The outer bark of some trees has such unusual configuration that the species can be identified from the bark alone.

Function of the stem. The functions of the stem, or trunk, of a tree are structural support, the conduction of water, food, and nutrients, and the storage of elaborated food materials.

One of the chief services rendered by the much-branched stem or trunk of a tree is support of the leaves in such a manner that they may derive maximum benefit from the direct action of the sun's rays. The importance of this role will become more apparent when the functions of the leaf itself are discussed. Let us remember for the present, however, that it is essential to the growth of the tree that as many leaves as possible be exposed to the rays of the sun. This exposure is accomplished by means of multiple branching of the tree trunk into limbs, branches, and finally twigs.

The trunk and its subdivisions, the branches, are the connecting links between the roots, where water and soil nutrients are absorbed, and the leaves, where food is manufactured. It is the function of these connecting links first to transport the water and inorganic salts in water solution from the roots to the leaves; and second, to distribute to every growing part of the tree, including every twig and root tip, the elaborated food products built up from raw materials.

In all deciduous trees, food materials manufactured in the leaves are gradually conducted to the twigs. These foods are partially assimilated for immediate growth. The surplus is stored in the twigs, in the trunks, and in the roots. Late in the season, when the leaves become less active in the manufacture of food, part of this surplus is reabsorbed for assimilation by the growing parts. In the spring, too, the initial twig and root growth,—which takes place before the leaves are expanded and active,—and frequently blossom production are entirely dependent upon food stored in the twigs, stems, and roots.

Structure of Leaves. If we look at the cross section of a leaf (Fig. 2) under a microscope, we see that the outer walls of the surface cells, or *epidermis*, are coated with a waxy covering or *cuticle*. This covering was once thought to be continuous except for the areas perforated with thousands of minute openings called *stomata* (singular *stoma*). Recent research has revealed, however, that instead of being a continuous coating over the surface, the cuticle is interspersed with pectin-like substances (Fig. 18). The cuticle and the epidermis help to prevent rapid drying out of the moist inner cells. Each stoma is surrounded by two kidney-shaped, movable cells, the *guard cells*. Through these pores moisture evaporates * from the leaf, thus allowing addi-

* The forces primarily responsible for the upward movement of fluids operate mainly in the leaf. Evaporation from the leaves sets up a pull on the countless columns of water in the xylem tissues. The tensile strength of the water itself holds these columns intact from the root to the leaf, thus aiding an influx of soil water at the roots proportional to the amount of moisture lost by evaporation.

tional water and nutrients to be drawn from the roots. The stomata also allow free interchanges of gases from within the leaf to the atmosphere surrounding it, and vice versa. The guard cells are very sensitive to light, opening in light and partially closing in the dark. Thus the loss of water and the gaseous exchange through the stomata are much greater during the day than at night.

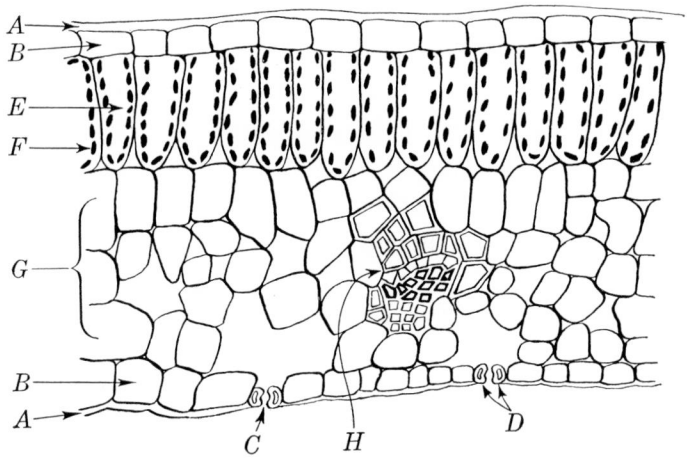

FIG. 2. Diagrammatic cross section of a leaf, showing *A*, the cuticle; *B*, the epidermis; *C*, the stomata; *D*, the guard cells; *E*, the palisade cells; *F*, the chloroplasts; *G*, the spongy parenchyma; and *H*, the veinal tissues.

Beneath the epidermis are one or more layers of long, cylindrical cells (the *palisade cells*). These contain many minute bodies known as *chloroplasts*, which in turn contain the green substance known as *chlorophyll*. Chlorophyll absorbs the light energy from the sun and transforms it into chemical energy, which acts upon water from the soil and carbon dioxide from the air to form sugars.

Numerous large, loosely arranged cells (the spongy *parenchyma*) lie between the palisade cells and the lower epidermis. The spaces between these cells facilitate the diffusion of gases between the stomata and the palisade cells.

Function of Leaves. The primary function of the leaf is the manufacture of sugars, the basic food materials used directly or after combination with other elements in various cells of the tree for root growth, maturation of woody tissues, formation of fruits and seeds, and storage. Green leaves, therefore, represent factories upon which the tree depends for its entire food supply. The foods used by trees for growth are the same carbohydrates, proteins, and fats used by animals, but green plants have the power to make their own food, whereas every other living thing is dependent upon green plants for its food, shelter, clothing, and fuel.

The process of food manufacture, *photosynthesis* (photo = light; synthesis = manufacture), is highly complicated and need be discussed here only in its essential details.

Food cannot be manufactured when a plant wilts and the stomata are nearly closed. But in a normal tree, carbon dioxide enters through the stomata of the leaf into the inter-cellular spaces, surrounds the cells, and is absorbed by the water solution there present. Having passed through the cell walls into the chloroplasts, the carbon dioxide and the water split up into their separate constituents. These recombine chemically to form a series of intermediate products, resulting eventually in sugars and starches, known collectively as carbohydrates. By the addition of nitrogen, small quantities of the element sulfur, and in some cases phosphorus also, these carbohydrates become proteins. Sugars may also be transformed into cellulose, lignin, fats, and oils.

During photosynthesis, oxygen is given off. Thus the process resolves itself into one of taking carbon dioxide from the atmosphere and returning oxygen to it. Animals bring about the reverse process (respiration) by utilizing the oxygen and returning carbon dioxide to the atmosphere. This is one of the interesting partnerships between plants and animals in which each helps contribute the elements necessary for the other's existence.

The leaves function somewhat differently when suitable plant foods are sprayed on them. The pectin-like substances previously mentioned have a high capacity for absorbing and releasing water and nutrients. They provide a direct connection between the outside atmosphere and the interior of the leaf via so-called *plasmodesma*. Hence, by spraying a balanced plant food on the leaves, one can supply nitrogen, phosphate, and potash, and certain minor elements to the plant. Most of these nutrients formerly were applied to the soil. More details on this method of supplying raw food elements to the trees are presented in the chapter on tree fertilization.

Structure of Roots. There are two principal types of roots, *tap* and *surface*, named according to their shape and depth of penetration. Examples of tap-rooted trees are hickory, walnut, tulip, and many oaks. Elm, spruce, and maple are surface-rooted species. Some trees have a combination of both types of root systems.

The root system of some trees is enormous. It may comprise one third to one half the entire volume of the tree. The total length of all roots of a large, spreading oak runs into hundreds of miles.

At the point of attachment to the trunk, the roots are relatively few and large. They divide and redivide, becoming progressively smaller in diameter, until extremely fine rootlets result at a distance from the trunk. The smallest of these strands are the absorption rootlets, which are covered a short distance back from the tips with numerous, very fine outgrowths of thin-walled individual cells, called the *root hairs*. Through these root hairs nearly all the water and nutrient salts are absorbed. As these minute appendages are very delicate, no single individual lives very long. The older ones are constantly dying, but new ones are continually being formed behind the growing tip. On certain rootlets more than 10,000 root hairs to a square inch have been found.

In advance of the root-hair region of rootlets is a section where primary elongation of the root takes place. This growing point is just behind the root tip and is protected from injury by a group of cells known as the *root cap*. The growing point is composed of a small group of rapidly dividing cells, which push the tip through the soil and lay down in their wake new tissues of the root.

Function of Roots. The functions of roots are primarily four: storage of elaborated foods, conduction, absorption, and anchorage. Tree roots, like those of other plants, store elaborated food materials, as do stems and twigs. The conducting tissues of the roots are, in form and function, essentially like those in the stem. Absorption and anchorage are so intimately associated with the medium in which roots grow—the soil—that they will be discussed in considerable detail in the following chapter.

2

THE SOIL AND ITS RELATION TO TREES

No book on the maintenance of shade and ornamental trees would be complete without some discussion of that important natural body, the soil. Usually considered by the layman to be an inert mass, soil is actually a dynamic entity in which physical, chemical, and biological changes are constantly occurring. A detailed treatment of these changes is beyond the scope of this book, but a brief presentation of the more important ones, particularly as they relate to trees, is essential.

Soil serves as a storehouse for plant nutrients, a habitat for micro-organisms, and a reservoir of water for the tree's growth. It also provides anchorage for the tree.

Soil Components

The soil is composed of three distinct parts or phases: gaseous, the soil air; liquid, the soil water; and solid, the soil minerals and organic matter. These parts are interrelated very closely.

The Soil Air. The pore space in soils amounts to 30 to 50 per cent of the volume. The quantity of air contained in the soil is governed primarily by the extent to which this pore space is filled with moisture. Ordinarily, air comprises about 20 per cent of the volume of a well-tilled soil.

Oxygen, one of the most important components of soil air, is used in the oxidation of toxins that otherwise would accumulate in the soil. It also supports the activity of numerous beneficial micro-organisms. Nitrogen, another component of the soil air, is used as a raw material by nitrogen-fixing bacteria to manufacture protein materials, which later decompose to yield nitrates that nourish the tree.

All tree roots and soil micro-organisms respire,* that is, they continuously consume oxygen and give off carbon dioxide. In poorly aerated soil, where gaseous exchange between the soil air and the atmosphere is retarded, the oxygen supply decreases, whereas the quantity of carbon dioxide increases. This unfavorable oxygen-carbon dioxide balance may retard or, in extreme cases, completely check root growth. Roots do not thrive in such areas. This, in fact, explains why the more active roots of city trees are often confined to the soil areas between the curb and the sidewalk, and to nearby lawn areas.

Although oxygen is absolutely necessary for a healthy root system, an insufficiency of this element for relatively short periods will not result in serious damage. Studies by Cornell University scientists revealed that the oxygen content of orchard soils in spring sometimes falls from the normal 21 per cent at the surface to as low as 1 or 2 per cent at a depth of 3 to 5 feet. The carbon dioxide content of the same soil increases to 2 to 3 per cent at the 1-foot level, and occasionally exceeds 12 per cent at 5 feet.

Flooding of soil in winter when roots are dormant is not so serious as it is during summer when the roots are making most active growth. Roots of apple trees can grow slowly with as little as 3 per cent oxygen, but they require at least 10 per cent for good growth.

Because the concentration of carbon dioxide must be very high, over 20 per cent, to cause direct injury to tree roots, lack of oxygen probably is the more common cause for restricted growth or death of roots.

Capillary and gravitational movements of water and the daily changes in temperature and barometric pressure are responsible for the aeration of soils. The oxygen supply is often renewed too rapidly in sandy soils and too slowly in heavy clay soils. In either case, the situation is improved by adding organic matter or by mixing the two classes of soil. Aeration of soils is also increased by structural changes caused by the use of lime and certain fertilizer salts. The channels remaining after the roots of plants decay and those caused by the burrowing of earthworms also aid in ventilating the soil.

Some trees can thrive under conditions of relatively poor soil aeration. Willows, for example, will grow along streams and ponds where the oxygen supply is relatively low. Some scientists believe that, in willows, oxygen produced by photosynthesis in the leaves is transferred internally to the root cells. The roots of one species of willow, *Salix nigra,* were found to be

* Actually, all living cells in the trunk and leaves also respire. Respiration, however, is of the utmost concern in roots, inasmuch as it is most often hindered in them.

capable of growing and absorbing nutrients in almost complete absence of oxygen. Cypress grows naturally in swamps where the soil is flooded for long periods. This tree has, on its roots, pyramid-shaped protuberances called 'knees' which extend as high as 10 feet into the air. The 'knees' are composed of light, soft, spongy wood and bark which permit air to reach the roots during the weeks and even months when the swamps are covered with water.

On the other hand, plants like Taxus and cherry thrive in soils that are well drained and well aerated. Such trees are killed or severely damaged if the soil is waterlogged for only a few days during the growing season.

Inadequate aeration of the soil may affect the roots in an indirect manner. It may change the kind of organisms in the soil flora and make the roots more susceptible to attack by parasitic fungi. For example, the fungus *Phytophthora cinnamomi* may become so abundant in a poorly drained soil as to cause severe damage to tree roots.

Then, too, certain organisms thrive under conditions of little to no oxygen. These produce compounds such as hydrogen sulfide and nitrites which may harm the roots.

Aside from direct injury to the roots, poor aeration also prevents the roots from absorbing water and minerals. A deficiency of oxygen and an excess of carbon dioxide reduce the permeability of the roots to water. They also prevent the roots from absorbing minerals from the soil solution.

Poor aeration of the soil results in development of roots near the soil surface. Such a shallow root system makes the tree more susceptible to toppling by winds and to droughts during the summer.

Soil Water. Trees depend on the water contained in the pore space of the soil in which they are growing. The roots absorb the water and pass it on to the leaves, where it functions both as a nutrient and in the absorption of carbon dioxide from the air. Water is an essential part of the protoplasm, or living matter, of every cell. It also functions in the soil as a solvent of the necessary mineral nutrients, and as a medium by which they enter the plant. Further, water transports other raw materials and elaborated food from one part of the tree to another.

Water may exist in any soil in the *hygroscopic, capillary,* or *free water* state. *Hygroscopic water* occurs in the form of a very thin film over the soil particles where the moisture in the soil is in equilibrium with that in the air. It acts both as a lubricant and as an attractant for *capillary water.* The latter exists as a thicker film, which is attracted to the particle by its own surface tension.

This tension is stronger than the pull of gravity. When two soil particles are in contact, and one has a thinner film of capillary water than the other, the drier surface has a greater attraction for the water. As a result, the water moves from the wetter to the drier surface, until both particles are surrounded by films of equal thickness. When a film of capillary water reaches the maximum thickness that can be maintained by surface tension, any additional water responds to the pull of gravity and flows downward into the soil to become *free water*. Free water occupies the spaces between the moist soil particles, just as the air does when only hygroscopic and capillary water are present. When free water occupies all the pore spaces in the soil or any horizon of the soil, its upper surface is known as the water table. Because of lack of air, roots cannot survive long in such a water-saturated zone (Fig. 3).

When soil water comes in contact with the complex organic compounds that comprise the cell walls of the root hairs, it is absorbed by *imbibition*. From this point the water now passes on, by a process known as *osmosis*, through the plasma membrane lining the cell walls and becomes part of the cell sap. The concentration of soluble substances in the cell sap is usually greater than that in the soil solution. The movement of water into the cell tends to equalize the concentrations on both sides of the cell wall. The pull exerted by the water lost from the leaves during transpiration further aids the movement of water from cell to cell and the influx of water from the soil into the root hairs.

When the soil surrounding the root hairs becomes very dry, as during a drought, the concentration of the solution in the soil may be greater than that within the cell. Then the reverse process takes place. Water moves out of the root hairs and plant cells, causing their collapse, and the continuous water column from roots to leaves cannot be maintained. Consequently, the plant wilts. Wilting also occurs if the quantity of water lost from the leaves is greater than that gained by the roots. Such reverses are often evident in shade trees during June, July, and August.

Trees vary with respect to their moisture requirements. Red maple, pin oak, sweet gum, and most species of willow prefer soils with a high moisture content, whereas red oak, chestnut oak, ash, and poplar grow best in soils containing smaller amounts of water. Spruces can tolerate wetter soils than can most other species of conifers.

A change in the water table often seriously affects a tree's health. Although not evident at the surface, a rise in the water table may cause drowning of

FIG. 3. These trees died because the high water table excluded air from the roots.

the deeper roots. This happened to London plane trees planted in Battery Park at the lower tip of Manhattan Island and to Hyssop crabapple trees planted in New York City's Stuyvesant Town several years ago.

The Soil Minerals. The solid body of the soil consists of varying proportions of coarse grains (sand), medium-sized grains (silt), and fine particles (clay), all of which are mineral particles originating from the disintegration of rocks. In addition, it contains varying amounts of humus substances formed in the bacterial decomposition of plant remains.

The smaller-sized particles of the clay portion of the soil are known collectively as colloidal clay. These particles aid greatly in the retention of water and plant nutrients in the soil. The working qualities of a soil depend largely on its content of colloidal clay and the condition in which this exists.

Colloidal clay, chemically, consists mainly of iron and aluminum silicates and oxides. The silicates are the seat of retentive or 'exchange' processes in soils. Surrounding each tiny fragment of the exchange mineral is a layer of adsorbed water containing both cations and anions.* The cations consist chiefly of hydrogen (H), calcium (Ca), magnesium (Mg), potassium (K), and sodium (Na); while the anions are hydroxyl (OH), phosphate (PO_4), nitrate (NO_3), sulfate (SO_4), and chloride (Cl) ions. Most of these ions are plant nutrients.

As a root hair enters the zone of adsorbed water, an exchange of ions takes place between this water layer and the root hairs. As a result, the mineral nutrients of the soil solution move into the plant, and the carbonic acid of the plant is excreted into the soil.

Soil Organic Matter. The second solid component of soil is usually concentrated in the upper layers. This organic matter is essential for the maintenance of soil micro-organisms. It also serves to make the soil porous and friable, to increase its water-holding capacity, and to improve its aeration. In its decay, organic matter yields up mineral nutrients and nitrogen for the use of the currently growing plants. In addition, organic matter has a high 'exchange capacity,' and acts much like the exchange minerals in colloidal clay.

Soil Reaction

Soils are acid, neutral, or alkaline in reaction. The degrees of acidity or alkalinity are expressed in terms of pH values, very much as temperatures are

* Soluble soil nutrients dissociate in water to form electrically charged particles known as cations and anions. Cations are positively charged (+), and anions are negatively charged (−).

expressed in degrees centigrade or Fahrenheit. Temperature scales center on the freezing point of water, whereas the acidity-alkalinity scale centers on a neutral point (pH 7). The latter scale is divided into 14 major units. Values between pH 7 and 14 are alkaline, and those from pH 7 to 0 acid. The intensity of the alkalinity increases tenfold with each unit increase in pH. For example, the alkalinity of pH 8 is 10 times as great as that of pH 7. Likewise, the acidity increases 10 times for each unit decrease in the pH scale. Thus the acidity at pH 6 is 10 times as great as that of pH 7, and at pH 5 it is 10 times as great as that at 6 and 100 times as great as that at pH 7.

The pH scale refers not to the total amount of acid or alkaline constituents in the soil, but to those portions of the acid and alkali that are in a free or dissociated state. A pH reading gives an indication of the relative number of free *acid* (H+) and *alkaline* (OH−) particles in the soil, and not of the combined or undissociated acid particles.

Most tree soils are acid in reaction, because a large portion of the alkaline exchange cations (Ca, Mg, K, and Na) have either been absorbed by the trees or lost through leaching, leaving acid exchange-minerals behind. In addition, the continued use of acid-forming fertilizers and sprays (especially those containing sulfur) tends to increase acidity. In cities, certain gases, such as ammonia and sulfur dioxide, which are carried down by the rain water, further tend to increase the acidity of the soil. The pH values of most Eastern tree soils lie between 5 and 6.5. They are seldom higher than 7, unless their content of limestone is naturally high, or large quantities of lime have been added to them.

Although soil acidity is not the only factor governing the availability of nutrients, it nevertheless plays an important part in the process. For example, an increase in acidity beyond certain limits means a decrease in the supply of certain plant nutrients, the hydrogen having taken the place of the basic cations in the exchange complex. The exchanged cations may have been removed by the trees, or may have been carried away by the drainage water. At the other extreme, as the alkalinity increases, the roots are less able to absorb iron, manganese, zinc, copper, and boron, because these minerals are insoluble under such conditions. A good case in point is that of the availability of iron salts, which are absorbed by the plant at a pH of 6.7 or below. If the tree is growing in an alkaline soil (above pH 7), there is a strong possibility that it will suffer from iron deficiency and that the leaves will become chlorotic (see Chlorosis, p. 196).

In general, most plants grow best on soils having a slightly acid reaction. A pH value between 5.5 and 7 seems most favorable for nearly all trees. Growth is best when the soils in which trees are planted and grown are maintained within the pH range to which they are adapted. Relatively few trees seem to prefer a distinctly acid soil. Among those that do are Silverbell (*Halesia*), Sourwood (*Oxydendrum*), Yellowwood (*Cladrastis*), and Franklin tree (*Franklinia*).

Some form of lime is usually recommended to sweeten excessively acid soils, and the use of aluminum sulfate and sulfur is suggested for overcoming alkalinity in soils. But sulfur should not be added to a soil that is wet and well supplied with organic matter in its deeper layers. In such a soil, it will form hydrogen sulfide, a chemical that is extremely toxic to roots.

Table I indicates the amount of ground limestone needed to increase the alkalinity to a 6-inch depth of the several types of soil.

Table I

LIME REQUIREMENT OF DIFFERENT SOIL TYPES OF SAME REACTION

Present pH of Soil	Pounds of Agricultural Ground Limestone Needed per 1000 sq. ft. To Raise:					
	Sandy loam soils		Silt loam soils		Silty-clay loam soils	
	to pH 6.0	to pH 6.5	to pH 6.0	to pH 6.5	to pH 6.0	to pH 6.5
6.0	none	23	none	41	none	58
5.5	23	46	41	83	55	115
5.0	46	69	83	124	115	173
4.8	55	78	97	138	138	196

County agricultural agents, agricultural experiment station chemists, and field representatives of agricultural chemical companies offer soil-testing services and they supply information as to the amount of one or the other of these materials that must be added to produce the desired change in pH. If preferred, inexpensive soil-testing kits can be purchased for individual use.

The listing of the most favorable pH range of most shade and ornamental trees and of a few of the shrubs will be found at the end of this chapter.

Table II indicates the approximate amounts of sulfur or aluminum sulfate per 100 square feet of area that are required to increase the acidity to a 6-inch depth, of a silt loam soil from:

Table II

		Sulfur (pounds)	Aluminum Sulfate (pounds)			Sulfur (pounds)	Aluminum Sulfate (pounds)
pH	8.0 to 7.0	2.0	4.5	pH	7.0 to 6.5	1.5	2.5
	8.0 to 6.5	3.0	7.0		7.0 to 6.0	2.0	5.5
	8.0 to 6.0	4.0	10.0		7.0 to 5.5	3.5	9.0
	8.0 to 5.5	5.5	13.5		7.0 to 5.0	5.0	13.0
	8.0 to 5.0	7.0	17.5		6.5 to 6.0	1.5	3.0
	7.5 to 7.0	1.75	3.5		6.5 to 5.5	2.5	6.5
	7.5 to 6.5	2.0	5.0		6.5 to 5.0	4.0	10.5
	7.5 to 6.0	3.5	7.5		6.0 to 5.5	1.5	3.5
	7.5 to 5.5	5.0	11.5		6.0 to 5.0	3.0	7.5
	7.5 to 5.0	6.5	15.5		5.5 to 5.0	1.5	4.0

Soil Temperatures

Old, well-established trees, as well as recently transplanted ones, are adversely affected by extremes in soil temperatures. Low temperatures retard or even prevent water and mineral absorption by the roots. If, at the same time, the air temperature is high enough to allow loss of water from the leaves and twigs, serious injury may result. The winter loss, by the leaves, of more moisture than can be replaced through the roots often results in drying and browning of both narrow- and broad-leaved evergreens (see Winter Drying of Evergreens, p. 169).

As soil temperatures normally drop slowly with the approach of fall and winter, heat can be conserved for a long period by early fall mulching. This practice prevents deep freezing of the soil and reduces the amount of winter drying of the trees. The mulching materials should be removed early the following spring; otherwise they will screen out the warm rays of the sun.

The Soil as an Anchor

By embedding themselves in the soil, the roots are able to hold the tree erect. Trees tend to develop anchorage where it is most needed. The tendency of isolated trees is to develop anchorage rather equally all around, with perhaps a slightly stronger development on the side toward the prevailing strong winds. The more a tree is protected from winds, the less is its dependence upon secure anchorage and the less it seems to provide for such anchorage.

The factors that are important to a tree from the standpoint of anchorage are the depth and the texture of the soil, the relative abundance of such mechanical allies as stones, and the roots of other trees and shrubs. Sufficient room for the roots to spread horizontally must also be available.

Trees growing in wet soils are more easily blown over during wind storms than are trees growing in well-aerated soils. One reason is the lower resistance resulting from the lubricating action of the wet soil particles. Another is the absence of deep anchor roots, which are unable to develop because of lack of air in the deeper layers. In wet soils most of the tree roots are confined to the upper 18 inches; in well-aerated soils the roots may penetrate much deeper.

Soil Improvement for Trees

Probably the best way to improve the physical condition of the soil is to incorporate large amounts of organic matter (Fig. 4). Such material not only improves the structure of the soil but increases its water-holding capacity as well. Disappointing or unprofitable results may ensue, however, if the organic matter is added to infertile, very wet, or extremely acid soils. Such adverse conditions must be rectified before much benefit from the organic matter can be expected. Wet soils, for instance, may be improved by the insertion of drain tiles to draw off the excess water, thus increasing the aeration. Shallow soils, likewise, will respond better to organic matter incorporations if any underlying hardpan or impervious subsoil is first broken up by the use of special digging tools or by blasting with dynamite. Incorporating ground limestone into the subsoil also markedly improves the physical condition of the soil, especially where the subsoil is very acid.

Of the many sources of organic matter, those most commonly used for shade and ornamental trees are stable manure and two distinct types of peat—peat moss and sedge peat. Manure that is to be placed in transplanting holes, or at any considerable depth in the soil around established trees, should be well rotted or past the stage of most rapid decay; otherwise, it may injure the tree. Peat, on the other hand, appears to decompose without the formation of harmful effects, when mixed to considerable depths in well-aerated and well-drained soils.

Peat moss, also known as sphagnum peat or peat, is very acid in reaction (pH 3.0 to 4.5) and is used principally for acid-loving plants, such as rhododendrons and azaleas, hemlocks, pines, and spruces. It is also used in back fill

FIG. 4. Effect of peat moss on root development of dogwood. LEFT: Tree grown in soil without the addition of peat moss. Notice the few scraggly roots. RIGHT: Tree grown in soil to which peat moss has been added. Notice the heavy mass of fibrous roots.

for almost all trees. It is fibrous in nature, light brown in color, low in nitrogen content (around 1 per cent on a dry basis), and has a high water-absorbing capacity. It should never be applied in the form in which it is purchased, but must be thoroughly soaked beforehand. This may take several days. When large quantities of this type of peat are used, some nitrogen in the form of nitrate salts should be applied from time to time throughout the following summer, to supply both the trees and the bacteria that decompose the peat. This practice will compensate for the leaching due to the frequent watering necessary for transplanted trees, and for the consumption by the bacteria.

Sedge peat, also known as reed peat or humus, may, in time, have more extensive use than peat moss. Large deposits of this type of peat are available for exploitation. As offered for sale, sedge peat is dark brown to black and relatively high in nitrogen (2.0 to 3.5 per cent). It is less acid in reaction (pH 4.5 to 6.8) and has a lower water-absorbing capacity than peat moss. It can be safely used around most trees.

Mycorrhiza

The roots of many trees are so intimately associated with strands of fungus tissue as to form compound organs known as mycorrhiza. There are several types of mycorrhiza, distinguished mainly by the location of fungus development on and in the root. The fungi on beech, elm, spruce, hickory, birch, oak, and red and white pine develop between the root cells, whereas the fungi on ash, maple, red cedar, sweetgum, and ericaceous plants develop within the individual root cells. The former are known as ectotrophic and the latter as endotrophic mycorrhiza.

The function of mycorrhiza is not clearly understood. In some instances, the intimate association between the root and the fungus is believed to be beneficial to the tree. In others, it is said to be harmful. The ectotrophic mycorrhiza are believed to play an important role in salt absorption in soils of low nutrient level. Among the endotrophic kinds a direct transfer of nutrients to the host takes place through the dissolution of the fungus strands.

The failure of some trees, especially beech, when moved away from their natural soil environment in the woods is attributed to the lack of mycorrhizal development around their roots in the new locality. Success in the transplanting of such trees has been reported by several tree specialists, when woods soil (containing the fungi) has been mixed with the soil in the new location. The greatest danger of such a practice lies in the possibility of introducing harmful species of fungi with the desirable ones.

SOIL REACTION ADAPTATIONS OF SOME TREES

Common Name	Botanical Name	Most Favorable pH-Range
Acacia	*Acacia*	6.5 to 7.5
Alder	*Alnus*, named spp.	6.5 to 7.5
Almond, Flowering	*Prunus glandulosa*	6.5 to 7.5
Apple	*Malus* spp.	6.5 to 7.5
Arborvitae	*Thuja* spp.	6.5 to 7.5
Ash	*Fraxinus* spp.	6.0 to 7.5
Bald Cypress, Common	*Taxodium distichum*	6.5 to 7.5
Beech	*Fagus*, named spp. & var.	6.5 to 7.5
Birch, Sweet	*Betula lenta*	4.0 to 5.0
Box, Common	*Buxus sempervirens*	6.5 to 7.5
Buckeye, Ohio	*Aesculus glabra*	6.5 to 7.5
Buckeye, Red	*Aesculus pavia*	6.0 to 6.5
Catalpa	*Catalpa*, named spp.	6.5 to 7.5
Cherry	*Prunus* spp.	6.5 to 7.5
Cherry, Choke	*Prunus virginiana*	6.5 to 7.5
Chestnut, American	*Castanea dentata*	4.0 to 6.0
Chinquapin	*Castanea pumila*	4.0 to 6.5
Cypress	*Chamaecyparis*, named spp. & var.	4.0 to 7.5
Dockmackie	*Viburnum acerifolium*	4.0 to 5.5
Dogwood, Flowering	*Cornus florida*	5.0 to 6.5
Douglas Fir	*Pseudotsuga taxifolia*	6.0 to 6.5
Elm	*Ulmus* spp.	6.5 to 7.5
Eucalyptus	*Eucalyptus* spp.	6.5 to 7.5
Fir	*Abies*, named spp.	4.0 to 6.5
Franklin tree	*Franklinia alatamaha*	4.0 to 6.0
Fringe tree	*Chionanthus virginica*	4.0 to 6.0
Hackberry	*Celtis* spp.	6.5 to 7.5
Hawthorn	*Crataegus* spp.	6.0 to 7.5
Hemlock, Canada	*Tsuga canadensis*	4.0 to 6.5
Hemlock, Carolina	*Tsuga caroliniana*	4.0 to 5.0
Hickory, Shagbark	*Hicoria ovata*	6.0 to 6.5
Holly, American	*Ilex opaca*	4.0 to 6.0
Hophornbeam, American	*Ostrya virginiana*	6.0 to 6.5
Hornbeam	*Carpinus*, named spp.	6.5 to 7.5
Horsechestnut	*Aesculus hippocastanum*	6.0 to 7.0
Juniper	*Juniperus*, many spp.	6.5 to 7.5
Juniper, Common	*Juniperus communis* & var.	6.0 to 6.5
Juniper, Creeping	*Juniperus horizontalis*	4.0 to 6.0

Common Name	Botanical Name	Most Favorable pH-Range
Juniper, Mountain	*Juniperus communis saxatilis*	4.0 to 5.0
Kentucky Coffee tree	*Gymnocladus dioicus*	6.5 to 7.5
Larch	*Larix*, named spp.	6.5 to 7.5
Linden	*Tilia* spp.	6.5 to 7.5
Loblolly Bay	*Gordonia Lasianthus*	4.0 to 6.0
Locust, Black	*Robinia pseudo-acacia*	5.0 to 7.5
Locust, Honey	*Gleditsia*, named spp.	6.5 to 7.5
Maidenhair tree	*Ginkgo biloba*	6.0 to 6.5
Magnolia	*Magnolia* spp.	4.0 to 7.0
Maple	*Acer*, many spp.	6.5 to 7.5
Maple, Mountain	*Acer spicatum*	4.0 to 5.0
Maple, Red	*Acer rubrum*	4.5 to 7.5
Maple, Striped	*Acer pennsylvanicum*	4.0 to 5.0
Mock-Orange	*Philadelphus* spp.	6.0 to 7.5
Mountain Ash, American	*Sorbus americana*	5.0 to 7.0
Mountain Ash, European	*Sorbus aucuparia*	6.5 to 7.5
Mountain Laurel	*Kalmia latifolia*	4.0 to 6.5
Mulberry	*Morus*, named spp. & var.	6.5 to 7.5
Oak, Black	*Quercus velutina*	6.0 to 6.5
Oak, Blackjack	*Quercus marilandica*	4.0 to 5.0
Oak, Chestnut	*Quercus prinus*	6.0 to 6.5
Oak, English	*Quercus robur* & var.	6.5 to 7.5
Oak, European Turkey	*Quercus cerris*	6.5 to 7.5
Oak, Overcup	*Quercus macrocarpa*	4.0 to 5.0
Oak, Pin	*Quercus palustris*	5.5 to 6.5
Oak, Post	*Quercus stellata*	4.0 to 6.5
Oak, Red	*Quercus borealis maxima*	4.5 to 6.0
Oak, Sand Blackjack	*Quercus catesbaei*	4.0 to 5.0
Oak, Scarlet	*Quercus coccinea*	6.0 to 6.5
Oak, Shingle	*Quercus imbricaria*	4.0 to 5.0
Oak, Shrub	*Quercus ilicifolia*	5.0 to 6.0
Oak, Southern Red	*Quercus falcata*	4.0 to 5.0
Oak, Swamp White	*Quercus bicolor*	6.5 to 7.5
Oak, White	*Quercus alba*	6.5 to 7.5
Oak, Willow	*Quercus phellos*	4.0 to 6.5
Oak, Yellow	*Quercus Muhlenbergii*	4.0 to 5.0
Paulownia, Royal	*Paulownia tomentosa*	6.5 to 7.5
Peach	*Prunus persica*	6.5 to 7.5
Pear	*Pyrus* spp.	6.5 to 7.5
Pine	*Pinus*, many spp.	4.0 to 6.5

SOIL REACTION ADAPTATIONS OF SOME TREES (*Cont.*)

Common Name	Botanical Name	Most Favorable pH-Range
Plane tree	*Platanus*, named spp.	6.5 to 7.5
Plum, American	*Prunus americana*	6.5 to 7.5
Plum, Beach	*Prunus maritima*	6.5 to 7.5
Plum, Common	*Prunus domestica*	6.5 to 7.5
Plum, Purple-leaf	*Prunus cerasifera pissardi*	6.5 to 7.5
Poplar	*Populus*, named spp.	6.5 to 7.5
Quince, Flowering	*Chaenomeles japonica* & var.	6.5 to 7.5
Redbud	*Cercis* spp.	6.5 to 7.5
Red Cedar	*Juniperus virginiana* & var.	6.0 to 6.5
Savin	*Juniperus sabina* & var.	6.0 to 6.5
Serviceberry	*Amelanchier* spp.	6.0 to 6.5
Silver-bell	*Halesia tetraptera*	4.0 to 6.0
Sourgum	*Nyssa sylvatica*	5.0 to 6.0
Sourwood	*Oxydendrum arboreum*	4.0 to 6.0
Spruce	*Picea* spp.	4.0 to 6.5
Spruce, Colorado	*Picea pungens* & var.	6.0 to 6.5
Stewartia	*Stewartia Malacodendron*	4.0 to 6.0
Storax	*Styrax americana*	4.0 to 6.0
Sweetgum	*Liquidambar styraciflua*	6.0 to 6.5
Sweetleaf, Horse Sugar	*Symplocos tinctoria*	4.0 to 6.0
Tree-of-heaven	*Ailanthus glandulosa*	6.5 to 7.5
Tulip tree	*Liriodendron tulipifera*	6.0 to 6.5
Walnut	*Juglans*, many spp.	6.5 to 7.5
White Cedar	*Chamaecyparis thyoides*	4.0 to 5.5
Willow	*Salix*, many spp.	6.5 to 7.5
Willow, Creeping	*Salix repens*	4.0 to 5.0
Yew	*Taxus* spp.	6.0 to 6.5

3

TRANSPLANTING TREES

Transplanting involves the moving of a tree from one location to another in such a manner that it will continue to grow. Complete success, however, should be measured not in terms of whether the tree merely lives or dies, but on the basis of minimum shock to the tree and ability to continue growth and development with the least possible interruption. The practice includes three distinct operations—digging or 'lifting,' moving to the new site, and replanting.

Importance of Good Transplanting

Except where trees are transported for long distances or where large trees are involved, the whole process of transplanting usually can be completed in a few hours or within several days. In any case, the time required is only a small fraction of the anticipated life span of the tree, which may vary from 30 to 300 years. Yet the future of the tree—its health, its resistance to insects and diseases, its beauty, and the cost of maintenance—can be greatly influenced by the methods used in digging, transporting, and planting. Faulty transplanting can result in considerable trouble and even in death of the tree, despite the use of good soil and a fine specimen.

A good transplanting job takes only a little more time, thought, and money than does a crude job and is always well worth the difference. Most of the trouble resulting from poor transplanting originates from the operator's thinking in terms of minimum requirements, that is, what the tree will stand rather than what it needs.

Before digging operations are begun, one must consider several points that will greatly influence not only the success of transplanting but also the future thriftiness of the tree.

Selection of the Tree

The selection of the tree is governed by soil type, exposure, elevation, and climate. A study of the relation of these factors to the future health of the tree requires considerable technical information not readily available to most individuals. Probably the most dependable criterion of what to plant in any locality, therefore, is *the selection of species that are already well established and thriving in that locality*. Any woodlot near the site picked for the new tree will reveal more pertinent information on the most suitable species than will the best scientific knowledge of the specific requirements of the tree. Trees in such areas are thriving because all the factors that combine to produce good growth are favorable.

Different tree species vary greatly in their reactions to being moved. As a rule, trees with compact, fibrous root systems are moved with greater ease and less shock than are those with large taproots or long scraggly root systems. Among the trees that are most easily moved and that appear to suffer least shock are ash, elm, ginkgo, linden, maple, pin oak, poplar, and willow. Trees moved with greater difficulty are apple, buckeye, dogwood, hemlock, and spruce. Those moved with most difficulty are beech, birch, hawthorn, hickory, magnolia, sourgum, sweetgum, and many species of oak.

Season for Transplanting

Although no blanket recommendation can be made for the best season to move all trees, most arborists agree that deciduous trees are best moved in the spring, fall, or winter. In northern states, however, it is sometimes difficult to move trees in winter, especially when the ground is frozen deeply, and some arborists now guarantee successful transplanting of fairly good-sized trees even during the summer. Extraordinary precautions are necessary in the latter case to reduce the shock of cessation of growth and food storage due to water loss and root injury. Some trees with fleshy root systems, such as dogwood, magnolia, yellowwood, and tulip, are more successfully moved in the spring than in the fall. Among other trees that are also moved more successfully in the spring are: American hornbeam, hickory, pecan, white fringetree, beech, butternut, goldenrain tree, sweetgum, tupelo, sassafras, and walnut.

A dry fall followed by a cold winter may result in the death of a high percentage of fall-transplanted trees. Because of this, many municipal tree men prefer to plant street trees in the spring.

With care, evergreens can be moved at any season of the year. Best success, however, follows their transplanting during August and early September. When evergreens are to be exposed to high winds, spring transplanting is better than fall. Among evergreen trees that are more successfully moved in the spring are: fir, American holly, and hemlock. Greatest success in transplanting evergreens is attained when the soil is warm and root growth continues immediately. Such conditions prevail in most parts of the country in early fall and late spring.

In general, evergreen trees as a group are best moved earlier in the fall and later in the spring than deciduous trees. The roots of evergreen trees cease to grow, or at least make very little growth, during dry summer periods. With the advent of late summer or early fall rains, the roots begin to grow again.

The roots of deciduous trees are capable of growing even when the soil temperature drops to as low as 45 degrees F. This means that deciduous trees transplanted before the soil gets cold can make some root growth in the fall and even during some of the winter months.

Preparing Trees for Transplanting

Nursery-grown Trees. Young trees develop few roots. These do not branch readily but spread through relatively large areas of soil and are thin and scraggly. Unless the growth of such roots is regulated by pruning, the trees can be dug and transplanted only with great difficulty. Properly grown nursery trees are root-pruned every few years, not only to confine the greater part of the root system within a small area but also to encourage the roots to branch and to develop a greater abundance of small roots and rootlets. Trees that have been properly root-pruned can be transplanted most efficiently and with maximum chances of success.

Collected or Wildling Trees. Many trees growing naturally in fields or woods without previous cultivation are transplanted to new sites to provide shade or ornament. These have fewer roots, many of which are long and scraggly. Trees growing in medium heavy or clay soil and in open areas are preferable as subjects for transplanting to those growing in sandy soil or in heavily wooded areas. A tree growing in the woods whose top has been protected by other trees and whose roots are shallow because of the protection by

FIG. 5. Method of root-pruning a wildling tree before it is moved. The soil is dug at sectors *A*, the exposed roots are pruned, and the soil is replaced after having been improved by fertilizing or manuring. One year later the same procedure is followed in sectors *B*. The tree should be moved a year after the pruning at sectors *B*. UPPER LEFT: Sketch shows the extent of the roots before root pruning. UPPER RIGHT: After root pruning.

shade and mulch is more apt to do poorly when moved out into the open and exposed to full sunlight and drying winds.

As a rule, very small collected trees are not root-pruned. Large trees, on the other hand, can be root-pruned to stimulate the formation of a compact root system and a greater number of fibrous roots. Best results are obtained when such trees are root-pruned, as shown in Figure 5, for two successive years before the actual moving operation. In the spring or fall, a ditch is dug with a spade at some distance from the tree,* and all roots are severed in this area. A year later, the remaining area is dug and the roots are treated similarly. The soil excavated each time should be mixed with manure, leaf mold, or a commercial fertilizer (see page 50) before it is replaced in the open trenches. The tree is ready for moving one year after the second series of trenches are dug and refilled. Trenching and root-pruning in this manner encourage greater root development within the cut area. They also spread the shock to the tree over a 3-year period rather than concentrate it in one severe shock, as is the case when trenching and moving are done at one time.

In commercial arboricultural practice, however, root-pruning a year or two in advance of moving is rare, principally because the client usually demands immediate delivery of the tree.

Digging Deciduous Trees and Evergreens

Proper digging includes the conservation of as much of the root system as possible, particularly of the smaller roots and finer rootlets. Any appreciable reduction in such roots, which are the most active in absorbing water and nutrients, results in a considerable checking of the tree's growth and reduces the speed of re-establishment.

Deciduous trees may be moved with bare roots or with balls of soil adhering to the roots (Fig. 6). The former method leads to greater transplanting shock and slower re-establishment; on the other hand, less equipment is required. Evergreens are always more safely moved with a soil ball than with bare roots (Fig. 7).

Before any digging begins it is advisable to tie in the tree's branches with heavy twine or quarter-inch rope. This will reduce the chances of damage or breakage to the branches.

* The distance varies with the type and size of the tree. For deciduous trees the radial distance from the trunk should be about 5 inches for each inch in trunk diameter. For example, the radial distance from the trunk to the point where the ditch is dug on a tree 4 inches in diameter at breast height should be about 20 inches.

FIG. 6. UPPER: Close-up showing a balled and burlapped deciduous tree ready for transplanting.

FIG. 7. LOWER: Balled and burlapped evergreens ready for transplanting.

Bare Root. In preparing trees to be moved free of soil, a trench is dug around the tree to a depth just below the greater part of the root system. With most trees this depth is usually 2 or 3 feet. The circumference of the trench should be just beyond the last root-pruning zone for well grown nursery trees, or just outside the root-pruning trenches for collected stock. If no previous root-pruning has been done, as is the case with most trees, the trench should be dug at a radial distance from the trunk of approximately 5 inches for each inch in trunk diameter. The roots are then combed by inserting a spading fork into the exposed face of the trench and pulling the upper end of the fork toward the operator. After the smaller roots are exposed, the combing is continued until the larger roots are cleared of soil to within a foot or so of the trunk. One part of the root system should be uncovered and freed of soil completely before another area is attacked.

As the roots are completely uncovered, they should be wrapped in wet burlap to prevent drying and mechanical injury. Each part of the root system should be uncovered and wrapped separately. If the soil is left intact close to the tree, sufficient anchorage is maintained to hold the tree erect until most of the roots have been uncovered and wrapped.

The tree is then bent over carefully to raise the roots, and the entire root system is tied in a large ball with wet burlap and rope. If the roots are to remain wrapped for several days, wet sphagnum or peat moss should be placed around them before they are enclosed in the burlap.

Soil Ball. Whenever possible, deciduous trees over 3 or 3½ inches in diameter should be moved with the soil adhering to the roots, the circumference and depth of the soil ball depending largely on the type of soil, the root habit and the type of tree, and several other factors. As a rule, trees with shallow roots require a flat ball, while those with deeply penetrating roots require a deeper ball with a smaller circumference. The main objective governing the size of the earth ball is the protection of the greatest number of roots in the smallest soil mass.

The soil ball for deciduous trees is usually 9 to 12 inches in diameter for each inch diameter of the trunk. As a rule, evergreens require a ball of smaller circumference.

Since most soils weigh about 110 pounds a cubic foot, no more soil than is absolutely necessary should be allowed to remain around the ball. A rapid method for computing the weight of a soil ball is to square the diameter of the ball in feet, multiply this figure by the depth in feet, then take two thirds of this total and multiply by 110. The resulting figure will give the approximate weight of the soil in pounds.

Digging should not be attempted unless the soil is moist. If the soil is dry it should be watered thoroughly at least two days before digging. Unless this is done the soil ball will break and the roots will be completely separated from the soil. Such a tree will have much less chance of surviving in its new location. If considerable time is to elapse before evergreens are to be replanted, they should be placed in shaded areas and the soil balls watered frequently.

The preliminary preparations for moving a tree with a soil ball are similar to those for bare-root trees. The trench is dug and the roots are combed out and then severed, both around and underneath, at approximately the predetermined size of the ball. The ball of earth is then held in place by wrapping it with a large piece of burlap or tarpaulin and securely tying this with rope. A tree thus prepared is known in nurserymen's catalogues as a balled and burlapped (B & B) tree. Because the soil ball is easily broken and the root system injured, only an experienced person should be entrusted with this task.

Table III

THE RECOMMENDED MINIMUM DIAMETER AND DEPTH OF SOIL BALLS FOR DECIDUOUS TREES OF VARIOUS SIZES

Trunk Diameter of Tree 1 Foot above Ground (inches)	Diameter of Ball (inches)	Depth of Ball (inches)
1¼-1½	18	14
1½-1¾	20	15
1¾-2	22	16
2-2½	24	17
2½-3	28	18
3-3½	33	20
3½-4	38	23
4-4½	43	26
4½-5	48	30
5-5½	53	31
5½-6	58	33
6-7	65	35

Several other methods of preparing and conserving the ball are used in different sections of the country. Where the soil is sandy and crumbles easily, the so-called box method is commonly employed. After a square trench is dug around the tree, four pieces of board are nailed together to form the sides of a box. The soil underneath the root system is dug out, and additional boards are slid under and fastened in place to complete the box.

Arborists interested in obtaining a detailed description of additional balling, wrapping, and roping methods are referred to the excellent treatment of this subject in *Tree Preservation Bulletin Number 1*, of the United States Department of the Interior, National Park Service. This publication is obtainable from the Superintendent of Documents, Washington 25, D. C., for 25 cents. Another comprehensive discussion on how to ball tree roots is presented in the *Proceedings of the 19th National Shade Tree Conference*, available for a small fee from Dr. L. C. Chadwick, Secretary of the Conference, Department of Horticulture, Ohio State University, Columbus, Ohio.

Container-Grown Trees

In recent years a new technique for handling trees and shrubs has proved popular with home gardeners and profitable for nurserymen. Trees are grown in containers just prior to sale. The practice, referred to as 'can culture,' originated in California, where the sandy soil in which trees and shrubs are grown does not lend itself to 'ball and burlap' handling because it breaks away from the roots.

The can culture method of growing and selling trees has several advantages over the ball and burlap method. First, it extends the planting season. Even in June and July, a gardener can plant a shrub or tree grown in a can. It is much like planting a pot-grown geranium or fuchsia in mid-summer.

A can-grown tree suffers far less transplanting shock than a field-grown balled and burlapped one. Few if any roots are lost from the former. Container-grown plants need not be planted immediately, as do bare-root or even balled and burlapped plants.

A nurseryman finds container-grown plants profitable because his selling season is extended, he can display more plants in a given space, his watering problem is less complex, and he can hold any surplus over for another season with less extra cost.

As with all other innovations, can culture has certain disadvantages. First, the size of tree is limited. Very large trees cannot be grown in cans, at least under present methods. For trees more than 10 feet high the cans would be too heavy and awkward to handle. Second, can-grown trees and shrubs held too long in their containers tend to become pot-bound. Unless such pot-bound plants are handled properly, they may not make satisfactory growth after they are removed from their containers and set into the garden. The roots of pot-bound trees should be spread out at planting time to reduce the possibility of girdling roots (p. 181) later on.

The metal containers used today are a great improvement over the original ones of a few years ago. They are so constructed that the gardener merely taps the edge of the can on a firm surface to release the plant. Both plant and soil then slide out as a unit.

Transporting

Small trees are transported very easily. Large trees usually require special types of transportation equipment. A sled with wooden runners can be used where the trees are to be transported for relatively short distances.

A number of well-designed and efficient tree-moving machines are now on the market. These are used mainly by commercial arborists, park superintendents, and other persons moving large trees. The machines are of two distinctly different types—trailer and automotive. The trailer types have 2, 6, or 8 wheels, and are drawn by trucks or tractors. Many of the trailer types of tree movers are designed to carry most of the weight on the rear wheels, the soil ball of the tree resting on a cradle just ahead of these wheels. The front wheels balance and guide the trailer. The automotive types of tree movers are mounted on trucks. At least five different types are in common use, one of which is illustrated in Figure 8.

The following firms manufacture one or more of the various types of tree movers: J. G. Butts Manufacturing Company, Cato, N. Y.; Gar Wood Industries, Inc., Detroit, Mich.; Charles F. Irish Co., 418 East 105 Street, Cleveland, Ohio; Williams and Harvey Co., 7313 Mission Road, Kansas City, Mo.; and Jeffrey Manufacturing Company, Falmouth, Kentucky.

The limit to the size of tree that can be moved is governed by the amount of road clearance over which the tree is to be transported, and the weight the roads and bridges will stand. The number of men needed for moving trees varies with the size of the specimens. As a rule, 3 to 7 men are needed for trees ranging from 8 to 24 inches in diameter.

Considerable injury may occur during the moving of large trees unless the following precautions are taken:

Pad unprotected areas with burlap, canvas, or some other material to avoid bruising and slipping the bark of the trunk and branches.

Carefully tie in all loose ends with a soft rope to avoid twig and branch breakage.

Keep the earth ball moist or cover bare roots with wet burlap or moist sphagnum moss.

Avoid excessive drying of the tops, especially evergreens, while the trees are in transit, by covering them with canvas or wet burlap, or spraying them with an anti-transpirant like Wilt-Pruf (p. 45).

FIG. 8. A modern tree-moving machine ready to place a large evergreen tree in the transplanting hole.

PLANTING

Preparation of the New Site. It has often been said that if a person has $10 to spend on a tree, he should spend $2 for the tree and $8 to prepare the planting hole. This advice may be slightly exaggerated but the fact remains that the care taken in the preparation of the new site is the major factor ensuring ultimate success in transplanting.

The hole should be dug sufficiently deep and wide to accommodate the full root system of the tree. In especially poor soils, it should be larger and deeper. As the hole is dug, the soil from the upper layers, which is richer, should be kept separated from the subsoil. The latter should be discarded and replaced with good topsoil, or improved by incorporating liberal quantities of well-rotted manure or peat moss.

Trees planted in heavy or poorly drained soils will not thrive unless some precautions are taken prior to planting. Properly installed 3-inch agricultural tile will help. But only a person skilled in the mechanics of laying tile drains and with a knowledge of subsoil conditions should be entrusted to a job of this sort. Another practice is to blast the holes with dynamite after all digging is completed. This fractures the deeper, more impervious layers and thus not only improves the drainage but allows deeper root penetration later.

Setting the Tree. The depth of planting must be close to the original as measured by the distance from the soil-stained ring at the base of the trunk to the base of the root bundle. Experienced arborists have found that if a tree is set a bit higher in its new location than in its former one, the chances of continued good growth are increased. It is a well-known fact that trees set too deeply will not thrive and may even die within a few years. Flowering dogwood, American beech, and narrow-leaved evergreens are especially sensitive to too-deep planting.

The soil to be placed around the roots should be neither too dry nor too wet. These extremes can be avoided by adding water if the soil is too dry, and peat moss if too wet.

Do not set the tree until the hole depth has been adjusted by adding or removing the soil and until the soil to be worked around the roots is in proper condition.

Because the roots of trees grow downward at a slight angle from the horizontal, it is wise to make a cone-shaped mound of soil at the bottom of the planting hole. The root crown of bare-root trees can then be set on this mound and the roots spread over and down the sides, thus assuring close contact with the soil along their entire length. The soil should be added gradually. Work the first lot in firmly at the base of the roots. Then add more soil and work it under and around the lower roots. The tree may be gently raised and lowered during the filling process to eliminate air pockets and bring the roots in close contact with the soil. Some water can be added also to eliminate air pockets. When the roots are covered, tamp the entire area firmly with the feet (Fig. 9). Heavy soils should not be tamped too firmly around the roots; otherwise, oxygen and water cannot reach the roots.

Balled evergreen and deciduous trees require somewhat different handling.

After they are set at the proper level in the hole, the rope holding the burlap in place should always be removed and the burlap loosened. If there is no danger of breaking the ball, the burlap should also be removed, or it may be loosened from the sides and left in the soil to decay. Soil is then added to half fill the remaining hole space, and firmly tamped. Water should be added to settle the soil further and to eliminate air pockets. When the water has drained off, the remainder of the hole is filled with soil (Fig. 10).

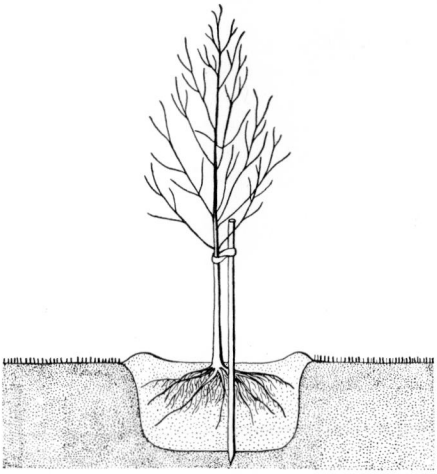

FIG. 9. LEFT: The correct placement of a tree moved bare-root.
FIG. 10. RIGHT: The correct placement of an evergreen moved with a soil ball.

The final level around any type of tree should be a few inches below the surface of the surrounding ground, to form a shallow basin for catching and preventing run-off of water. The ground should be leveled off, however, before winter sets in, since water collecting in the basin may freeze and injure the trunk.

Supporting. Newly planted trees need some artificial support to prevent excessive swaying by the wind, which would disturb the roots and hinder their normal functioning and development.

For trees up to 20 feet in height, 1 or 2 strong stakes or poles, 6 to 8 feet in length, can be driven 2 feet or so into the ground and 6 to 12 inches from the trunk. A short length of old garden hose with a wire run through it, is then placed around the trunk to protect the bark, and the wire is twisted around the stake in the form of a figure 8 to complete the support (Fig. 11).

Two Methods of Supporting Newly Transplanted Trees

FIG. 11. UPPER: Wire run through a piece of garden hose and twisted around the trunk and the supporting stake in the form of the figure 8.

FIG. 12. LOWER: Paul's Tree Brace is adjustable and quickly set in place.

A wide cloth tape may be substituted for the wired hose. This is wound around the tree and the loose ends nailed to the supporting stake. A patented brace, Paul's Tree Brace (Fig. 12), is available from Colonial Gardens, Delaware, Ohio.

Guy wires or cables must be used to support larger trees. They should slope from about halfway up the trunk to the ground at an angle of about 45 degrees. The upper ends of the wires are attached with lag screws or by encircling the trunk with a loop encased in a piece of garden hose or other protective cover. The lower ends of the wires are anchored to stakes driven deeply into the soil. From one to four wires or cables are usually placed around a large tree.

Guy wires or cables are rarely placed around trees set along the city streets or in public parks. Unless protected or plainly marked so as to be readily visible, especially at night, they are a constant source of danger to passers-by.

Pruning. The tops of transplanted trees are pruned to compensate for the loss of roots and thus to maintain a more natural balance between the two parts. Pruning should be heavier on collected than on nursery-grown trees. The amount of pruning * depends not only on the condition of the tree at transplanting time, but also on the care the tree is to receive while it is becoming firmly re-established. Less pruning is needed when most of the roots have been conserved during digging, and when adequate water and after-care are assured.

Pruning should be confined to the secondary branches in order to maintain the natural shape of the tree. The growing tip of single-stemmed shade trees should not be cut back. Twigs and buds along the branches and upper part of the trunk should not be removed until the tree is well established, for these tissues manufacture much-needed food and, by providing shade, help to reduce scalding of the trunk and branches.

Little or no pruning of balled evergreens is needed except to remove broken or severely injured branches. If the leading shoot is broken during moving, one of two practices may be followed: (1) one of the next nearest lateral branches may be bent upward and held in place by tying it with raffia to a small piece of wood; or (2) all the nearest laterals but one may be re-

* Some municipal tree men prune transplants drastically. For example, all branches except the leader on a tree 2 to 2½ inches in diameter are pruned, so that only stubs ranging from 4 to 12 inches remain on the bole. In general, as much as 30 to 40 per cent of the leaf-bearing wood may be removed.

moved. The lone lateral will tend to grow upright and will soon replace the broken leader.

Watering. The soil around the tree must be watered thoroughly after the tree is set in place. Additional water must be applied from time to time until the tree is firmly established. The frequency of watering depends on the type of soil, the size of the tree, and the amount of rainfall. Deciduous trees moved during late fall or winter require very little water at least until new leaves emerge the following spring.

The water must be allowed to soak into the ground slowly until the soil is moistened down to the roots. This can best be determined with a soil auger or by digging with a narrow spade. The tree benefits little from frequent light waterings which moisten the soil to a depth of only a few inches. If water must be applied with a pail, it is best poured on 2 or 3 pailfuls at a time and allowed to soak in completely before more is added.

One common error of inexperienced gardeners is to overwater the soil around the newly transplanted tree. This drives out the oxygen from the soil, an element that is essential to normal root development, and it also favors the development of root-decaying fungi. The soil around newly planted trees should be kept relatively dry in spring until root growth starts and the leaves begin to emerge.

The soil ball around a large newly transplanted tree, however, can dry out in a surprisingly short time. Summer showers cannot be depended upon to supply sufficient moisture to keep the soil ball moist. Hence an occasional good soaking with the garden hose is necessary. To get good penetration into the ball itself, it may be necessary to punch holes into the ball with a crowbar or to bore holes with a soil auger at varying depths before applying the water.

Frequent syringing of the leaves of newly planted evergreens on cloudy days helps to cut down water loss by the leaves, and in some cases washes off spider mites and soot.

Mr. Edgar Rex of the New Jersey Department of Agriculture and Mr. C. N. Keyser of the Bartlett Tree Expert Company recently devised a method for facilitating the watering of newly planted shade trees. It involves the establishment of two vertical porous cores containing ½- to 1-inch mesh dolomitic limestone or crushed stone. Through these cores rainwater or applied water can more easily reach the critical root zone. Details of installing the Rex-Keyser method of subsurface irrigation are illustrated in Figure 13.

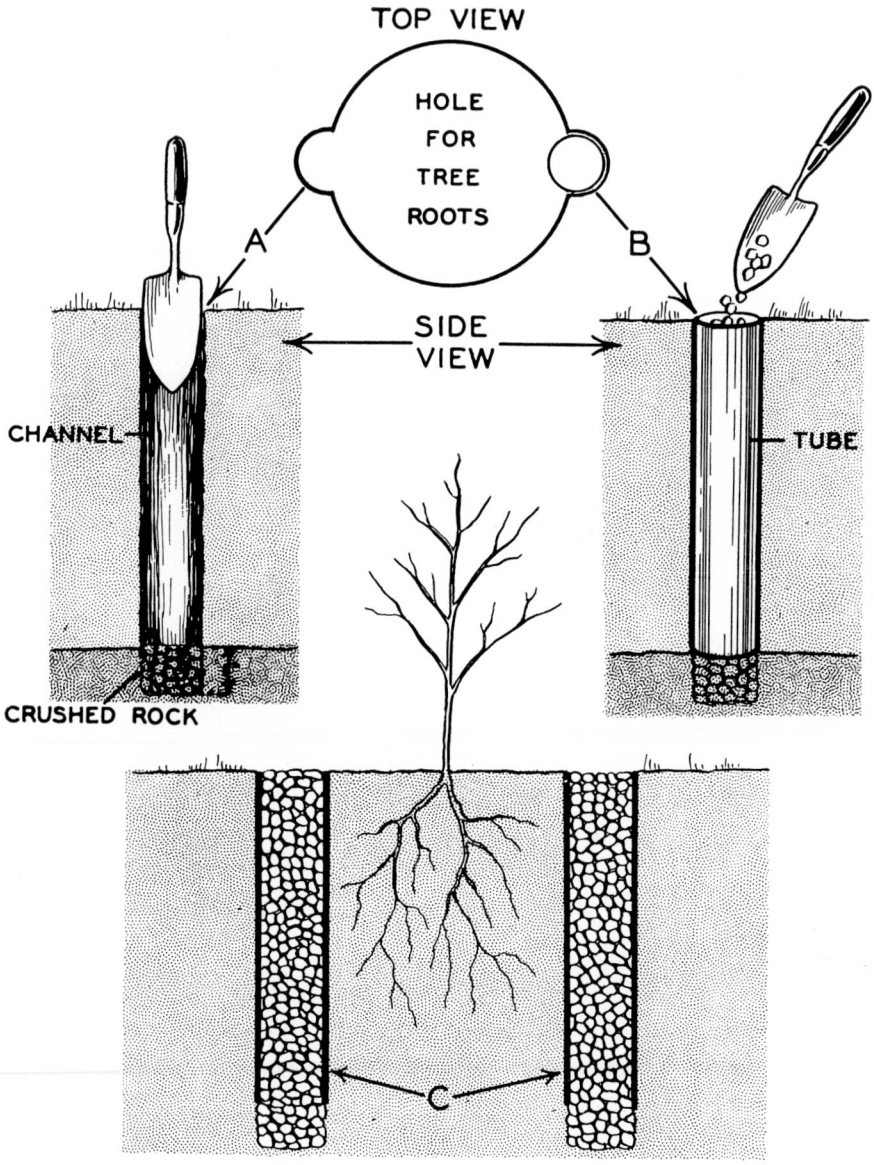

FIG. 13. Diagrammatic sketch showing Rex-Keyser method for facilitating watering of newly planted trees: After the hole is dug, semi-circular areas are cut into the sides of the hole. Cylindrical cardboard tubes approximately 18" x 3" x .070" are inserted above a few inches of crushed rock or dolomitic limestone, with the upper ends flush with the soil surface. The tubes are filled with the crushed stone and the top of the tubes are covered until the tree is planted in the normal manner. The lower sketch (C) presents a cross section of the completed installation.

Mulching. With particularly valuable trees, especially evergreens planted in the fall, a good practice is to mulch the area above the roots with straw, leaf mold, or well-rotted manure. Such mulches prevent wide fluctuations in soil temperature, keep the soil warmer, and help to conserve moisture. They are usually left on over winter and either removed or worked into the soil in spring. A newly planted deciduous tree collected from the woods should be mulched to a depth about equal to that of the mulch in its former location.

Sand is sometimes used as a mulch in the depressed area around small newly planted trees along streets. Water is thus absorbed more rapidly and is held longer. The sand also keeps the area free of weeds and is said to increase the chances of the tree's survival.

Wrapping. The trunk and larger branches of a transplanted tree should be wrapped with burlap, specially prepared crepe paper, or some other material to prevent sunscald and drying of the bark and to reduce the possibility of borer infestation. A protective covering is especially desirable on trees collected from shaded woods, since the bark of such trees is extremely susceptible to sunscalding. The covering should be securely fastened in place with cord and left on two years or until it falls off.

Wrapping the trunk of a newly transplanted tree has one disadvantage. In rainy seasons the trunk is kept unduly wet beneath the wrapping, a condition which fosters the development of fungus cankers. The author has found that pin oaks are especially sensitive to such cankers. Perhaps a coating of some fungicide on the trunk prior to wrapping might help to reduce such infections.

Any wires that hold labels should be loosened or removed so that they do not girdle the trunk or branch as it increases in girth.

Fertilizing. As a rule, dry commercial fertilizer should not be applied until new roots are formed and can absorb it. It is best to wait until the end of the first growing season before applying dry fertilizer. However, a dilute solution of a completely balanced chemical fertilizer can be added to the soil the first summer after transplanting, for by that time the tree will have formed sufficient roots to absorb the nutrient solution. A more immediate response to nutrients can be obtained by spraying the leaves with a plant food, such as Ra-pid-gro or Agrico Water Soluble Plant Food, which is formulated especially for foliar applications. Start as soon as the leaves are half grown, and repeat 4 or 5 times at 10-day intervals.

Anti-transpirants

Loss of water from evergreen trees or deciduous trees in leaf can be partially overcome by spraying the trees with Wilt-Pruf just before moving them. This product is a liquid plastic which is diluted in water and then applied over the leaves and stems with a pressure sprayer. It dries in 20 minutes or so to form a protective film that helps to reduce the loss of water from the leaves.

Trees sprayed with Wilt-Pruf must still be watered in their new location, although perhaps not so frequently as untreated trees. Treated trees become established in their new locations more quickly than untreated ones.

For use on evergreen and deciduous trees, one part of Wilt-Pruf is mixed in 4 to 6 parts of water. The diluted mixture is then applied as a fine spray at air temperatures above freezing. According to the manufacturer, the material needs to be applied to the upper surfaces of the leaves only.

Regular spraying equipment, except that with gear pumps, can be used to apply Wilt-Pruf. After use, the sprayer should be rinsed thoroughly with soapy water. Otherwise, the Wilt-Pruf will harden and clog the nozzle and other parts should the gardener forget to rinse out the sprayer. A special solvent is available to clean clogged parts. Unused Wilt-Pruf must be stored where it will not freeze.

According to the manufacturer, Wilt-Pruf has not been known to injure humans and pets. Nor is it likely to cause foliage burn when used as directed. Many professional arborists and nurserymen now use Wilt-Pruf whenever they move trees out-of-season. They find that it largely eliminates the gamble of moving large trees in leaf.

Although the use of this product assures more successful transplanting of trees, it should not be considered a substitute for good gardening practices. Carelessly handled trees and those with poor root systems will not survive transplanting shock nearly so well as those that are properly handled and have an adequate root system.

4

FERTILIZERS AND THEIR USE

Many abnormalities in trees are closely associated with plant-food deficiencies. Professional arborists recognize the benefits trees derive from periodic feeding and for this reason always stress the importance of such a practice. They also appreciate that although fertilization does not constitute the whole of soil improvement, it is an important part of any successful tree maintenance program.

Many laymen are unaware that trees should be fed in much the same manner as smaller plants and vegetables. Woodland trees thrive despite the absence of artificial feeding, but many of our shade and ornamental trees do not grow in such a favorable environment. Their leaves are gathered and burned in the fall, thus destroying an important food source. Concrete or other impervious pavements usually cover a considerable area that otherwise might be good hunting ground for hungry roots. Water often becomes a limiting factor for good growth, even where trees are growing on open lawns. The heavy sod uses much of the available water and food that might otherwise be utilized by the trees. Moreover, lawns inhibit good soil aeration.

The effects of undernourishment are usually evident in the gradual decline in the vigor of the tree. This decline, which at times may be very rapid, is hastened by the more serious effects of fungus and insect attack.

Until recently the problem of feeding shade and ornamental trees was not so clearly understood as was the fertilizing of fruit trees and farm crops, principally because less research has been done. Within the last 20 years, however, agricultural experiment station workers and some of the larger tree companies have investigated the problem more carefully, with the result that general recommendations on shade and ornamental tree fertilization can now be made. Nevertheless, many questions are still unanswered, and considerable work has yet to be done.

The application of fertilizers improves the appearance and condition of trees and in some cases enables the trees better to withstand future attacks by parasites. Recovery from mild cases of Verticillium wilt of maples (see p. 252), in which the fungus invades the vascular tissues, has been repeatedly observed following liberal applications of high nitrogenous fertilizers. It should be clearly understood, however, that fertilization exercises no 'cures' upon trees affected by fungus leaf spots, heartwood rots, or bark and cambium diseases.

Not all trees respond to fertilizer to the same degree; oak, elm, walnut, and basswood are among those which usually show the quickest response.

How Fertilizers Act. Contrary to the general belief, tree roots do not obtain food for growth directly from the fertilizers added to the soil. The food for any part of the tree is obtained from the leaves. Fertilizing materials supply elements necessary for the maintenance of the bodies that constitute the green coloration in leaves. As already discussed in the first chapter, these green bodies (chloroplasts) have the ability to change the light energy from the sun into the chemical energy of foods. Obviously, a dark-green leaf can manufacture and store more food than a yellow one or one attacked by fungus or insect pests.

To cope with the problem of feeding trees, it is necessary to understand not only the fundamental requirements of plants but also the influence of the various elements upon the development of plant tissues. With such knowledge one can better adapt the fertilizer recommendations to the soil and other environmental conditions in which the particular tree is growing.

The Essential Elements for Food Manufacture

Plant foods and tissue-building compounds of plants are manufactured from the following raw materials: carbon, oxygen, hydrogen, nitrogen, phosphorus, sulfur, potassium, calcium, magnesium, iron, and several so-called trace elements, such as boron. Of these, only carbon, oxygen, and hydrogen are essential in the first step of food manufacture. As already discussed in the first chapter, carbon and oxygen are provided by the carbon dioxide in the atmosphere, which enters the leaves through the stomata or breathing pores. Hydrogen is obtained from water absorbed through the roots and carried to the leaves through the water-conducting system. These three elements are then united by means of energy obtained from the sun by the chloroplasts in the leaf. The results of this reaction are expressed in the simple equation:

$$6\,H_2O + 6\,CO_2 = C_6H_{12}O_6 + 6\,O_2$$

6 parts water plus 6 of carbon dioxide gives 1 part sugar and 6 of oxygen

This equation represents mainly the gaseous exchange involved, and not the complex nature of the process.

Nitrogen is taken out of the air and fixed or assimilated by certain kinds of soil bacteria. Trees in turn obtain nitrogen from decaying organic and inorganic nitrogen in the soil.

All other elements required by trees are taken from the soil. These are absorbed by the roots only in solution in the manner already described in Chapter 2. Obviously plants will absorb many elements from the soil solution that are not used for food manufacture and tissue formation.

From the standpoint of their relative value as raw materials for plant food, the elements in the soil may be divided into three classes: nonessential; essential and abundant; and essential and critical.

The nonessential elements include silicon (Si), aluminum (Al), sodium (Na), and some of the rarer elements. These elements are believed to have no role in the nutrition of plants.

The second group includes iron, calcium, magnesium, and sulfur, all essential to plant growth but usually present in soils in ample amounts. One or more of these elements may occasionally be deficient, however. For example, pin oaks in alkaline soils often develop yellow or chlorotic leaves because of a deficiency of iron or the inability to absorb it. Such a condition can be corrected by the addition to the soil of iron salts and sulfur or by injections of certain iron salts into the trunk and by the use of iron chelates.

The third class includes nitrogen, phosphorus, and potassium, and in some cases magnesium, manganese, and boron. These elements, which are essential for plant growth, are usually present in the soil in only small quantities and are often the limiting factor in plant growth. This is especially true of nitrogen. Nitrates, the source of nitrogen, are soluble in water and are readily leached by drainage water. During very rainy seasons considerable nitrogen (as nitrates) is lost in this manner, whereas in dry seasons this material accumulates. All commercial fertilizers have as their main constituents the first three elements in this important group. The percentage of these elements in any commercial fertilizer is expressed by formulas as 12-8-6, 6-8-4, or 10-8-6, the first number denoting the percentage of nitrogen; the second, the phosphoric acid; and the third, the potash. Recently many fertilizer companies have added small amounts of the rarer but essential elements such as boron, magnesium, and manganese to their standard formulas. These are then sold as complete fertilizers with 'minor elements' added. Only an expert is capable of diagnosing troubles due to deficiencies of the minor elements. Descriptions of some of the deficiency symptoms are available in experiment station publications to persons interested in pursuing this subject further.

The Influence of Essential Elements on Growth. The essential elements exert various influences on plant growth. Some of these influences are not clearly understood, and others are difficult to detect, but the knowledge available will give some basis for fertilizer recommendations and some indication of what may be expected from these applications. Only the main functions of the various elements will be considered here.

Carbon, hydrogen, and *oxygen* are constituents of all sugars and starches. They, of course, are not added in fertilizers but are obtained from the air or through the soil.

Nitrogen is a constituent of all proteins that are active chemical parts of protoplasm. Lack of nitrogen results in small yellow foliage. This element is the principal fertilizer constituent involved in the production and maintenance of the green color in leaves. If in proper balance with other nutrients, it will produce abundant twig as well as leaf growth. Overdoses of nitrogen may damage roots, especially when the soil is rather dry.

Nitrogen is obtained from sodium nitrate, calcium nitrate, ammonium sulfate, and such organic materials as cottonseed meal, tankage, dried blood, and activated sludge. The organic fertilizers can supply nitrogen only after they are decomposed by the soil organisms (principally bacteria), and for this reason they are the more slowly available but longer lasting sources of nitrogen. The new ureaform nitrogens also supply nitrogen at a very slow rate.

Orchard soils do not usually show a great shortage of nitrogen, since cultivation and turning under of cover crops tend to increase the supply. In contrast, nitrogen is usually deficient in the conditions under which shade and ornamental trees grow. Nitrates are present in the surface layers of the soil, but only if there is no competition from the sod or small shallow-rooted plants. Lawn clippings should be left where cut, since they supply some food to trees growing in lawns.

Phosphorus has a less obvious role than nitrogen. It stimulates root growth, hastens the maturation of tissues, stimulates flower and seed production, and is an active part of certain proteins. In addition, it helps to make stored carbohydrates, such as starch, available for plant growth in spring. Phosphorus, as used by plants, is in the form of phosphates, which penetrate into the soil very slowly when applied to the surface. Consequently fertilizing materials containing phosphates should be worked into the soil or inserted into the zone occupied by the roots. Phosphorus is most readily available from soils containing abundant amounts of organic matter. It becomes less available in highly acid or highly alkaline soils.

Potassium or potash must be present for the manufacture of sugars and starches and for the reactions that are necessary before these materials can move from one part of the tree to another. It helps to overcome succulence and brittleness, hastens maturity of the plant, and intensifies the color of flowers.

Calcium exerts an influence both in the plant cells and in soil. In the plant it is essential for growth and cell divisions. In the soil it overcomes harmful effects of excessive amounts of aluminum and iron salts. In addition, it favors development of beneficial bacteria and aids in the liberation of nitrogen, phosphorus, and potassium.

Magnesium aids in the transportation of phosphorus and is a chemical constituent of chlorophyll. Without it, leaves become yellow.

Iron is essential to the formation of the all-important chlorophyll. Though this element is not a constituent of chlorophyll, as is magnesium, plants lacking it have yellowed foliage.

Sulfur is a constituent of plant proteins.

Molybdenum acts as a catalyst in enzyme systems that function in reducing nitrate to ammonium in preparation for the synthesis of amino acids and proteins.

Vanadium is the latest element to be found essential for normal leaf growth.

Fertilizing Materials and Formulas

One of the best materials for small trees is well-rotted stable manure. This is not used extensively, however, because its application is limited to open areas and because it is not available in large quantities. Commercial fertilizers are more readily available and can be used under most conditions.

No general agreement exists among scientists or commercial arborists as to the best formula for tree fertilizer. Some claim a tree cannot be over-fertilized, whereas others claim that severe injury may result from too much fertilizer. Virtually all agree on one point, however, that a complete mix, containing the three critical elements, nitrogen, phosphorus, and potassium, should be used.

There is much disagreement on the amount and source of nitrogen. Some workers maintain that one third to one half of the nitrogen must be derived from organic sources, such as cottonseed meal, tankage, or dried blood. Others claim this is unnecessary, that the nitrogen can come entirely from inorganic sources, such as ammonium sulfate, nitrate of soda, or calcium nitrate.

Most of the recent research on shade-tree fertilization indicates that fertilizers with the approximate analysis 5-10-5, which are commonly used for vegetable and other commercial crops, are not entirely suitable for shade trees. These investigations reveal that formulas with higher amounts of nitrogen—for example, 10-8-6 or 10-6-4—are more effective. Some of the larger tree companies have especially prepared formulas which they feel are most effective.

It should be borne in mind that most ornamental trees grow naturally in an acid soil. For this reason, commercial fertilizers tending to leave an alkaline residue should not be used. If the soil reaction needs a change toward the alkaline side, this can best be accomplished by adding lime.

A SUGGESTED FORMULA FOR MIXING A 10-8-6 FERTILIZER

Materials	Amount Required (lbs.)	Plant Nutrients in 2,000 Pounds		
		Nitrogen (lbs.)	Phosphoric Acid (lbs.)	Potash (lbs.)
Nitrate of Soda, 16% N	600	96	..	..
Sulfate of Ammonia, 20.5% N	280	57.4	..	..
Urea, 48% N	40	19.2	..	..
Tankage 6% N, 2% P_2O_5	470	28.2	9.4	..
Superphosphate 42%	360	..	151.2	..
Muriate of Potash, 48%	250	..	..	120
TOTAL	2,000	200.8	160.6	120
ANALYSIS	..	10	8	6

The mere insertion of fertilizer into the soil around the tree is no guarantee that the tree will benefit. The fertilizer must be placed where the tree's roots can use it. The reason is that the various components of fertilizers do not move readily through the soil. Unless they are placed in the immediate vicinity of roots, they will do little or no good. Nitrate nitrogen is perhaps the only ingredient in a complete fertilizer that will move to areas beyond its application. The other major essential nutrients such as calcium, magnesium, phosphorus, and potassium are absorbed on the surface of the soil particles, and the plant roots must come in close contact with these particles before the nutrients can become available.

Rates of Application

Opinions differ on how much commercial fertilizer should be applied. Recommendations range from 1 to 8 pounds for each inch in diameter of the trunk at breast height. Probably a figure about midway between these extremes is the best general recommendation. With this quantity, there will be no danger of root injury, and if it does not produce the desired results, an additional application can be made in a year or two.

A safe method to follow, then, is to use about 2 to 4 pounds of complete fertilizer for each inch in diameter of the tree trunk at breast height. Small trees (less than 6 inches in diameter) should have half this dosage. For example, a tree 10 inches in diameter should receive 20 to 40 pounds (2 to 4 pounds per inch diameter) of commercial fertilizer, whereas a tree 4 inches in diameter should receive 4 to 8 pounds (1 to 2 pounds per inch). When grade changes or building operations have restricted the feeding area, most arborists recommend slightly larger amounts. For large trees, as much as 4 to 5 pounds of 10-6-4 per inch in trunk diameter may be used.

In determining what amount of fertilizer to apply, one should consider the age of the tree and what one desires to achieve. If the tree is young and the owner wants it to grow as rapidly as possible, then extra food is called for. On the other hand, relatively less fertilizer would be applied to an older tree that must be kept in good vigor but with further growth restricted.

Fertilizing Narrow-leaved Evergreens

At one time commercial fertilizers were not generally applied to evergreens because they were supposed to be extremely toxic. It is now known that good growth responses can be obtained with these materials. The only precaution necessary is to avoid excessive applications.

Since small evergreens may be occasionally injured by commercial fertilizers, it is probably safer to apply nitrogen in some organic carrier at first. For small narrow-leaved evergreens, about 5 pounds of tankage or cottonseed meal should be applied to each 100 square feet of bed area in the spring. The fertilizer should be hoed or watered in, so as to mix readily.

For large narrow-leaved evergreens, commercial fertilizers can be used with safety. From 2 to 4 pounds of 10-6-4 fertilizer should be applied to every 100 square feet of bed area, if the trees are in a group. For large evergreens standing in open areas, about 2 pounds of this fertilizer can be applied for each inch in diameter of the trunk.

Fertilizing Broad-leaved Evergreens

The problem of fertilizing broad-leaved evergreens is more complex. Such plants as azalea, rhododendron, laurel, and leucothoe require a distinctly acid soil. For this reason fertilizing materials that tend to sweeten the soil should be avoided. Greatest success with broad-leaved evergreens is obtained when liberal quantities of humus are incorporated in the soil and applied as a mulch. Acid peat moss and rotted oak-leaf mold are two of the best materials. As a rule they provide enough nutrients and at the same time help to keep the soil acid. In poor soils, where additional nitrogen fertilization is necessary, cottonseed meal or tankage may be added at the rate of 5 pounds per 100 square feet of bed area. An excellent formula for broad-leaved evergreens, developed by Dr. R. P. White, is made up as follows:

	lbs.
Tankage or Cottonseed Meal (6% N)	650
Sulfate of Ammonia	200
Ammoniated Superphosphate (1.88-17.5%)	800
Muriate of Potash	200
Magnesium Sulfate or Magnesium Oxide	150
Total	2,000

This should be applied at the rate of 1,000 pounds an acre for large plantings or 1 pound to 50 square feet for small plantings. It contains all the ingredients required by broad-leaved evergreens and has an acid reaction. The magnesium sulfate (epsom salts) in this formula aids in the production of deep-green leaves.

Fresh manure, lime, or wood ashes should not be used to fertilize rhododendron, azalea, or related plants, since these materials will neutralize or sweeten the soil.

Methods of Applying Fertilizers to Trees

Obviously the area to be fertilized is governed by the location of the feeding roots. In this respect there is considerable variation among species of trees

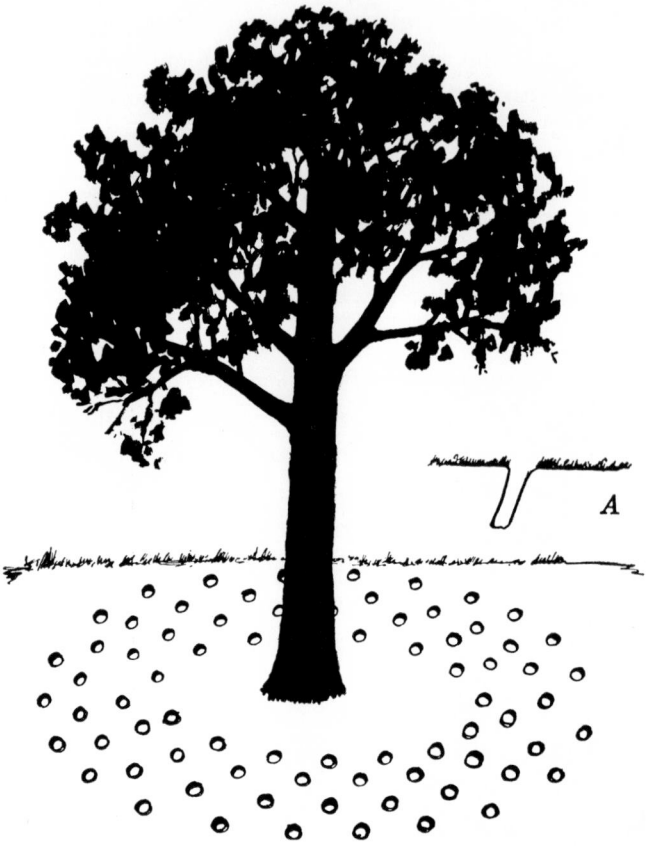

Fig. 14. Fertilization of shade trees, showing the location of the holes to which fertilizers are added. The holes should be 18 to 24 inches deep, about 2 feet apart, and slanted toward the tree, as shown in diagram A.

and in different soil types. But the feeding roots usually occupy the outer band of a circular area whose circumference lies just beyond the spread of the outermost branches. The width of this band is equal to about two thirds of the radius of the circle. With most trees, few feeding roots lie within the center third of the circle (Fig. 14). A rule-of-thumb method for determining the region in which most of the roots occur is as follows: the radial spread of the roots in feet is equal to the diameter of the tree (1 foot above the soil) in inches. For example, a tree whose diameter 1 foot above the ground is 9 inches, will have most of its roots within a 9-foot radius and its feeding roots in the outer 6-foot wide band.

Another point to be borne in mind: it is the roots of smaller diameters, 1/16 inch or less, that function in absorption of nutrients.

Surface Applications. Fertilizers may be broadcast under small trees growing in open soil. Satisfactory results are obtained only when such fertilizers are hoed or watered into the soil thoroughly. As already mentioned, most of the fertilizer elements, especially phosphorus and to a lesser extent potash, remain where they are placed and do not move through the soil readily. When a fertilizer is spread on the surface for a period of years, the tree tends to develop roots near the soil surface. If the tree is growing in a lawn, these roots not only interfere with mowing, but they affect the growth of the grass beneath the trees. Moreover, a tree with a shallow root system is more apt to suffer from dry spells than one with a deeper root system. Elms and maples normally produce roots near the soil surface. Application of fertilizers to the soil surface merely aggravates that condition.

No dry commercial fertilizer should be applied within one foot of the trunk, since injury to the root collar and trunk base might result.

Trenching. The trench method was developed in an effort to place nutrients nearer to the root system. A trench about 2 feet wide is dug in a circle beneath the outer spread of the branches. This area is then filled with rich composted soil to which fertilizers have been added. The disadvantages of this method are obvious. First, the soil is enriched in only a small part of the total area covered by the roots; second, many roots are injured when the trench is dug; and third, large amounts of sod must be lifted and replaced where trees are growing in lawns.

Punch-bar Method. The most practical method for large trees and those growing on lawns is the punch-bar method, because it distributes the fertilizer throughout the root area. Holes, spaced about 2 feet apart and about 18 inches deep are made with some implement throughout the area to be fertilized (Fig. 15). The correct amount of fertilizer is evenly distributed in the holes, which are then filled completely with loosely packed peat moss, shredded manure, or topsoil and closed by a push with the heel of the shoe.

FIG. 15. The author's son Joseph is shown punching holes with a special apparatus known as Pok-a-hol, prior to inserting fertilizers near the tree's roots.

An ordinary crowbar is the implement most readily available to home owners for making the holes. Obviously the small, closely compacted hole made with such an implement will keep the fertilizer localized. For this reason it is essential that a large number of holes be made, about 10 to 20 for each inch in diameter of the tree. A soil auger, which bores into the soil without compacting the sides of the hole, is a more efficient implement. A post-hole digger is recommended by some authorities, but the holes made by this implement are much too large to be satisfactory for fertilizing trees growing in lawns.

Many commercial arborists and park and cemetery superintendents have discarded punch-bars in favor of more modern equipment, such as electric and compressed air drills or augers. With these, a great number of holes can be bored to the desired depth in a very short time.

Holes should be made in a slanting direction (Fig. 14) rather than vertically, regardless of the implement used. This increases the fertilizable area to a small extent.

If no response becomes evident from fertilizing applications, one should check by probing in a few spots to see whether the plant food was placed near roots. Some investigators have found that some trees have no feeding roots as far out as the branch drip area. In some trees the bulk of the feeding roots are within 6 to 8 feet of the trunk.

Compressed-air Feeding. A method invented by a well-known commercial arborist in the Middle West, and one that represents a great improvement over previous methods, involves the use of compressed air. Holes are made with compressed-air drills or augers as outlined above. A pipe that tapers toward the lower end is inserted in the hole, and compressed air is forced into the soil to drive off accumulations of soil gases and to fracture the soil. (This treatment alone often helps to stimulate tree growth.) Dry fertilizer is then placed in the hole, and compressed air is again forced in.* The fertilizer is thus distributed as well as possible with the least disturbance to the sod. Fertilizing with compressed air is most effective in homogeneous soils. It is not so effective in soils made up of several different layers.

Feeding Needles. Several types of feeding needles or guns are available. With these, fertilizers either in dry form or in water solution are placed in

* To prevent the air and fertilizer from blowing back into the operator's face, a metal cover resembling an inverted funnel is attached to the pipe.

the holes. The addition of fertilizers in water is especially helpful when the soil moisture supply is low and water is a limiting factor. Aside from supplying much-needed water, the method ensures almost instant availability of the nutrients, since these must be in solution before the roots can absorb them (Fig. 16). The soluble fertilizer elements are dissolved in water in a tank and are then pumped through the needle into the soil. Among the more commonly used soluble fertilizers are Ra-pid-gro—discussed later in this chapter under Foliage Feeding—Heller-Gro, and Agrico Water Soluble Plant Food.

FIG. 16. Professional arborists apply nutrients with a tree-feeding needle attached to a pressure sprayer.

Another type of gun includes a hopper, holding about a pound of fertilizer. The water from the spray tank or main is turned off, and a valve is opened to release the charge of fertilizer from the hopper. Another valve then permits water to enter the gun and force the fertilizer down into the soil.

These latest improvements in feeding methods are now used by many of the larger tree preservation companies and have proved more effective than the simpler methods outlined earlier in this chapter.

Tree owners who want to feed their trees with an apparatus resembling that used by professional tree men can use the Ross Root Feeder (Fig. 17), manufactured by Ross Daniels, Inc., 207 Court Avenue, Des Moines, Iowa. A plant food cartridge is placed in the chamber at the top of the apparatus, the garden hose is attached at the proper end, and the pointed rod end inserted into the soil to the proper depth. When the water is turned on, it dissolves the plant food and delivers the solution to the root area.

FIG. 17. The Ross Root Feeder can be used by home owners who want to feed their own trees.

Fertilizers in liquid form have the following advantages over the dry fertilizers. They are immediately available and are more uniformly distributed through the soil. Their application requires less labor, makes an additional use for the sprayer, and gives the tree a good watering.

Among the disadvantages are: they do not last very long, and some of the constituents leach out rather quickly; they do not work so well in heavy, poorly drained soil; they offer no opportunity to improve the soil by incorporating organic matter as is the case with dry fertilizer applications.

When and How Frequently Trees Should Be Fertilized

Trees may be fertilized either in late fall or in early spring. Fall applications should not be made until all possibility of top growth is past. In the New York City area, the first of October is a safe date to start such treatments. Fall applications have an advantage in that some nutrients will enter the roots immediately and in very early spring. The remainder will be available later. Tree roots always become active long before the buds begin to swell, and fertilizers already in place can be more quickly utilized. Fertilizer applications in the spring may be made from the time the frost is out of the ground until May 1 around New York, but the later the application the less will be the returns as measured in root and top growth.

Summer applications are not recommended, as such treatments stimulate the production of soft succulent tissues, which do not harden properly and are therefore subject to winter injury. A starving tree, however, should be fed regardless of the season.

The frequency of the fertilization treatments depends on the kind of tree involved, the growth response, and other factors. It may vary from an annual treatment to one made every 3 or 4 years.

Cost of Fertilizers

Price alone should not be the deciding factor in the purchase of fertilizers. Some fertilizers costing $40 or more a ton are worth no more than some costing only $10. Others, manufactured by reliable firms that honestly attempt to include materials known to produce the desired results, may cost more but are well worth the additional price. Some tree fertilizers are more expensive because they contain more nitrogen, some of which may be in the more expensive organic forms such as cottonseed meal, tankage, or dried blood. The synthetic ureaform fertilizers are even more expensive.

Home-mixing of fertilizers as outlined on page 53 is not recommended unless a substantial saving would be realized over the cost of ready-mixed formulas.

Feeding Above-Ground Parts of the Tree

Foliage Feeding. Interest in foliage applications of nutrients has increased by leaps and bounds during recent years. If one individual could be selected as the moving force behind this method of feeding plants, it would be Thomas

Reilly of Dansville, New York, a member of one of America's early tree-growing families. Mr. Reilly was an early advocate of high analysis, completely soluble plant foods which could be applied as foliage sprays. His product, Ra-pid-gro, was the first complete chemical fertilizer to be made commercially available for spraying on trees, shrubs, flowers, and food plants. Ra-pid-gro is a 23-21-17 completely soluble chemical fertilizer with minor elements added. The nitrogen is obtained from crystal urea, monoammonium phosphate, and potassium nitrate; the phosphoric acid from monoammonium phosphate and monopotassium phosphate; and the potassium from potassium nitrate and monopotassium phosphate.

Scientific corroboration of Mr. Reilly's early observations that plants could benefit from foliage applications of nutrients has appeared in recent years. Most notable was the work with radioisotopes at the Michigan Agricultural Experiment Station under professors Tukey, Wittwer, Bukovac, and others. These men were among the first to report that leaves are not completely covered with cutin (except for the stomatal openings), as was believed formerly.

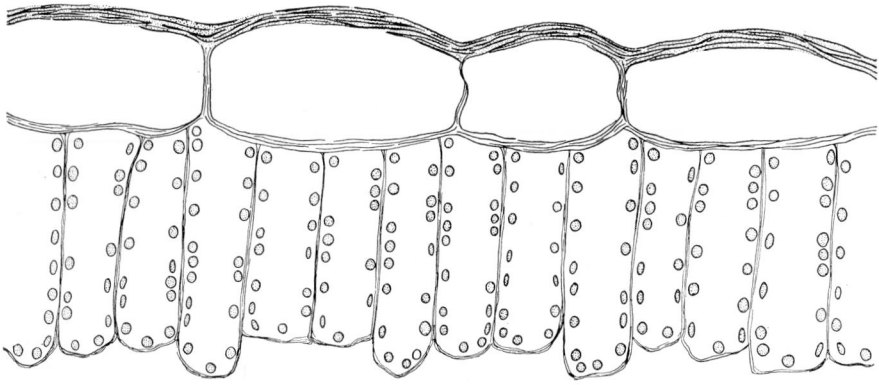

FIG. 18. Highly magnified cross section of leaf surface illustrated in Figure 2, which shows that the waxy surface does not consist of a solid, impervious layer of cutin, but is actually interspersed with layers of pectin-like substances which allow nutrients to penetrate the leaf surface.

It is now known that both the upper and lower surfaces of leaves consist of intermittent parallel layers of pectin-like substances interspersed with the cutinized layers (Fig. 18). The pectinaceous substances have a great capacity for absorbing and releasing water and nutrients. Leaf surfaces, therefore, are no longer considered to be relatively impervious to materials in solution.

Formerly it was thought that the lower leaf surface was more efficient in absorbing solutions because of the greater number of stomata on that surface. This apparently is not the case, because the upper surface of some leaves can absorb as much phosphate as the lower surface, even though it has only one seventh as many stomata.

It is not easy to appreciate the potential absorbing surfaces involved. The leaf area of a 12-year-old apple tree, for example, is about 10 times the land area is occupies. And the ratio between leaf area of a shade tree growing along a city street and the constricted land area it occupies is even greater.

The absorption of nutrients by leaves is generally greater during daylight hours.

Among other factors which favor absorption and transport of nutrients are high humidity, temperature, and incident radiation; a carbohydrate supply; and vigorous growth. Chemical and physical properties of the nutrient spray such as pH, carrier ions, and surface active agents also exert a great influence.

Factors inherent in the leaves themselves also influence nutrient intake. Among these are thickness of the cuticle, number of discontinuities (interspersed pectinaceous areas), hairiness or smoothness of leaf surface, and other surface properties that influence adherence.

Young, rapidly expanding leaves absorb more nutrients per unit area than do fully mature ones. Generally speaking, the more vigorous the plant, the more nutrient it can absorb via the leaves.

Not all plants have the same capacity for absorbing nutrients via foliage sprays. Apple leaves, for example, can absorb more than peach leaves can.

Nitrogen applied in the form of urea is the most readily absorbed of the major elements. In fact, the entire nitrogen requirements of some crops can be satisfied by means of foliage sprays. After the urea in solution is absorbed into the leaf, it is split up into ammonia and carbon dioxide by means of the enzyme urease.

Sodium and potassium are two other elements readily absorbed by leaves, and they are among the most highly mobile once inside the leaves. Phosphorus, chlorine, sulfur, zinc, copper, manganese, iron, and molybdenum are less mobile in the order given. Calcium is absorbed but does not move out of the leaves. Despite this fact, calcium and its close relative magnesium can be applied successfully to the leaves of some plants to overcome a deficiency of those elements.

The rate of movement of nutrients such as nitrogen, phosphorus, and potassium absorbed by leaves has been estimated at 1 inch in 5 minutes.

The leaves are not the only tree parts capable of absorbing nutrients. The bark, buds, petioles, and flowers are also capable of some absorption.

Phosphorus and potash applied to the bark of apple and peach trees moved into the branches 18 to 24 inches from the points of application within a 24-hour period. Most absorption via the bark occurs when the nutrient is applied just before the buds begin to swell in early spring. And the nutrient solution applied to the bark can be ten times as strong as that applied to the leaves, without causing damage to the tree.

This method makes it possible to supply important nutrients, before the start of the growing season, to trees that have suffered from some winter injury or that may have a nutrient deficiency.

Foliage sprays are particularly effective in maintaining good growth in dry seasons because the tree can start using the evenly distributed sprays immediately. Dry fertilizers placed in holes around the tree are not available to tree roots in dry spells.

Foliage-nutrient sprays should be applied either with a pressure sprayer or with one of the specially designed fertilizer spray guns available in seed stores. The latter have a siphoning device and can be attached to the garden hose. As a rule they cannot reach high enough to leaf-feed tall trees.

The amount of high analysis all-soluble plant food to use will vary with the grade or formula. Generally speaking, no more than the following amounts should be dissolved in each 10 gallons of water used: 5 ounces of a 12-12-12, 13-13-13, or 13-26-13; 4 ounces of a 15-15-15, 15-30-15, or 16-16-16; 3 ounces of an 18-18-18, 20-20-20, or 23-21-17.

The tree should be sprayed until the nutrient solution begins to drip heavily from the leaves.

Tree owners are cautioned that not all fertilizers that are quickly soluble in water are safe to spray on the leaves. The manufacturer will indicate clearly on the label whether his product can be used safely in foliage sprays.

Most of the concentrates that are safe to use as foliage sprays can be combined with the newer organic pesticides, such as ferbam, DDT, and malathion. This enables the tree owner or commercial arborist to feed the tree and control pests and diseases in one operation. Here, too, the manufacturer of the plant food can advise the gardener as to which pesticides are compatible with his product.

Leaves act as a two-way avenue for plant nutrients. Not only are nutrients taken in, but nutrients absorbed either through roots or leaves can be leached out of the leaves during rains.

In the not-too-distant future, large areas such as forests will be fertilized by applying the plant food by airplane. Early experiments revealed this method to be quick and relatively inexpensive, and the forest trees responded by making good growth within 2 years of the applications.

Tree Injection. Direct injection of essential nutrients into the trunk has been moderately successful. But the method cannot yet be recommended as a standard commercial practice. Probably the only exception is the injection of iron salts into chlorotic trees that are known to suffer from iron deficiency. Details of this treatment are given on page 198.

Though fertilization by injection into the trunk or branches is still in the experimental stage, some recent work indicates that it is feasible, inasmuch as favorable responses have been obtained in some instances. Apple trees have been injected with complete fertilizer solutions containing 0.25 per cent potassium phosphate plus 0.25 per cent urea, at rates varying from $\frac{1}{30}$ to $\frac{1}{6}$ pound per tree. Sufficient quantities of these materials were absorbed in 24 hours to bring about an increase in growth equivalent to that resulting from heavy applications to the soil. Moreover, the injection of urea at the rate of 1 gram of the salt to each inch of the branch diameter produced no injury and increased the nitrogen content of the tissues of some trees.

The greatest danger in tree injections is the possibility of injuring the tree tissues. Another danger is that of introducing wood-rot fungi in the bored holes.

For the present, injections should be attempted only where single salts are needed to overcome a particular deficiency, or where soil treatments are not easily applied because of pavement, walks, and other obstructions over the root area.

5

PRUNING TREES AND TREATING WOUNDS

The pruning of trees and the subsequent treatment of wounds are probably the most important of all tree maintenance practices. Proper and systematic pruning helps trees better to withstand adverse environmental conditions. In addition, properly pruned trees require less fertilizing, bracing, and spray materials to keep them healthy. Correct wound treatments reduce losses due to wood-decay fungi. The two practices are so closely related that they must be considered together.

Need for Pruning

Trees are pruned principally to preserve their health and appearance, and to prevent damage to life and property.

Pruning for Health. Broken, dead, or diseased branches are pruned to prevent decay-producing fungi from penetrating into the part of the tree to which these branches are attached. Live branches are removed to permit penetration of sunlight and circulation of air through the canopy or to compensate for loss of roots. Overlapping branches and those that interfere with utility wires are removed to prevent rubbing and eventual decay. Branch stubs are removed to promote proper healing.

Tops are sometimes cut back to revitalize the remaining parts of the tree. Such pruning, if properly done, increases the general vigor of the tree by supplying additional nourishment to the parts that remain. Drastic pruning involving the removal of much potential leaf surface may, however, seriously affect the food supply and weaken the trees. The removal of a large portion of the top may also favor fungus and insect attack, and increase the possibility of scalding of the bark by suddenly exposing tender tissues to the direct rays of the sun. The removal of many small branches rather than a few large ones not only reduces the amount of sun scald, but also makes the work less conspicuous and facilitates maintaining the desired shape. In addition, the many small wounds will heal more rapidly and completely than the few large ones.

Since the removal of dead or dying branches constitutes one of the major objectives of pruning, it is well to list some of the principal causes of dieback. These include: lack of adequate or proper plant food; damage to the root system; a deficient or overabundant moisture supply; poisonous elements in the atmosphere or soil; inadequate aeration of the root system owing to grade changes; excessively dense crown; fungus, bacterial, or insect attack; mechanical injury to the trunk and branches; and lightning injury. While it is imperative that all dead wood resulting from any of these causes be removed, it is extremely important that the causal factors be eliminated or corrected whenever feasible.

Pruning for Appearance. Only in formal plantings or to restore the characteristic form to a badly misshapen tree is pruning for appearance warranted. An appreciation of landscape values and some idea of the normal form of the tree being pruned are essential for success in pruning trees in formal plantings. Certain branches on many street trees, particularly Norway maples, tend to develop faster than others. These should be pruned back to preserve the tree's form.

Occasionally it is desirable to keep normally large-growing shade or ornamental trees within restricted bounds of small properties. Judicious pruning of the new growth during summer will produce the desired dwarfing effect, with no harm to the tree.

Pruning for Safety. Dead, split, and broken branches are a constant hazard to life and property. Danger from falling limbs is always greatest in trees along city streets and in public parks. Low-hanging live branches must be removed to a height of 10 to 12 feet when they interfere with pedestrian and vehicular traffic. This phase of pruning is probably the most important of all street tree maintenance practices.

Pruning Equipment

Adequate equipment (Fig. 19) is an essential of good pruning. Equipment must be maintained in first-class condition both for the safety of the operator and the good of the tree. The following chart lists the most essential pieces of pruning equipment and their use.

Equipment	Specifications	Use
Hand Saw	at least 6 teeth to an inch	for normal pruning cuts
Crosscut Saw	3 feet long	for large cuts
Pole Pruner Pole Saw	10-12 feet long	for removing small limbs not reached by hand saw
Manila Rope	150 feet, ½ inch in diameter	for climbing
Manila Rope	100 feet, ¾ inch in diameter	for lowering large limbs
Pruning Shears	hand operated	for cutting back small trees and pruning roots
Ax	small size	for removing short stubs that cannot readily be removed with a saw
Mallet and Chisel	small size	to point and shape final cuts
2 Belt Snaps		to hold saw and paint cans
Shellac	see page 80	to paint cambial region of cuts
Wound Dressing	see page 80	to paint cuts
Safety Belt		to fasten operator securely

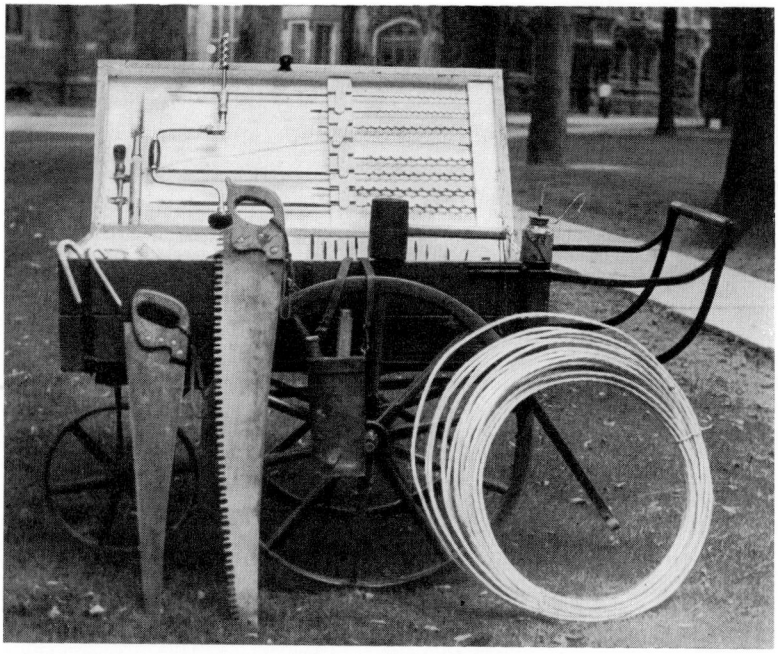

Fig. 19. Some of the equipment essential for pruning, bracing, and cavity work.

The knack of using ropes properly in climbing and working in a tree can be developed best through visual instruction and practice; consequently, a discussion of this subject is beyond the scope of this book.* Men most skilled in the use of ropes have been trained principally by the staffs of large tree companies. The use of ropes in tree work contributes safety, comfort, and dexterity to the climber. The amateur would best use saws and pruners that he can operate from the ground for most of the work, and employ experienced tree climbers to remove high limbs in older trees.

Among the well-known manufacturers or distributors of tree-pruning tools at present are Bartlett Manufacturing Co., Detroit 2, Mich.; Henry Disston and Sons, Philadelphia 35, Pa.; A. M. Leonard & Son, Piqua, Ohio; Fred Marvin and Associates, Akron 12, Ohio; H. E. Muench, Stamford, Conn.; H. K. Porter, Inc., Somerville 43, Mass.; Seymour Smith & Son, Oakville, Conn.; J. B. Sebrell Corp., Los Angeles, 13, Cal.; and Wachtel Supply Co., Wauwatosa, Wis.

World War II speeded the development of one of the most useful of arborists' tools—the *power* chain saw. This saw is designed to give high-speed cutting of trees. One type can fell a hardwood tree 15 inches in diameter in less than a minute and a softwood tree of the same size in a quarter of that time.

The chain saws are simple to operate and light in weight. They reduce limbing, felling, or bucking time to a minimum. Some types can be operated by one man; others require two. The power to run these high-speed saws is usually generated by a light-weight gasoline engine.

Among the manufacturers of chain saws are the Disston Company, mentioned above; Homelite Corp., Port Chester, N. Y.; Lancaster Pump and Mfg. Co., Lancaster, Pa.; Mall Tool Co., Chicago 19, Ill.; McCulloch Motors Co., Los Angeles 25, Cal.; Monarch Silver King, Chicago 35, Ill.; and Porter-Cable Co., Syracuse 8, N. Y.

The McCulloch Company has an excellent booklet entitled *How To Use a Power Chain Saw*.

Among the newest types of power saws recently introduced to professional tree men is the Wright Rebel, which saws back and forth like a handsaw,

* *Tree Preservation Bulletin Number 4* of the U. S. Dept. of Interior, National Park Service, presents a complete discussion of the use of ropes in climbing and working in trees. This is available for 15 cents from the Superintendent of Documents, U. S. Government Printing Office, Washington 25, D. C.

but at 166 strokes a second. It is manufactured by Thomas Industries of Louisville 3, Ky.

To facilitate pruning and line clearing jobs in trees growing along streets and highways, engineers have perfected hydraulically operated aerial lifts to which are attached one or two crows' nests or baskets. From these the operator can reach any point in the ordinary city tree. The apparatus is powered by a conventional power take-off from the truck engine.

Among the manufacturers of such equipment are Maxwell Equipment Co., Milford, Conn., manufacturer of the Hydro Skylift; Emhart Manufacturing Co., Milford, Conn., manufacturer of the Tree King, Mobile Aerial Towers, Inc., Fort Wayne, Ind., manufacturer of Hi-Ranger.

How To Prune

There is no standard method of pruning, but certain procedures and precautions generally are followed by competent arborists to ensure best results. In general, it is best to start pruning operations in the upper parts of the tree and work downward. This facilitates shaping the tree properly and saves time in clearing the tree of pruned branches, which become lodged in the lower branches as they fall. All dead, broken, fungus- and insect-infested, and interfering branches are removed. Small branches that may prove undesirable within a few years should be treated as interfering branches. Clean cuts should be made as nearly flush as possible with the branch that is to remain. Dead branches should be cut back to a healthy crotch, so that healthy tissue surrounds the final cut.

The treatment of bark on the trunk and branches is included in the pruning operation. All dead bark areas should be cut to healthy tissue and old wounds not healing properly should be recut, and then shellac and a wound dressing applied.

Shaping the Final Cuts

All final cuts should be made as close as possible to the remaining part. Stubs or protruding 'lips' should never be left, since proper healing is inhibited and water pockets that promote wood decay may develop. The branches of some trees, such as the Norway maple, are so shaped that wounds resulting from their removal are flat at the top or bottom. Such wounds do not heal properly because the sap flow fails to reach some of the bark. Wounds of this type should be shaped, as shown in Figure 20, by pointing the bark above and below the circular cut, and by removing the protruding wood with a chisel. As will be discussed in Chapter 6, sap movement runs parallel to the length of the branch, and abrupt cuts across the conducting vessels result in a decrease or stoppage of the sap flow to the bark and in the drying and death of these unfed areas. Wounds made by removing branches at sharp crotches should also be pointed with a chisel to facilitate proper healing.

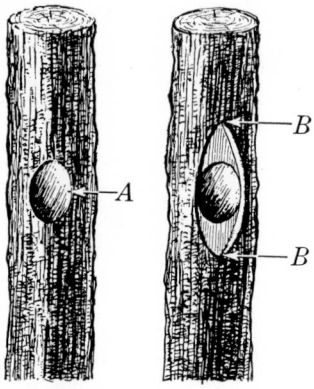

FIG. 20. How to shape pruning cuts. LEFT: Branch removed at *A*. RIGHT: The bark is cut away and pointed at *B*.

Theoretically, all wounds, regardless of size, should be painted with a dressing. In actual practice, however, only wounds 2 or more inches in diameter usually are painted. One should bear in mind, however, that decay may readily develop in smaller untreated wounds, especially in trees low in vigor, in which healing progresses slowly.

REMOVING LARGE BRANCHES

A good-sized saw with teeth so set as to make a wide cut is best for removing large branches. The procedure is the same as that followed in removing smaller branches, except that additional precautions must be taken to prevent injury to the tree itself.

The tree will suffer considerable damage if only a single pruning cut is made in a large branch, as shown in Figure 21. As the cut deepens, the remaining wood will become too weak to support the weight of the limb, and much of the bark below will be ripped off. The proper way to remove a large branch is shown in Figure 22. About a foot beyond the proposed final cut, a preliminary undercut is made by sawing until the blade begins to bind. On the upper side of the branch an inch or so beyond the first cut, a second cut is made to sever the branch. The short stub remaining can then be removed by making the final cut as nearly flush with the main branch as possible. The stub should be held in place by the operator's free hand, or by rope, to prevent tearing of the bark. When an entire branch is small enough to be held firmly in place by ropes or by hand, the first and second cuts my be omitted.

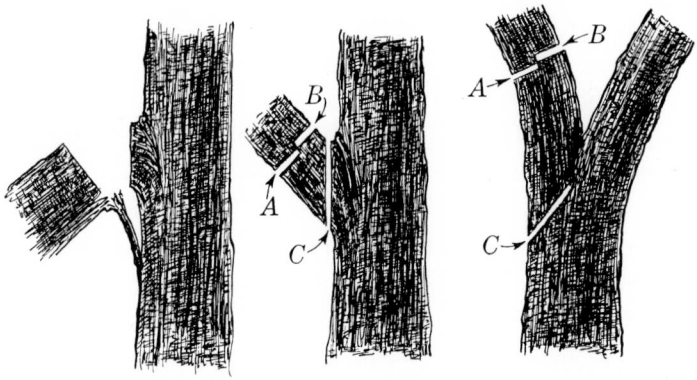

FIG. 21. LEFT: The wrong way to prune a large branch. A single cut close to the main stem may result in tearing of the bark.

FIG. 22. CENTER: The correct way to prune a large branch. A preliminary undercut is made at *A*; a second cut is made at *B*, to sever the main part of the branch. The remaining stub is removed by cutting at *C*.

FIG. 23. RIGHT: Removal of V-shaped crotch. A preliminary undercut is made at *A*, and the second cut at *B* to remove the major part of the branch. The final cut, starting at point *C*, slants upward to the point of union of two limbs.

Large final cuts cannot always be made as close to the remaining branch as necessary. In such cases, the cut surface must be smoothed over with a chisel.

Removing large limbs from trees along city streets, in parks, and in cemeteries entails an additional precaution. In many instances, such branches must be lowered to the ground in pieces or as a unit, to avoid damage to electric wires, private property, monuments, or shrubs and lawns. This is usually accomplished by means of two or more heavy ropes which lower and guide the branches in their descent.

REMOVAL OF V-CROTCHES

The removal of one or two limbs that have developed adjacent to each other and are connected by a V-shaped crotch is commonly necessary. The actual point of union of the wood of these two limbs may be several feet below the visible joining of the bark. Only careful cutting at the right point will ensure proper healing. Preliminary cuts, as described for removing large branches, are made a foot or so above the crotch, to remove the upper part of the branch. A final cut starting from the main trunk and slanting upward to the point of union of the two limbs completes the operation (Fig. 23). After this stub is removed it may be necessary to finish the cut with a chisel, to form a pointed wound that will heal rapidly.

Dehorning

The practice known as dehorning involves the drastic cutting back of the larger limbs of a tree. The operation may completely disfigure the tree by removing the normal canopy. Moreover, it may eventually lead to serious branch decay beginning at the severed ends, and to the scalding of the bark, which is suddenly exposed to the full rays of the sun. This practice must be employed, however, where there is considerable dying back of the branches due to severe insect or fungus attack, or where a reduced root area cannot sustain a large canopy. It may also be necessary as an invigorating measure, as a means of eliminating interference with overhead wires or other objects, and as a safety measure along city streets.

Certain precautions will help to reduce many of the detrimental after-effects. The branches should be removed with a slanting cut starting just above a vigorous bud or shoot, and running back and across the limb at an angle of about 45 degrees. The bud or shoot must always be at the peak of this cut and should point in the direction where new limb development is desired. It is well to remember that a cut made at right angles to the long axis of the branch is not likely to heal, while one made at too sharp an angle will weaken the elbow and eventually result in breaking or splitting of the new limb, when it has attained sufficient size. All cut ends should, of course, be properly painted.

As a rule, elm, linden, locust, plane, and silver maple can be cut back severely without seriously reducing the vigor of the tree, but hickory cannot withstand such drastic treatment. Many Carolina poplars and willows growing along city streets are cut back every few years to prevent their canopies from becoming too large and unsafe. Some attention must be given to pruning suckers or small twigs that usually develop at the severed ends.

Though dehorning is justified under certain conditions, it would be unnecessary in many instances if the tree had received early and judicious pruning.

Pruning Suckers or Waterspouts

Adventitious buds often give rise to thin slender twigs that grow nearly parallel and closely appressed to the branches or trunk. The presence of a large number of such shoots, or suckers, is usually an indication of structural injuries, fungus or insect attack, changes in environmental conditions, or excessive or improper pruning.

The number and size of suckers, the species of tree on which they form, their position on the tree, and the cause responsible for their formation determine what disposition is to be made of them. The presence of sprouts on the upper branches of an elm, for instance, is a natural phenomenon with this species.

Suckers at the base of the trunk on any ornamental or street tree should be removed as they develop, since they not only detract from the appearance of the tree, but are also a nuisance.

When the top of a tree has been cut back severely, the suckers that form should be allowed to remain to provide protection from sun scalding of the bark.

Pruning Storm-damaged Trees

Drastic pruning is sometimes necessary after a hurricane or after ice and wind storms. Where large branches or the entire tree is damaged severely enough to endanger life and property, the pruning should be done as quickly as possible. Except for emergencies, however, pruning can be delayed until mild weather comes along.

Damage to large branches or to uprooted trees is best handled by a professional arborist. The tree owner can undertake repair jobs on smaller trees whose damaged branches are relatively close to the ground. Trees should be examined carefully to determine the extent of injury. If this is not too extensive and the tree is worth saving, a definite plan of rehabilitation should be worked out. Broken branches should be cut back to the next lower crotch if possible. In some instances, especially where a main branch is involved, it may be desirable to leave a considerable part of the stub. Treatment in such cases should follow that outlined under Dehorning. Many limbs or crotches that are split can be saved by artificial bracing with bolts. Wounds made by splitting limbs should be cleaned of splinters, the wood smoothed, and the bark traced to the proper shape to ensure rapid healing. Unbroken branches should be left untouched: they should never be headed back to conform with cuts made on broken branches. Though the immediate effect of long and short branches in the canopy may appear odd, additional judicious pruning within a few years will help to fill the wide openings and will eventually result in a well-shaped tree.

Pruning for Disease Control

Very often it is necessary to prune trees that are partially affected by diseases, such as fire blight of apple, pear, and mountain ash; sycamore anthracnose; Cytospora canker of blue spruce; Diplodia tip blight of conifers; wilt of maples and elms; and twig blight of chestnut oak. In such cases, the infected twigs or branches should be removed at least 3 inches below the point of visible infection.

Extreme caution must be exercised so as not to transmit the causal organism from a diseased to a healthy tree by means of the pruning tools. After use on a tree suspected of being diseased, the cutting edges of all tools should be thoroughly disinfected by dipping them in, or wiping them with a cloth saturated with, 70 per cent denatured ethyl alcohol. Furthermore, pruning when the leaves are wet should be avoided, since parasitic organisms are easily spread under such circumstances.

When To Prune

Though there are advantages and disadvantages in pruning during certain seasons, trees may be pruned at any time of the year, and the selection should be governed generally by practical considerations.

Trees can be best pruned to the desired shape when they are in foliage. It is easier to see weakened or dead branches then.

Since most rapid healing occurs on wounds made in early spring, this season is also a good time for pruning. But some trees, like maples and birches, bleed so profusely when cut in spring that it is better to delay pruning them until summer, when the sap runs less freely. Street and park trees, except maples, are usually pruned during the winter, to distribute tree maintenance work more evenly over the entire year.

Pruning Evergreen Trees

Evergreens do not usually require the periodic pruning necessary for deciduous trees. Dead branches in evergreens rarely give rise to such extensive cavities as are found in shade trees, partly because decay is checked by the presence of resins in the wood. If the removal of large branches is necessary, however, the methods previously discussed should be followed.

The main objective in pruning evergreens is to produce a more bushy or compact plant. Pruning the growth at the ends of young shoots forces the plant to make new growth along the branches.

Not all evergreens can be pruned in the same manner or at the same time. The method depends on the species and the type of growth.

Pines and spruces make only one growth during the year. They can be trimmed at any time, but best results are obtained when the new growth is soft. In pruning spruces, especially the blue spruce, the long ends should be removed at the point where side shoots are formed. This will check the terminal growth for a year or two and will enable the branches to fill out. Most pines are best pruned before the new needles start to unfold, that is, when the new growth still gives the appearance of candles. No damage is done to the needles, and the tree will retain its original healthy appearance.

Most of the other evergreens, including arborvitae, juniperus, yew, cypress, and chamaecyparis, continue to develop during the entire growing season. These may be trimmed up to the month of August in the New York City area, although June or July pruning is preferable, since new growth will cover the cuts before the end of the growing season.

Winter pruning is sometimes preferable when an evergreen has become straggly as a result of long neglect. The long branches are cut to encourage new bud development early in the next growing season. When pruning is postponed until summer, new buds will not develop until the following spring, and in the interim the plant will appear unattractive.

Evergreens with a distinctive shape should be so pruned as to retain the desirable effect. Only the long shoots should be removed. Formal evergreen hedges are best pruned with hedge shears. Hemlocks may well be pruned to increase their branching and fullness. In the New York City area late June appears to be the best month, since new growth soon starts and continues throughout the summer. By early fall, the effects of pruning are well hidden by the new growth.

Pruning for Line Clearance

Friction is bound to arise when any two objects attempt to occupy the same area. Such a situation exists in the case of trees growing along highways and city streets, and light, power or communication wires. Tree lovers argue that the wires should be set underground or moved out of the way, while the public utility officials often feel that the trees should go. Both trees and wires are essential to the health, prosperity, and happiness of communities, and the only immediate remedy when such an unhappy situation exists is intelligent trimming of trees.

Many utility companies employ a reliable tree-trimming organization for this type of work, as the use of inexperienced help often results in complaints by tree lovers of unnecessary butchering of the trees.

There are four types of pruning usually employed in the vicinity of overhead wires: topping, side pruning, directional pruning, and drop-crotching.

Topping merely involves the removal of the ends of the branches much as a hedge is sheared. It is usually practiced when young trees are growing directly below the wires. This method spoils the natural shape of the tree.

Side pruning is the removal of the side branches of large trees when these interfere with the wires. Usually the opposite side of the tree is pruned to preserve the symmetrical shape.

Directional pruning involves the opening of paths for the wires through the trees by removing smaller branches. Branches are also trained to grow away from the wires before the two meet. An expert pruner can anticipate the direction of future growth and by early and correct pruning reduce the necessity of later pruning. While this method is more expensive than the others, it results in least disfigurement of the tree and gives the most lasting benefits (Fig. 24).

Drop-crotching is the cutting back of large branches to laterals, in order to encourage growth to the side and away from the wires. This method provides a longer duration of clearance, but it may have disastrous effects on the tree because of the danger of decay at the cut ends of the branches.

Disposing of Prunings

The disposal of tree prunings in cities has always been a problem. It has been aggravated recently by the passage of laws, as in New Jersey, that curtail the burning of such material in city dumps because of the resulting increase of air pollutants.

One remedy for municipal arborists, at least insofar as disposing of the branches and twigs is concerned, is the so-called chipper. The pruned branches are fed into the revolving blades of a chipping machine and are broken up

FIG. 24. Trees can be pruned so as not to interfere with power lines.

into small pieces or chips. Four truckloads of brush (branches, twigs, and leaves) can be reduced to one load of chips. Best known among professional arborists is the product made by the Asplundh Chipper Company of Jenkintown, Pa.

The wood chips, incidentally, can be used in many ways. They make fine soil conditioners after being composted for a year or two, and they make excellent mulching material. On farms they can be used as poultry house litter and as animal bedding. Recreational departments are using the chips as a cushion under playground swings and slides.

Safety Hints to Pruners

Determine the general condition of the tree. Greater precautions are necessary when working in an old or weakened tree than in a sound young one.

Examine pruning equipment often for safety and efficiency.

Know the type of wood in the tree. Greater precautions are necessary in pruning weak-wooded trees, such as poplars, silver maple, willow, and tulip, than in pruning elm, oak, hickory, and plane, which have strong flexible wood.

Danger is greatest when branches are wet and when temperatures are low.

Bark peeling and fungus growths are signs of dead and dying branches. Limbs showing these symptoms should never be depended upon for support.

When electric wires run through a tree, remember that the danger of electric shocks is increased when the tree is wet.

Do not allow tools to come in contact with wires, even though the wires are supposed to be insulated.

Most important of all, *always have a safety rope properly attached.* Any branch, no matter how sound in appearance, may give way under the weight of the climber.*

TREATMENT OF WOUNDS

A major cause for decay and premature death of trees can be traced directly to neglected wounds made years earlier. The bark of trees, like the epidermis of humans, acts to protect tissues beneath. Once the protective cover is broken, the parts below are subject to infection by fungi, bacteria, and other parasites.

* Tree Preservation Bulletin No. 2 *Safety for Tree Workers*, revised 1956, is obtainable for 20 cents from the Supt. of Documents, U. S. Govt. Printing Office, Washington 25, D. C.

Wounds made by a tree owner's pruning saw in removing a dead, broken, or unwanted branch, by a skidding automobile, or even by a mischievous boy's ax, should be treated as soon as possible after the damage is done. The sooner the treatment, the less chance of invasion by wood-decaying fungi or insect pests.

A tree wound heals by the formation of successive annual layers of callus tissue around the edge of the wound (Fig. 25). The callus tissue averages about ½ inch in thickness each year. Consequently, the smaller the wound, the sooner the healing. When a branch is cut very close to the trunk, the wound is much larger than it should be. On the other hand, if the branch is cut so that a stub remains, callus tissue must develop up along the sides of the stub before it can cover the cut surface. The proper cut should be in between the two extremes.

FIG. 25. Callus tissue has completely covered the wound on this tree.

All large wounds should be smoothed over by removing stubs or lips of wood, and then shaped into a pointed ellipse to aid rapid callus formation. Following this, the cambial region, from which regenerated tissue (callus) will develop later, should be immediately painted with shellac. When the shellac has dried, a wound dressing should be applied over the entire surface exposed by pruning or by injury to the bark.

Wound dressings on living trees help to protect exposed surfaces and to facilitate healing. These are best applied when the wound is thoroughly dry.

Bark-deep wounds, in which only the bark is damaged and the cambium layer is undisturbed, will heal more rapidly if the damaged area is covered immediately with new burlap or polyethylene film. Mr. Carl Fenner, nationally known city forester of Lansing, Michigan, found that on trees such as elms, maples, and hackberry prompt shading of damaged bark promoted excellent healing. The shade is provided by a burlap or canvas cover placed on supports of wood or steel frames about six inches from the tree. Just before the damaged area is shaded, the wound is shaped with a sharp knife, as described under Shaping the Final Cut, earlier in this chapter, and the cut edge painted with orange shellac. There is no need for painting the exposed cambium area inside the margins of the wound. Four to eight weeks of shading is usually sufficient to ensure the survival and recovery of the injured cells.

Mr. Fenner's method of handling bark-deep wounds arose from his observation that small wounds of this nature on the north or shaded sides of trees usually healed without a tree dressing or any special care.

Wound Dressing

An ideal wound dressing would disinfect the area treated, prevent entrance of wood-rotting fungi and checking of wood, stimulate callus formation, and be toxic to parasitic organisms with which it comes in contact. It would be easily applied, sufficiently porous to allow excess moisture to evaporate from the wound underneath, and not crack on drying or weathering.

No single dressing has yet been developed to meet all these qualifications, but the following wound dressings are most commonly used:

Orange Shellac. One of the least harmful of all materials available, orange shellac, which is sometimes used as an undercoat for spar varnish or ordinary house paint, is most commonly used to cover traced bark and the adjacent sapwood (Fig. 26). While the shellac is not specifically a wound dressing, the alcohol in it is a good sterilizing agent. Shellac applied alone is not very durable.

Asphaltum Paints. Various tree paints that contain asphaltum as the principal ingredient are available. These may be obtained from manufacturers and horticultural supply houses. They may be made up as follows: For each pound of solid asphaltum melted over a slow fire, add about a quart of turpentine, or petroleum oil, or one of its refined fractions, mineral oil. When cooled, the resulting mixture is ready for use. This type of asphaltum paint is more injurious than those made up of water emulsions, but it dries out and weathers more slowly.

The newest asphaltic-based tree dressing to appear on the market, Arbor-Aid 101, comes nearest to meeting the specifications of the ideal wound dressing. This dressing is said to contain a substance, A-X-100, which helps speed callus tissue development. At present Arbor-Aid 101 can be obtained directly from the manufacturer, Arbor-Aid Associates, Box 48, Villa Park, Ill., in 6-ounce tubes and in half-gallon and full-gallon, and larger, cans.

FIG. 26. LEFT: Applying orange shellac to freshly made wound.
FIG. 27. RIGHT: Tree wounds should be inspected periodically and recoated with tree paint.

Creosote Paints. These are best for painting large surfaces in the interior of a cavity where fungi may have become established. They should be used with extreme caution in the vicinity of fresh wounds when rapid healing is desired.

Ordinary commercial creosote is one of the best materials for destroying and preventing the growth of wood-decay fungi. Most paints will not stick to creosote, but coal tar or hot asphalt will adhere readily. Creosote, like creosote paint, is toxic to live tissues and should be used principally on dead heartwood.

A mixture of equal quantities of creosote and asphaltum is also satisfactory and apparently not so injurious as creosote alone.

Grafting Waxes. Prepared with alcohol, these are satisfactory for treating small surfaces.

House Paints. Paints prepared for outside use, containing oxides of lead and zinc mixed with linseed oil, are fairly satisfactory. They are not so durable as asphaltum paints and will injure tender tissues unless the tissues have been previously painted with shellac.

Lanolin Paints. Tree paints containing lanolin as the major ingredient have also found wide acceptance in recent years. By protecting the cambial and bark tissues, this ingredient allows wound callus to develop unchecked.

Workers at the Northeastern Forest Experiment Station at Philadelphia have reported that mixtures of lanolin, rosin, and crude gum promoted wound healing and provided a greater degree of protection to the most crucial zone, the wound edges, than did either shellac or asphaltum.

A suggested mixture of these ingredients is lanolin 10 parts by weight, rosin 2, and crude gum 2. The ingredients are blended simply by melting them and gently stirring together.

The author did considerable research with lanolin mixtures while employed at the New Jersey Agriculture Experiment Station. He found that a mixture of 2 parts of lanolin, thinned with 1 part of raw linseed oil to which .25 per cent of the total mixture of potassium permanganate solution was added, gave better results in wound callus development than any of several commercial tree paints used in the tests. Because the permanganate is not soluble in the lanolin-linseed-oil mixture, it must be dissolved in a small quantity of acetone before being stirred into the mixture.

Varnish-Mercury Paint. For use on plane trees growing in areas where cankerstain (canker, p. 357) disease is prevalent, a special paint has been developed by Dr. James Walter, formerly forest pathologist for the United States Department of Agriculture. The base of this paint is gilsonite varnish fortified with 0.2 per cent phenyl mercury nitrate.

The mercury is included in this paint to eliminate the possibility of spreading the causal fungus from tree to tree via the paint during after-pruning operations. Its presence has been found to retard only slightly the development of wound callus on planes. In Dr. Walter's tests it did not interfere with good callus development when experimentally applied to wounds on elm, tulip, Norway and silver maples, white and pin oaks, and apple.

The mixture is poisonous and must be handled with care. If any gets on the skin, it should be wiped off immediately, otherwise it will cause irritation. The W. J. Sutcliffe Company, East Rutherford, N. J., manufactures a paint that conforms to Dr. Walter's recommendations.

Bordeaux Paste. This is prepared by slowly stirring raw linseed oil into dry bordeaux powder until a thick paint is produced. It has antiseptic properties but hinders callus development the first year after it is applied. In addition, it weathers rapidly and must be replaced often.

Commercial Tree Paints. Ready-made tree paints are also available. While many of these can be depended upon to give satisfactory results, a few are known to be incapable of killing parasitic organisms that get into them. Diseases may actually be spread through use of infested tree paints.

Tree paints are now available in aerosol cans which merely require pressing a button to release the material on the tree wound. This avoids messing up a brush, clothing, and hands. The du Pont Company manufactures such a product in which 'Freon' is the propellant for the dressing. Just recently Gar-Prod, Inc., Queens Ave., Lindenhurst, N. Y., placed its 'Tree Healing Dressing' aerosol on the market. The active ingredients are asphaltum, petroleum oils, and phenyl mercury oleate. The latter has fungicidal properties.

Regardless of the wound dressing used, best results are assured only when the dressed surfaces are inspected periodically and recoated once or twice a year. This is especially necessary when the dressing blisters, cracks, or peels (Fig. 27). When preparing old wounds for recoating, it is best to clean the surface with a stiff wire brush to remove all blisters and loose flakes. The new callus growth should not be covered with the dressing.

Wounds callus over more rapidly when no dressings, with the possible exception of shellac and certain asphalt emulsions, are used. There is some evidence that wounded trees form certain gums that aid in healing and help to retard decay. Since the rapidity of healing is of secondary importance to the entrance of wood-rotting fungi, it is best to continue the use of some good dressing.

Wounded or pruned branches may continue to bleed for a long time. This condition is discussed in detail under Slime Flux on page 165.

Cracks in the wood of large flat wounds constitute important foci for wood-decay fungi. These can be largely prevented, especially when wood is sound at the start, by covering the freshly dressed wound with heavy cloth. The cloth should be cut smaller than the scar, so as to allow room for callus formation. It should then be tacked firmly to the wood, and dressed with the same kind of material previously applied over the wound.

It was formerly believed that trees which produce considerable amounts of resin, such as conifers, are not so subject to heartwood decay as are most deciduous trees. The resins which exude from wounds of some pines and spruces do not have fungicidal properties and hence wood-decaying fungi can enter through the wounds. Large wounds on conifers should be protected with wound dressings and should be inspected regularly as are those on shade trees.

6

CAVITY TREATMENTS

Filling cavities in the trunk and branches is the most spectacular of all tree maintenance practices. The layman has the impression that this is also the major phase of tree care. Although cavity work has a place in any balanced tree program, it does not rank in usefulness with such practices as fertilizing, pruning, and spraying. Its importance in the mind of the layman probably arises from the similarity to the filling of dental cavities. Oddly enough, though only dentists treat cavities in teeth, anyone having access to some cement feels capable of filling a cavity in a tree (Fig. 28). Treating cavities in trees, however, involves more problems than appear on the surface and should be entrusted only to a reliable arborist.

History of Cavity Work

The impulse to beautify unsightly objects is, perhaps, as old as man. Conceivably, even in prehistoric times a decaying cavity in an otherwise beautiful tree would have been a good subject for improvement. Attempts to cover cavities with mud and stone are recorded in many ancient writings. In his book, *Enquiry in Plants*, written about 300 B.C., the Greek scientist Theophrastus described 'plastering wounds with mud' to prevent trees from decaying. Within the last 200 years many investigators have written on this subject and have advocated the use of wood, brick, clay, mortar, lime, and many odd and 'secret' mixtures. The last 50 years have witnessed greater advance in cavity treatments than all previously recorded history, primarily because of the impetus furnished by several tree-preservation companies in the United States. Within this period, cement, cork products, wood and rubber blocks, asphalt compounds, and various kinds of sheet metals have attained extensive popularity. In addition, several patented mixtures have been developed that are definitely superior to any of the materials used before the start of the present century. Without doubt, more satisfactory materials will be discovered in the future.

FIG. 28. LEFT: An amateurish job of cavity filling on horsechestnut. Notice how cement is being pushed out of the cavity.

FIG. 29. RIGHT: How cavities start. As a result of improper treatment after pruning, the wood has begun to decay. The upper part of the wound is not healing because the bark has not been properly treated.

How Cavities Start

All cavities originate from neglected bark injuries (Fig. 29). Sound bark forms an effective protective barrier for the more susceptible tissues beneath it. The sapwood dries out shortly after an opening is made, either mechanically or by some natural agent. When the tree is growing vigorously and the injury is not extensive, the open area is covered by a layer of healing tissue (callus) within a year or two, and little damage follows. Where considerable injury occurs (Fig. 30), or where a branch stub remains after faulty pruning (Fig. 31) or breakage by wind, healing proceeds slowly or not at all. In such cases, wood-rotting fungi and boring insects have ample time to enter and

FIG. 30. UPPER: Extensive bark injuries near the trunk base serve as entrance points for wood decay fungi.

FIG. 31. LOWER: Stubs such as the one shown not only prevent callus formation, but furnish ideal points of entrance for wood decay fungi.

cause decay. The action of these organisms in turn prevents additional healing, and in time a large cavity is formed. Where the heartwood is exposed by some injury, decay progresses more rapidly, because of the formation of so-called checks or deep fissures in the wood.

A cavity in the heartwood does not seriously impair the vigor of the tree, as is commonly believed. Its presence, however, weakens the tree structurally, making the parts above more susceptible to breakage by wind and ice storms. It also serves as a breeding place for ants, boring insects, and vermin. An extensive opening in the trunk may reduce the vigor of the tree indirectly, because the loss of bark, cambium, and sapwood (which normally would occupy this area) greatly decreases the food transportation and tissue-building facilities.

Decay by most wood-rotting fungi progresses rather slowly, the rate being about equal to that of the annual growth of the tree. For this reason, a vigorous tree may attain great size despite the presence of an extensive cavity. Dr. Ray Hirt, plant pathologist, recently retired from the New York State College of Forestry, found that the fungus *Fomes igniarius*, growing in aspen, travels 23 inches in 10 years or slightly over 2 inches each year. The same fungus in maples moved about 18 inches in a 10-year period. A few highly virulent fungi, however, may cause extensive decay within a short time.

Some trees, like cherry and willow, decay more rapidly than other trees. The older a tree, the more susceptible it is to decay.

Major cavities occur in trunk, branch, root, crotch, and base. Trunk cavities usually originate from mechanical injuries and broken or improperly pruned large branches; branch cavities, from broken or diseased twigs or from rubbing of interfering branches; crotch cavities, from splitting of the main branches; base cavities, from mechanical injuries, fire, root diseases, and feeding of animals; and root cavities, from mechanical injury and from animal, fungus, and insect attack.

Objectives of Cavity Treatments

The primary objectives of cavity work are to improve the appearance of the tree by removing decayed and insect-riddled wood and to eliminate a breeding place for boring insects, ants, mosquitoes, termites, snails, and rodents.

Many persons have the impression that cavity work includes several other objectives—the principal one being to eradicate and cure heartwood decay. Cavity treatment will not cure decay; and rarely will it check further spread, unless a very small area is involved. As a rule the fungus threads extend from a few inches to several feet beyond the rotted zone, into apparently sound wood. Although the gouging out of such wood is theoretically possible, it is neither economical nor practicable. Extensive laboratory work must be conducted to determine whether all fungus-infected wood has been removed. Moreover, the extensive removal of all wood containing fungus tissue would so weaken the tree that it would be a menace to life and property. Occasionally trees are killed outright when extensive removal of heartwood exposes the sapwood, which then dries out. Such trees might have lived for many years if only badly decayed portions had been removed.

Method of Treating Cavities

Though greatest emphasis in the past was placed on filling cavities with some solid material, the tendency in more recent years has been to leave the cavity open, once it has been cleaned out, sterilized, and painted. This is especially true in the tree programs of large public enterprises, such as city and county park systems, whose limited budgets do not permit extensive work on individual trees.

The principal steps in treating a cavity according to the standardized methods are cleaning, shaping, bracing, and sterilizing and dressing.

Cleaning Out the Cavity. To remove the decayed and insect-riddled wood, the following tools are needed: a wooden or composition-rubber mallet, a chisel, several gouges with curved edges, and a knife. The knife is used on the bark and cambium, and the other tools on the sapwood and heartwood. For large-scale cavity operations, chisels, gouges, and bits operated by compressed air or electricity are available (Fig. 32).

All discolored and water-soaked wood should be removed from small cavities, even though the discolored wood appears relatively sound, for such wood harbors the most active stages of the wood-rotting fungi. Discretion must be used in treating extensive cavities, however, since the removal of much discolored wood may so weaken the tree structurally that it may become dangerously liable to breakage.

As a rule, old neglected cavities have such an extensive callus growth over their edges that the opening may be nearly completely closed, making it almost impossible to clean out the rotted wood (Figs. 33 and 34). The removal of this callus from the sides may result in considerable weakening, since it will eliminate important food-conducting tissues of the tree. Judgment must determine whether the cavity is better left alone or whether it

FIG. 32. Removing decayed wood with an automatic gouge.

should be widened to permit easier access to the decayed wood. If a considerable amount of callus is cut away, the foliage of the tree should be reduced proportionately by pruning to compensate for the loss in water- and nutrient-conducting tissues. The lengthening of the cavity interferes little with the function of the tree.

Shaping the Cavity. The interior of the cavity should be so shaped that no water pockets remain. When the cavity penetrates downward to form a pocket in the trunk or branch, the outer shell should be cut away to eliminate the pocket. If cavities are so shaped that deep water pockets cannot be avoided, some provision must be made to drain off the water. A hole should be bored from the outer bark below the pocket, into the lowest point of the depression, and a drainpipe placed at the opening. Because of the presence

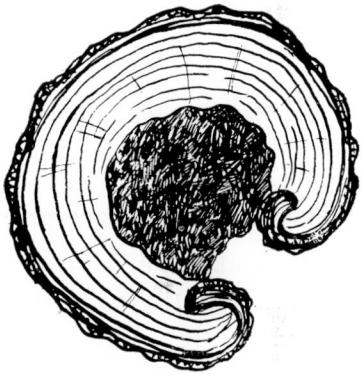

FIG. 33. LEFT: An old cavity nearly closed by callus growth. Such a tree might best be left untouched, because removal of the callus tissue incident to clearing out the decayed wood may weaken the tree considerably.

FIG. 34. RIGHT: Cross section of a trunk showing callus development over a cavity.

of sap and moisture, which collect near the opening, drainpipes are excellent foci for development of yeast, fungi, bacteria, and insects, and should be avoided wherever possible. A drainpipe cannot be inserted when a cavity extends below the soil surface. In such cases, the lower part, after being properly cleaned and dressed, should be filled with a satisfactory solid material to a few inches above the soil level, and the upper surface of the filling sloped to allow water to drain out of the cavity.

The treatment of the edge of the cavity unquestionably requires more care than any other phase of cavity shaping. Proper and vigorous healing will not take place unless the bark and cambium are healthy and sufficient sapwood is left under these tissues to prevent drying out. The bark should be cut only with a sharp knife to ensure smooth, clean edges, and orange shellac applied to the cut edges immediately to prevent drying of these tender tissues. The bark on the sides of the cavity should be shaped as nearly parallel to the line of sap flow as is practicable, and the upper and lower ends should depart from this line so gradually that they form a long tapering point. When the conducting channels in the bark and sapwood are cut across abruptly, as in round or slightly oval wounds, the sap does not move laterally with sufficient ease to promote healing or prevent death of unfed tissues.

Bracing the Cavity. Many large cavities require some type of support to keep the sides rigid or to hold some solid material more securely, if a filling is to be used later. Only an arborist with a good knowledge of the mechanics and physiology of the tree should be allowed to install the supports.

Bolts are placed in the tree by any of the methods shown in Figures 35 and 36. A strip of healthy bark and cambium, at least 2 inches wide, should be left between the edge of the cavity and the point where the holes are bored into the bark with a sharp bit. Bolts (made by a mechanic or obtained from an arboricultural supply store) of the proper length and diameter are inserted through the holes, oval iron washers placed in position, and the nuts screwed on. The washers and nuts should be countersunk into the wood deep enough to allow callus to form over them. All cut surfaces are then covered with tree paint.

So-called screw rods are commonly substituted for bolts, because they provide more support and eliminate the need for washers and nuts. Briefly the installation of a screw rod is as follows: A hole is bored through the trunk at the proper point with a bit $\frac{1}{16}$ inch smaller in diameter than the screw rod that is to be inserted. The openings in the bark are then reamed down to the wood and the cut surfaces shellacked. The length of the bore is carefully measured and marked on a piece of screw rod. At this point the rod is cut about two thirds through with a hacksaw. It is then screwed into the hole with a pipe wrench or some other suitable implement. When the rod is completely in place, the projecting end can easily be snapped off at the point previously weakened.

The importance of bracing long cavities by inserting bolts or screw rods in the sound wood above and below the wound (Fig. 37) should not be overlooked. Such a practice minimizes the possibility of the cracking of the wood by frost.

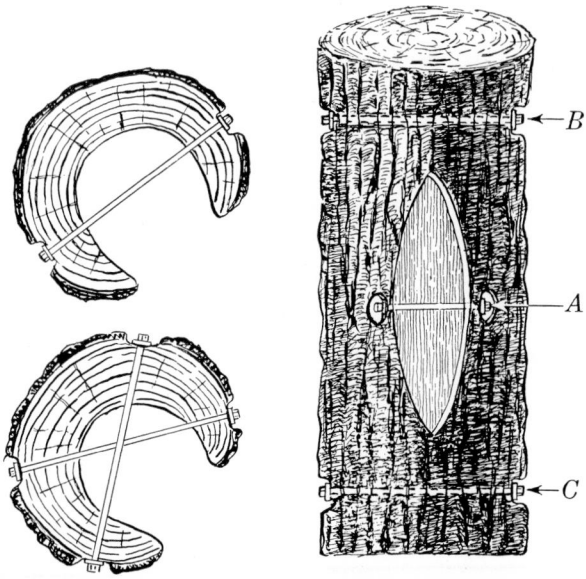

FIG. 35. UPPER LEFT: Cross section of trunk with extensive cavity, showing the installation of a single bolt to keep the sides rigid.

FIG. 36. LOWER LEFT: Cross section of the trunk showing the installation of double bolts to reinforce the thin sides of the tree.

FIG. 37. RIGHT: Bracing a long cavity. Aside from bracing at A, it is also essential to brace at B and C to minimize the possibility of cracking of the heartwood by frost action.

Additional mechanical support is also needed when treating split or crotch cavities. Before the holes are bored and the bolts set in place, the split sides should be drawn together by means of a block and tackle in the main branches arising from the split crotch. At least two bolts must be used above the crotch, and wire cables should be placed higher in the same branches, so that all parts will sway as a unit and thus prevent separation of the split sides.

Sterilizing and Dressing. The final essential operations in cavity work include sterilizing and dressing. Sterilization is accomplished by painting all exposed parts of the wood with creosote, copper sulfate solution (1 pound of copper sulfate crystals dissolved in 4 gallons of water), or bichloride of mercury solution (1 ounce in 7½ gallons of water). Copper sulfate and bichloride of mercury should be dissolved in wooden containers or crocks. These materials are poisonous to human beings and animals when taken internally, and extreme caution must be exercised in handling them.

After sterilization, a dressing of some ready-made tree paint should be applied over all exposed wood, and also over the bark and sapwood which were previously shellacked.

Open Cavities

The judgment of the arborist and the desires of the tree owner should determine whether a cavity treated in the manner described is to be left open or is to be covered or filled with some suitable material. If the cavity is left open, it should be inspected frequently to determine whether additional dressings are necessary or new rot has developed, in which case prompt attention should be given. The newly developing callus should also be inspected periodically and additional bark traced or cut back wherever healing is not progressing satisfactorily.

Covered Cavities

Occasionally, very old trees are so weakened structurally by decay that no extensive excavating, bolting, and reinforcing can be done, nor can such trees withstand the excessive weights of solid fillings. In other cases, the tree owner may decide that the open cavity is unsightly or otherwise undesirable and may wish to cover the opening. The so-called shell-filling can be used in such circumstances.

After the removal of decayed tissues, the cavity walls are sterilized and dressed as already described. A half-inch strip of bark is cut around the entire opening, leaving a ledge of wood exposed. The cut area is immediately shellacked. A large sheet of paper is then placed over the opening, and a pattern is made to fit the opening and overlap the half-inch ledge of wood. Sheet tin or copper is cut to the exact size of the paper pattern, the back is painted with asphalt or tar, and the metal sheet is nailed to the ledge of wood. Finally the entire surface is waterproofed with some dressing.

Decay continues in many hollow shells, despite most careful cleaning of the cavities. This method, however, is inexpensive, keeps out vermin and decaying leaves, and is a quick, simple way to cover a cavity.

Filled Cavities

Though the various individuals, firms and institutions interested in tree preservation are agreed that the practices already discussed—that is, cleaning, shaping, and sterilizing—should be employed in treating any cavity, they disagree about whether such a properly treated cavity should be filled or left open. The author does not intend to take sides in this controversy but will attempt to present both points of view. Here too, the judgment of the arborist and the desires of the tree owner should determine subsequent treatment.

It is commonly assumed that the principal objectives of filling a cavity are prevention of further decay, addition of mechanical support, provision of a foundation for healing callus, and improvement in appearance of the tree.

The first of these objectives has already been discussed, but some points merit repetition. The evidence at hand today indicates that when decay is well established in the heartwood, it is economically impossible by methods now in use to eradicate all the fungus tissue responsible for decay. Where decay is confined to the outer wood and sapwood, or to small cavities in small branch stubs, however, proper treatment is effective.

Advocates of cavity fillings argue that since normal callus growth has been repeatedly observed to seal small wounds perfectly and to check further decay, similar checking should also occur if an opening can be sealed perfectly by artificial means. Perfect closing of a wound, however, can take place only when the materials used as fillers do not become defective under the stresses set up by the swaying of the trees, and when the contraction and expansion of the wood are about equal to those of the filler. Some tree men argue that no filling is completely effective in sealing wounds or in preventing reinfection, since microscopic openings are present along the edge of the filling and beneath the callus and surface of the filling. Such openings appear constantly during the growing season and even during cold winter days. Normal trees shrink in diameter during the day while most fillings expand, and the reverse occurs during the night. Thus perceptible openings may be present several times each day between the filling and the tree. These openings allow tiny fungus spores to enter and cause infection.

Whether or not filling a tree with any material adds structural strength is another disputed point. Solid materials such as cement may actually weaken the tree by their excessive weight or by creating a fulcrum over which the part above the filling may snap off during wind or ice storms. When the cavity is left open, the normal rolling over of successive layers of callus provides extra support to the tree. The principle involved is much the same as that used in making T- or H-shaped structural steel beams. A thick rolling callus over an open cavity imparts considerably more strength to a tree than does a thin flat one growing over a filling. However, long narrow cavities close much more rapidly when a firm base is provided for the callus.

A large mass of concrete in a cavity will not bend with the swaying of the tree. As a result, either the concrete cracks or crumbles, or the wood wears away at the point where friction occurs. Some patented fillers are more flexible than concrete, thus minimizing the latter possibility.

A properly filled cavity improves the appearance of a tree and imparts the feeling to passers-by that the tree is receiving proper care. It also provides a considerable amount of personal satisfaction to most owners of trees.

Many trees are well worth the effort and expense required to excavate and fill the decayed portions, but before much money is spent, the following points should be carefully considered.

Factors Determining Whether To Fill a Cavity

Size of Cavity. The more extensive the cavity, the more difficult it is to remove wood-rotting fungi. The greater the size of the wound opening, the more difficult it is to maintain an effective and permanent barrier.

Age of the Tree. As a rule, callus formation proceeds very slowly in old trees, thus leaving a relatively large area exposed to the danger of reinfection for some time. Old trees are also subject to other troubles.

Vitality of the Tree. Trees that have been struck by lightning, injured by manufactured illuminating gas, covered with a grade fill, or otherwise weakened should not be filled until the condition has been remedied and the trees have regained their vigor through pruning and fertilization.

Longevity of the Tree and Susceptibility to Virulent Fungus or Insect Attack. Extensive fillings are not justified on such short-lived or 'weed' trees as the tree-of-heaven. As a general rule, cavities should not be filled in any of the following trees: chestnut, black locust, boxelder, white birch, cultivated cherries, dogwood, yellowwood, Lombardy poplar, redbud, mountain-ash, goldenrain, and most deciduous magnolias. Neither should trees already infected by some wilt fungus, such as Verticillium, be filled. Cavities may

Fig. 38. An extensive cavity ready for filling.

be filled in the following, but only if the specimen is in otherwise excellent condition: apple, catalpa, black oak, mimosa, paulownia, and willow.

Filling the Cavity

The preliminary procedure for filling a cavity should always include excavating, shaping (Fig. 38), sterilizing, and dressing as already described, with one modification. To hold the filling in place more securely, the sides of the cavity should be cut (undercut) as shown in Figure 39. This cut must not be made too close to the outer edge, since the wood between the cut area and the edge may be too thin and may dry out the sapwood. As a substitute for undercutting, strips of wood or lath can be nailed along the edge of the cavity, just below the cambial region. These should be water-proofed with tar or asphalt.

Where a cavity is to be filled with a material such as concrete, some provision must be made for draining sap and moisture, which may collect behind the filling. Drainage grooves, resembling in form the veins of a leaf, are cut in the back and sides of the cavity; that is, a center groove corresponding to the midrib, with side grooves leading into it, as do the lateral veins. A hole is then bored from the outer bark up into a pocket just below the lower end of the main drainage groove, and a drainpipe inserted. After being sterilized and painted, the entire inner face of the cavity is lined with 3-ply roofing paper held in place with tacks. This paper will prevent clogging of the drainage grooves and will eliminate seepage of the liquids to the surface of the filling. Nails may be driven halfway into the sides and back of the interior of the cavity to provide additional support for the filler (Fig. 39).

The surface of a filling should never extend above the level of the cambium. The filling of cavities above the level of the inner bark not only reduces callus development but in many instances inhibits its growth completely. As a guide to determine the proper level of the filling, the bark should be cut around the edge of the cavity, exposing a ledge of wood $\frac{1}{8}$ to $\frac{1}{16}$ inch wide. The surface of the filling should be level with that of the exposed wood.

Extreme care must be exercised at all times to prevent bruising of sound bark. An accidental blow from the mallet or the careless use of the chisel or knife may severely injure the cambial layer. In such cases, callus formation must start farther away from the edge of the cavity. This leaves the wood beneath the injured area subject to decay fungi for one to several years longer than normally.

Specifications for a Good Filler

Before beginning the filling operations, select a good filling material. The following are the specifications for a good filler:

Durability. The material should not decompose, crack during sharp temperature changes, or melt during the summer heat.

Safety. The filler should not contain harmful chemicals likely to injure the tree.

Flexibility. It must be sufficiently flexible (either by itself or because of the way it is placed in the cavity) to allow for the swaying and twisting to which the part filled is subjected during windy or stormy weather. Without flexibility, the filling breaks or crumbles or grinds away the margin of the wood adjacent to it.

Plasticity. A good filler should be plastic enough to conform to the outline of the cavity.

Water-proofness. The filler must not absorb moisture, since a moist environment adjacent to the wood is conducive to decay.

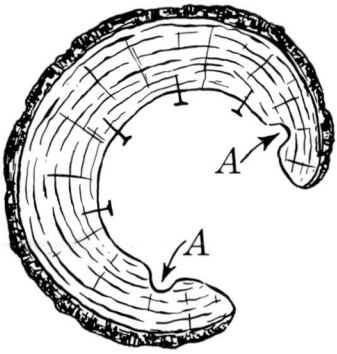

Fig. 39. left: Cross section of the trunk showing insertion of nails to help hold the filling in place. The wood is undercut at A to further help hold the filling.

Fig. 40. right: Filling the cavity with cement.

Filling Materials

Special Fillers. Several patented materials used by some of the larger tree companies meet the specifications for a good filler. They cannot be used, however, without the permission of the patent holders.

Concrete. Probably the cheapest and most readily available filling material is cement mortar or concrete. Two parts of clean sharp sand are mixed dry with one part of cement, and sufficient water is added to produce a stiff mortar. The mixture is then placed in the cavity with a trowel and tamped firmly, especially near the walls of the cavity. Large fillings should be placed in layers or sections not over 6 inches wide to prevent cracking from expansion and contraction and from twisting and bending when the tree is swayed by the wind (Fig. 40). Three-ply roofing paper or some other suitable material is used to separate the layers. In wide cavities the sections should have vertical as well as horizontal layers with the roofing paper in between. After the surface is smoothed, the concrete should be painted with a water-proofing material. Cement fills are most satisfactory in small cavities or in large root or trunk-base cavities, where no twisting occurs when the wind sways the tree.

Asphalt Mixtures. Several asphalt mixtures make better fillers than cement, but are more difficult to prepare and put in place. A commonly used mixture is prepared as follows: Melt asphalt in a container over a fire and slowly add dry hardwood sawdust, excelsior, or wood shavings. Continue adding and stirring the solid material as long as it is blackened by the asphalt. The pasty granular mixture formed can then be put in the cavity and tamped into place. When the cavity is very large, or the mixture is too soft, the filling will slump or flow out of the base of the cavity. Canvas, burlap, or thin strips of lath can be fastened over the surface to hold the filling in place. After the filling has hardened, the barriers are removed and the face is smoothed with a hot iron and dressed with tar, asphalt, or paint.

The principal disadvantage of asphalt mixtures is their tendency to soften and slump when exposed to hot summer sunlight. They are much more flexible than concrete, however.

Other Cavity Fillers. Wood blocks, cork products, rubber blocks, and other materials are commonly used for filling cavities. Most of these have some outstanding advantages over the concrete and asphalt fillers.

Quality Cavity Fills. In the various methods outlined, no attempt has been made to discuss many of the details that might arise during the operation.

FIG. 41. A properly filled trunk cavity. Notice the thick roll of callus one year after filling and the points where bolts were inserted to reinforce the sides of the tree.

Actual practice, rather than reading, makes the tree expert. For that reason, the filling of a particularly valuable tree should be entrusted only to a reliable commercial arborist who thoroughly understands not only the mechanics of tree structure but also the materials and their use (Figs. 41 and 42). The skilled arborist takes greater pride in filling cavities than in any other phase of his work. Several companies are so confident of their results that they guarantee their cavity work. A reliable tree expert will not promise to cure a tree of heartwood decay by some secret preparation. He will not undertake extensive fillings if the tree is not vigorous enough to stand the treatment. Above all, he realizes that a good tree poorly filled will suffer far more in health and appearance than one left untouched.

FIG. 42. A properly filled basal cavity.

ADVICE TO TREE OWNERS

A written contract covering certain phases of the work is usually advisable when a tree owner employs a commercial arborist to take care of his trees. Any reliable concern will readily agree to the following specifications:

1. The type of cavity work to be done—that is, whether the cavity is to be filled, covered, or left open—should be agreed upon.

2. All badly diseased, punky, or worm-eaten wood should be removed.

3. Orange shellac or some other efficient and harmless covering should be applied to the cut edges of the sapwood, cambium, and bark immediately after the final tracing cut, or as soon afterward as the cut surface becomes sufficiently dry to ensure a good coverage.

4. All other cut surfaces should be painted with some disinfectant, followed by some effective and more or less permanent tree paint or dressing.

5. Climbing spurs should not be used on any part of the tree, and only soft-soled shoes should be worn by workmen.

6. A guarantee to repair free of charge any defects of workmanship appearing within a reasonable period (a year or so) should be gvien.

7. Regular inspection of the trees every year should be provided unless the owner prefers to do this himself.

7

BRACING AND CABLING

The importance of supplying artificial support to structurally weak or injured trees has long been recognized. Some trees, such as oak, beech, and locust, rarely need artificial support because they consist of tough wood fibers and usually have a single major stem from which branches arise at approximately right angles. Despite the absence of tough wood fibers, even pine, spruce, and fir seldom need support because of their single stem and type of branching. Other trees, such as elm, maple, and poplar, which have inherent structural weakness resulting from the manner in which their branches arise, often require considerable mechanical support.

The provision of some support by means of iron braces or wire cables is an inexpensive form of tree insurance and is an important phase of any complete tree maintenance program. Artificial supports prevent such injuries as crotch-splitting and branch breakage, both of which eventually lead to wood decay. They also prolong the life of many trees after decay or extensive cavities have already developed. It is well to bear in mind that the principal function of the heartwood, or center of the tree, is to provide mechanical support, and that a tree can live for many years despite the presence of a cavity or extensive rotting of this tissue. In such cases, most of the sapwood is free of decay, and together with the cambium and bark above it, can continue to function in the tree's metabolic processes for a long time.

When Is Artificial Support Justified?

In many situations the proper installation of braces and cables proves to be extremely helpful in prolonging the life of a tree. This is especially true of trees having tight V-shaped crotches, split crotches, or inherent susceptibility to breakage.

Tight V-Shaped Crotches. These occur commonly in certain trees in the course of normal development. In others, they result from lack of proper pruning of the young trees. A brief discussion of how V-shaped crotches develop is essential for a better understanding of preventive and subsequent treatments.

In the narrow angle between two sharply forking major branches, the bark and cambium are hindered from developing normally or may even be killed by the pinching action of the branches as they increase in size. This results in a weak union, of the type that readily splits when the branches are subjected to heavy wind or ice storms. In contrast, from trees with a single main trunk, the lateral branches arise more nearly at right angles, forming a U crotch (Fig. 43), which does not split off so readily in times of stress.

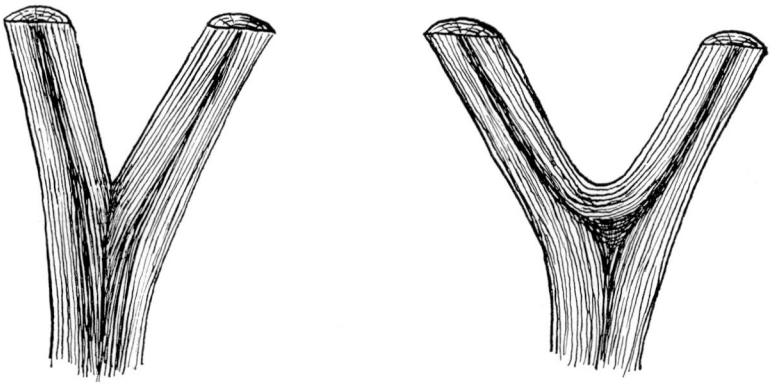

FIG. 43. LEFT: A V-shaped crotch results in a weak union. Branches that form such a crotch are easily broken during storms. RIGHT: Branches joined by a U-shaped crotch do not split or break off so easily.

Split Crotches. The factors favoring ease of splitting are present very early in the tree's life. In many cases, early and judicious pruning by an expert will eliminate the possibility of a split crotch later. As a rule, the arborist finds it difficult to convince the tree owner that preventive practices should be applied when the tree is young, and because of improper early care, considerable work is necessary in repairing split crotches on older trees.

Split crotches should be supported and treated not only to prevent further splitting and complete loss of one or both branches, but also to check wood-decay, which is sure to follow.

Susceptibility to Breakage. Because of such inherent properties as heavy foliage production and brittleness of the wood, some species of trees are more susceptible than others to breakage by wind and ice storms.

Among the trees that are quite susceptible to breakage by winds are: chestnut oak, honey locust, horsechestnut, mimosa (silk tree), poplars, maples (red, silver, and sugar), sassafras, Siberian elm, tulip tree, and willows.

The following trees are more resistant to wind damage, although they too can lose branches and be toppled by winds of very high velocity: American beech, American elm, ginkgo, hackberry, little-leaf linden, London plane, Norway maple, red oak, sweetgum, and white oak.

Among other conditions that govern the need for supporting a tree are the following: weakening of the tree as a result of removal of some roots during wall or curb construction or grade changes; decay of large areas in the trunk or roots; removal of nearby trees, which previously helped to support the tree with their intertwining root systems or to shield it; and exposure to high winds and sleet storms.

Types of Artificial Supports

Despite the fact that bracing and cabling of trees is an old practice, real improvements in materials and methods have been made only in rather recent years. The iron collars, rods, and chains formerly used were expensive, unsightly, inefficient, and often harmful. The iron collar, especially, caused considerable damage by preventing normal sap flow in the limb it was supposed to help. Today improved and standardized materials are readily available at tree-surgery supply houses.

Modern tree-supporting technique is divided into two main types: rigid and flexible bracing. Rigid bracing involves the use of bolts and threaded rods for supporting weak or split crotches, long cracks in the trunk or branches, and cavities. The types for cavities are discussed in Chapter 6. Flexible bracing involves the installation of wire cables high in the tree to reduce the load on weak crotches and long arching limbs, without hindering the normal branch sway. Both rigid and flexible braces are commonly installed in the same tree.

Rigid Bracing

Prepared lengths of steel or duralumin rods with lag threads at both ends (screw rods) are commonly used in rigid bracing. The holding power of such rods is similar to that of ordinary wood screws fastening two boards together. Rods or bolts with machine thread, when used, are held in place by washers and nuts of appropriate size. Though these do not supply so much support as do lag thread rods, they must be used in the following instances: when more than half the wood in the area where they are to be inserted is decayed; where only a thin shell of sapwood remains around large cavities; and where future decay may develop at the point of insertion. All rods should be covered with a tree wound dressing before they are inserted, and all exposed metal portions painted with a metal preservative paint.

Rigid braces are employed to support crotches, to hold split limbs or trunks together, to support cavities, and to prevent two branches from rubbing against each other.

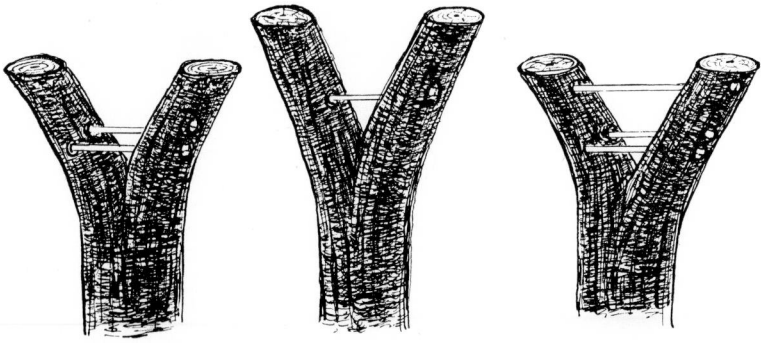

FIG. 44. LEFT: For larger limbs, two parallel screw rods provide more support than a single rod.

FIG. 45. CENTER: Installation of a single screw rod to support a V-shaped crotch.

FIG. 46. RIGHT: For very large limbs, the installation of a third rod above the parallel rods is suggested.

Crotch Bracing. The installation of artificial support is justified on sound trees with weak crotches, and when crotch cavities have already developed. To provide additional support, as a rule, wire cables must be installed (see p. 110) higher in the tree whenever the branches extend more than 20 feet above the crotch. In small trees or in secondary crotches of large trees, a single rod may be inserted through the center of the two limbs above the crotch (Fig. 45). Larger limbs may need the support of two parallel rods (Fig. 44). These are usually placed above the crotch at a distance twice the diameter of the limbs at the point of rod insertion, and horizontally separated by a distance equal to about half this diameter.

Split crotches between large limbs should first be closed by pulling the

FIG. 47. The three rods sketched in Figure 46 are shown installed in an American beech.

limbs together with properly placed rope and tackle blocks high in the tree. When the crack is closed, a hole is bored through the split parts with a bit 1/16 inch larger than the bolt to be used. The bolt is then inserted and held in place with washers and nuts. Additional support is provided by inserting a second bolt a few feet above in the same branches, as shown in Figures 46 and 47.

Split Limbs or Trunks. Screw rods may be used to hold split limbs together or to support the sides of long splits in the trunk. They are usually placed about a foot apart and are staggered so as not to be in the same direct line of sap flow.

Cavity Bracing. Cavity bracing is described in detail under cavity treatments.

Rubbing Limbs. Limbs that rub against each other can be braced together or kept separated by proper bracing. A single rod screwed through the center of each limb at the point of contact will hold the limbs rigidly together and thus eliminate rubbing (Fig. 48). If it is desirable to keep such limbs separate,

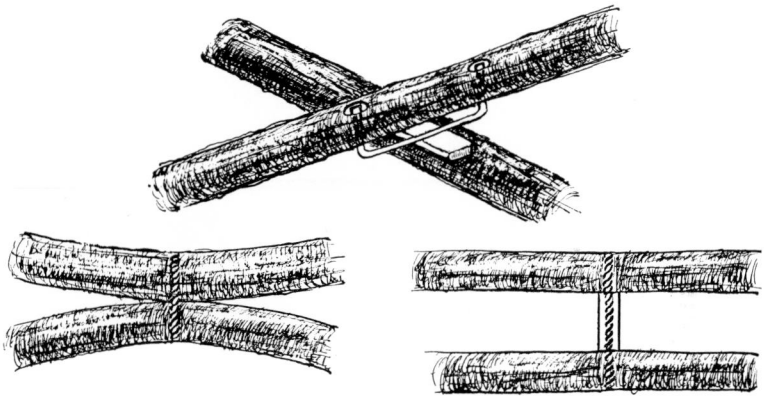

FIG. 48. LEFT: Method of inserting a screw rod to hold two limbs together and thus prevent rubbing.

FIG. 49. RIGHT: Separating two branches by means of a screw rod and a short piece of pipe.

FIG. 50. ABOVE: Method of installing a U-bolt to prevent rubbing of two limbs and at the same time allow free movement of the branches.

a length of pipe equal to the desired space between the two limbs can be placed over the rod before the branches are fastened together (Fig. 49). If the limbs are to be separated and free movement between them is desired, the installation of a U bolt in the upper limb, which rests on a wood block attached to the lower limb, will prove satisfactory (Fig. 50). In time the wood block will need replacement because of wear.

Flexible Bracing

Wire cables are used to brace trees together and strengthen weak crotches. These are usually placed rather high in the tree to provide maximum support by the use of materials of minimum size and cost. Copper-covered steel wire, although more expensive than ordinary galvanized wire, is stronger, size for size, and does not rust.

Flexible bracing systems commonly employed are of several types. In the simple direct system, a single cable is installed to support two limbs arising from a single crotch (Fig. 51A). In the box system, cables are attached to all

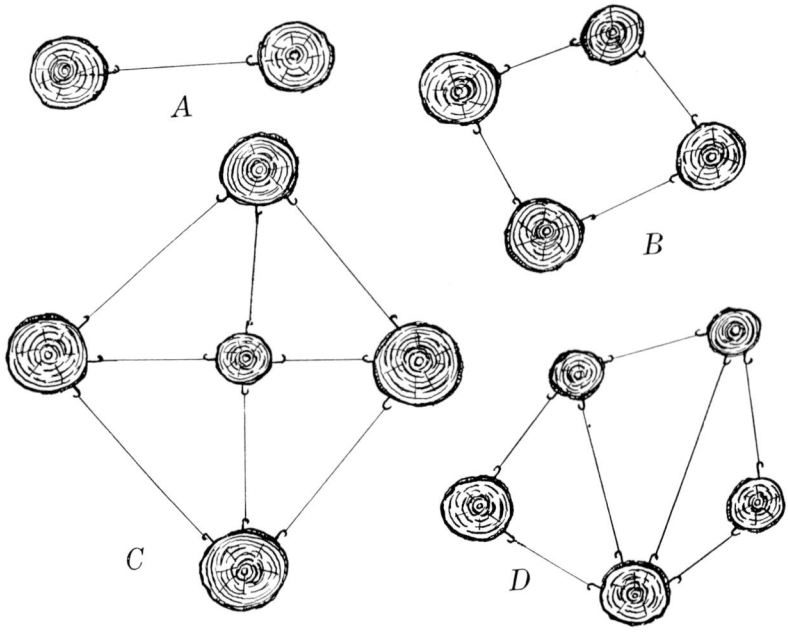

Fig. 51. Flexible bracing systems. *A*, Simple direct: a single cable supports two limbs arising from a single crotch. *B*, Box: cables attached to all large limbs in a rotary manner. *C*, Hub and spoke: cables radiate from a limb or metal ring in the center of the main branches. *D*, Triangular: cables placed in triangular form.

large limbs of a tree in a rotary manner. This arrangement provides lateral support and maximum crown movement, but little support to weak crotches (Fig. 51B). The hub and spoke system, in which cables radiate from a limb or a metal ring in the center of the main branches (Fig. 51C), provides little direct or lateral support. In the triangular system cables are placed as shown in Figure 51D to provide direct support to weak or split crotches and lateral support to the branches. Triangular bracing is probably the most efficient of the various types described.

Materials and Installation Procedure

The wire cables are usually placed at a point approximately two thirds the distance from the crotch to the top of the tree. They should be so placed as not to rub against each other.

Lag hooks, hook bolts, bent eyebolts, and other materials are most often used to anchor the cables. Such anchors should be well separated, when more than one is used on a single branch, and should be placed in a straight line with the cable and as nearly at right angles to the branches as possible.

Lag Hook Installation. Cadmium-plated lag hooks (½ to ⅝ inch diameter and 6-inch thread) are recommended as anchors for most branches. The exact size selected is governed by the amount of stress and the size of the branches. Holes about 1/16 inch smaller than the hooks are bored at the proper anchorage points, and the hooks are screwed in just far enough to allow a cable to be slipped over the open ends. After the lag hooks are in place, the limbs are pulled together sufficiently so there will be no slack in the cable after it is hooked on. A thimble is inserted over the hook and the cable is then passed through it and spliced to form a permanent union between the two anchors. All spliced areas should be coated with a metal preservative paint.

Eyebolt Installation. Where wood decay is present or considerable stress occurs, eyebolts or hook bolts are generally used as anchor units. The hole for these is drilled completely through the wood, the bark is countersunk to a depth below the cambium, and the bolt, washer, and nut are put in place. The cable ends are then inserted through the eyes and properly spliced.*

* The reader desiring more detailed information on the engineering principles involved in the installation of cables and braces is referred to the excellent bulletin *Tree Bracing*, prepared as *Tree Preservation Bulletin* No. 3 of the National Park Service, United States Department of the Interior. The Superintendent of Documents, U. S. Government Printing Office, Washington, D. C. will send copies at 20 cents each.

8

TREES SUITABLE FOR VARIOUS LOCATIONS

Tree species vary greatly in their ability to grow in different environments. Before selecting a tree, therefore, one should know something about the different kinds of trees, the conditions under which they are to grow, and what to expect of them.

When the author wrote his first book on trees about twenty years ago, the consensus among shade-tree commissions, park department officials, arboricultural firms, and other agencies empowered to regulate the selection, planting, and care of trees on public property was that tall-growing trees such as elms, oaks, planes, and maples were best suited for planting along city streets.

Times have changed and so have ideas on the proper kind of trees to plant. Many officials now realize that it is best to select trees that will give a minimum of trouble and maintenance costs in future years. While many of the tall-growing or standard shade and ornamental trees have long been favorites on wide city streets, they are rapidly losing their popularity because of high maintenance costs. Such trees are more expensive to spray and prune, and during severe wind, ice, and snowstorms or hurricanes cause great damage to power and telephone wires. Moreover, tall-growing trees have a greater tendency to crumble curbstones or push them out of line, crack and raise sidewalks, and clog sewer pipes. Tall-growing trees, discussed later in this chapter, should be planted where they have lots of room to grow and where they will not interfere with public and private utility services.

The modern city planner realizes that trees should be fitted into the available space. This idea developed from a street tree program started under Edward H. Scanlon for the city of Cleveland about twelve years ago.

People acquainted with tree raising realize that the tall-growing trees such as elms, pin oaks, and planes were popular in the past for several reasons. They are easy to reproduce either from seeds or from cuttings, grow rapidly, transplant readily, and require relatively little after-care.

The low-growing trees, on the other hand, require more skill to grow and maintain. For the first four or five years after being transplanted they must be helped more, and perhaps sprayed and pruned more. Although these practices are more expensive, the small type trees are well worth it, for they usually serve their purpose two or three times as long as the tall-growing types, without the many disrupting habits of the latter. And should they have to be removed for any reason, the cost of removal is only one quarter to one sixth that of tall trees.

Not only are most of the small type trees suitable for streetside planting, but they are ideal for modern ranch-type houses.

Proponents of the tall-growing trees claim that low-growing trees have several serious drawbacks: their relative scarcity as compared with tall-growing kinds; their low-hanging branches, which interfere with vehicular and pedestrian traffic; their inability to provide as much shade as tall-growing kinds.

In time, as more nurserymen grow the low-growing forms, the first drawback will be overcome. The second problem can be overcome if low-growing trees are grown properly with a central leader; for example, with Kwanzan flowering cherry, the graft is made 6 to 7 feet above the ground level. The drawback of low-hanging branches can also be overcome if communities permit the planting of small-type trees inside the property line instead of between the street and sidewalk. The third drawback is not solved so easily, but actually shade is not so important as it was formerly. Street beauty and stabilization of property values are more important today. Where more shade is desired, close planting of low-growing trees is suggested.

The ideal small type tree should grow rapidly; it should reach a useful size soon after being transplanted to its permanent site and should then abruptly slow down. It should be reasonably free from insect pests and diseases and should have pleasing growth habits.

The initial cost of such trees of streetside planting size is much greater than that of the standard tall-growing types, but, as has been said, maintenance and removal costs, when necessary, are much lower.

Low-growing Trees for Streetsides and Lawns

Many of the following low-growing trees can be used for streetside planting where their low height will not interfere with traffic or where children are not apt to break branches for their showy flowers. All of them are ideally suited for lawn planting on small properties or near ranch-type houses. A few, like *Magnolia Kobus* and *M. grandiflora,* may grow to considerable size, however.

Ashes

Flowering ash, *Fraxinus Ornus,* is a low-growing, round-headed tree with fragrant, whitish flowers.

Modesto ash, *F. velutina glabra,* is a fast-growing, lovely tree which reaches a height of 25 feet. It has a smooth green trunk and bright crisp green, slightly pendulous branches.

Sorbus alnifolia is a handsome small mountain ash from Asia. Bright green leaves turn orange and red in autumn, and deep pink berries persist well into the winter months. *Sorbus Aria* is another pleasant small mountain ash. Its leaves, bright green above and white-tomentose below, contrast with the large orange-red to scarlet fruit. The variety *aurea* has yellow leaves; those of *chrysophylla* are even deeper yellow.

Other desirable ashes are the Manchurian ash and the blue ash.

Cherries

Varieties of the oriental cherry *Prunus serrulata* produce an unsurpassed display of flowers in early spring. They should be planted in full sun and in well-drained fertile soil. The most popular variety, Kwanzan, is the showiest, hardiest, and best of all flowering cherries for street tree use (Fig. 52). When grafted on 6½ foot stems, its upright branches do not interfere with vehicular and pedestrian traffic as do those of other varieties. More than a hundred Kwanzan cherries are planted in the gardens of the United Nations Headquarters in New York. Other good cherries being grown for streetside planting are Akebono, Daikoku, and Shirotae.

Yoshino cherry (*P. yedoensis*), another excellent cherry, grows rapidly when young, then slows down. It is covered with pendant clusters of fragrant pale pink to white flowers in spring. It and Kwanzan are the major components of the famous cherry display at the Tidal Basin in Washington, D. C.

Other excellent flowering cherries are Sargents, *P. Sargentii,* and double-flowered Mazzard (*P. avium plena*).

FIG. 52. Kwanzan Flowering Cherries grafted on 6½ foot stems make excellent streetside trees. These, growing in Portland, Oregon, are about 30 years old and 24 feet tall.

CRABAPPLES

Crabapples, *Malus* species, can be both ornamental and utilitarian. Some produce bright-colored fruits that make an excellent jelly in addition to the truly magnificent flowers in early spring. Others are grown primarily for their showy flowers.

Among the best known that serve the dual purpose are Arnold crab, *M. Arnoldiana,* and Aldenham crab, *M. purpurea aldenhamensis.* Two recent introductions that fall in the same category are Redfield and Redford. Both are hardy and produce good-sized fruits with flesh that is red throughout. Other good varieties are Dolgo and Young America. The former is a Russian variety introduced into America by the South Dakota Experiment Station, while the latter is an old variety of unknown origin. Fruits of both make excellent jelly.

Among the older varieties of crabapples grown primarily for their flowers are Bechtel's crab, with fragrant double pink blossoms; Carmine crab, with large rose-madder flowers; and Dorothea, with bright tyrian rose flowers. Newer ornamental crabapples are Van Eseltine, which produces double pink blossoms and retains its small fruits during the winter, thus providing food for birds; and the very hardy Crimson Brilliant, with semi-double, vivid crimson flowers. Other excellent crabapples are Shakespeare, with pink flowers; Beauty, with white flowers; and Frau Louise Dittman, with bright pink flowers.

Dogwoods

The dogwoods are among the most popular of the low-growing trees. Best known and most widely planted is the flowering dogwood (*Cornus florida*) (Fig. 53). It does best in the eastern United States in well-drained soils. It

Fig. 53. The flowering dogwood is one of the most beautiful low-growing trees.

flowers either in full sun or in partial shade. The large white bracts surrounding the flowers are very conspicuous in spring, and its leaves and fruits are colorful in autumn. It is most successfully transplanted in spring immediately after flowering. There are a number of pink and red strains of the flowering dogwood. Latest to appear is the Prosser flowering dogwood, which has deeper red bracts than the older red-flowering kinds.

A close relative of our native dogwood, the Japanese dogwood (*C. Kousa*) grows about 20 feet tall and blooms about three weeks after the flowering dogwood. Its small yellow flowers, which last about a week, are surrounded by four conspicuous white bracts which are pointed at the tips and not notched and rounded as are those of the flowering dogwood. Because the flowers and fruit are borne on the upper side of horizontal branches, the tree is best planted where it can be seen from above.* The Chinese dogwood (*C. Kousa chinensis*) usually has larger bracts than the Japanese species.

Dogwoods are extremely sensitive to droughts. Hence, water should be supplied to the soil around dogwoods when no rains have fallen for a week or so during the summer months. Otherwise, the leaves will begin to wilt at the tips and turn red, then brown, long before they ordinarily would.

Magnolias

The outstanding magnolia for northern areas, except for the extremely cold regions, is the star magnolia (*M. stellata*). It produces semidouble white flowers more than 3 inches across. The variety *rosea*, with pink buds that open to pinkish flowers, is even more striking. A newer form, *M. stellata rubra*, has dark pink or purplish-red petals. Waterlily is a white hybrid with slightly smaller flowers but more compact growth than the species.

Star magnolia does best in rich, porous soil, although it tolerates clay soils. It withstands city conditions rather well. It definitely dislikes lime and fresh manure. It is most successfully moved with a soil ball in spring after growth starts.

The saucer magnolia (*M. soulangeana*) is best known for its large pink to purple flowers, which are 5 to 10 inches across in early spring. It also does well under city conditions.

The two best magnolias for streetside plantings are *M. Kobus* and *M. Loebneri*.

* A pink-bracted form of Japanese dogwood (*C. Kousa rubra*) is available at a few nurseries.

Southern magnolia (*M. grandiflora*) is a magnificent tree with large evergreen foliage and fragrant white flowers up to 8 inches across. It thrives in hot, humid climates and will tolerate some cold but not prolonged freezing. Although a few old specimens are growing as far north as Long Island, this tree does best in the southern part of the United States. Sometimes used as a street tree, it is more effective in lawn plantings.

Maples

There are a goodly number of low-growing maples suitable for streetside plantings, for lawns, and for small properties. Many of these are of Asiatic origin; a few are native species.

Amur maple (*Acer ginnala*) does best in the northern United States in rich, well-drained soil. It will thrive either in full sun or in partial shade. It has excellent foliage and will grow in dry spots. It is most effective when planted in groups of two or three.

Hedge maple (*A. campestre*) is an unusual low-growing tree that thrives in the eastern United States in well-drained, deep, fertile soil. Its growth resembles that of the Japanese maple, and its foliage is yellow in fall.

Japanese maple (*A. palmatum*) is a fine-textured, graceful tree with deeply lobed leaves. Some varieties, such as *atropurpureum,* have beautiful red leaves which retain their color throughout the season; others, such as *Oshiu-beni,* have 9-lobed red leaves in spring which turn green when mature.

Mountain maple (*A. spicatum*) grows well in the northern states on rocky hillsides. It is a very hardy tree with light green leaves that turn orange and scarlet in fall, and with fruits that are bright red in summer.

Two outstanding low-growing maples are David's striped-bark maple, and the Paperbark maple. The former is an upright grower of great value for streetsides. The latter has cinnamon-brown bark which exfoliates in paper-thin strips similar to that of certain birches. The peculiar bark is noticeable at some distance, which makes the tree outstandingly interesting throughout the year.

Other low-growing maples are the Trident maple (*A. buergerianum*); Coliseum maple (*A. cappadocicum*) and its varieties *aureum* and *rubrum;* and the Montpelier maple (*A. monspessulanum*).

Among other excellent low-growing maples, which unfortunately are not yet widely available, are *A. truncatum, A. diabolicum purpurascens,* and *A. mandshuricum.*

Miscellaneous Trees

Crape myrtle (*Lagerstroemia indica*), though usually considered a shrub, can become a tree reaching a height of 25 feet in the South (Fig. 54). Its showy, lilac-like clusters of flowers, which vary from deep red to purple, pink, lavender, and white, appear from late summer into early fall. Although re-

Fig. 54. Crape myrtle is a low-growing lovely tree in the warmer parts of the country.

liably hardy only as far north as Baltimore, good-sized specimens are to be found in sheltered spots in New York City and along the coast to Cape Cod. It does best in a hot, dry climate and in a fertile, well-drained loam.

The *dove tree* (*Davidia involucrata*) is a small to medium-sized tree valued for its unusual flower, which is a yellowish-tan ball surrounded by white bracts.

Franklin tree (*Franklinia alatamaha*), a native of the southeastern United States, is now known only in cultivation. Its beautiful flowers, up to 3 inches across with yellow stamens in the center, appear in late summer and early fall when most other flowering trees are through for the season. It prefers a moist but well-drained soil and is more difficult to establish than most trees.

Fringe tree (*Chionanthus virginica*) grows best in the East and in Gulf Coast states in moisture-retentive sandy loam soils. It is fairly tolerant of smoky city conditions. In late May and early June it forms numerous white blossoms. Later, purple fruits resembling grapes appear. These are devoured by birds.

Goldenchain tree, or Waterer laburnum (*Laburnum Watereri*), an upright, rather stiff tree, bears many pendulous clusters of bright yellow pea-like flowers. It does best in moist soils with a limestone base but will tolerate a wide variety of soils. It is reliably hardy only in the southern parts of the country. *L. Vossi,* a hybrid between the common laburnum and the Scotch, is said to be even more desirable than the Waterer laburnum.

Goldenrain tree (*Koelreuteria paniculata*) grows almost anywhere but is most common in the Midwest and does especially well in dry soils. It is drought-resistant and has average resistance to windstorms and ice. It forms showy panicles of yellow flowers in summer, followed by conspicuous bladder-like pods that are handsome in autumn.

Hawthorns (*Crataegus* species), like the crabapples, are especially attractive when in flower and fruit. They do best in the northern and eastern states, in well-drained, fertile soils. Nursery-grown trees are transplanted easily. Hawthorns are excellent trees to plant around modern ranch-type houses. Best of the older kinds are Paul's Scarlet thorn (*C. oxyacantha pauli*), with clusters of half-inch blossoms of deep rose color, and the Washington thorn (*C. phaenopyrum*), with white blossoms in May followed by scarlet berries and brilliant foliage in autumn. Lavalle hawthorn (*C. lavallei*) is considered the best of the white-flowering kinds. The newest thorn to be made commercially available by Edward H. Scanlon and Associates, Olmsted Falls, Ohio, the Glastonbury thorn or the Holy Thorn of Glastonbury (*C. monogyna biflora* or *praecox*), has had long association with early Christian history.

The *American hophornbeam* (*Ostrya virginiana*) is a low-growing tree with very hard wood. A mysterious fungus disease has been killing this tree in recent years.

A low-growing graceful tree, *Yeddo hornbeam* (*Carpinus tschonoskii*) does well in the eastern United States from New York southward. It makes a compact head and has up-reaching branches which lend it an illusion of tallness. Its leaves are an attractive reddish-bronze color in the fall, and best of all, it has no serious faults. Two other excellent members of this genus are *Japanese hornbeam* (*C. japonica*) and upright *European hornbeam* (*C. betulus pyramidalis*).

Katsura tree (*Cercidiphyllum japonicum*) forms numerous upward-branching trunks somewhat like the American elm. If the tree is allowed to grow with a single trunk, its habit is columnar like the Lombardy poplar. It grows best in fertile, moist soil in eastern and northern United States. Its rounded leaves, nearly 4 inches long when mature, are markedly resistant to insect pests and fungus parasites.

Eastern redbud (*Cercis canadensis*) may be grown throughout the South and in the eastern United States as far north as Boston along the coast. It prefers a rich soil and a sunny or slightly shaded area. Its pea-shaped purplish-pink flowers are very showy in early spring. It is used as a street tree in the South and as a lawn specimen in the North. A white-flowering form is also available. The Chinese redbud (*C. chinensis*) is less hardy than the ordinary species and it produces dense clusters of flowers that completely hide the twigs which bear them.

Russian olive (*Eleagnus angustifolia*), one of the hardiest of the low-growing trees, withstands extremely low temperatures far up into Canada. It has beautiful gray-green to near-silver leaves, which are very striking when ruffled by the breeze. Unfortunately it does not transplant easily, and hence large specimens must be moved with a soil ball.

Silk tree or *mimosa* (*Albizzia julibrissin rosea*) has leaves composed of many small leaflets which give the tree a delicate lacy appearance. The fern-like leaflets curl up at night. Round clusters of dainty pink flowers resembling fluffy balls 1 to 2 inches in diameter cover the tree in midsummer. This tree will grow on almost any soil, including poor, dry rocky ones. It is not reliably hardy north of Washington, D. C., although large specimens are growing along the east coast as far north as Boston. It is a fast-growing tree, and its wood is so brittle that heavy rain and wind storms frequently cause branch breakage. Although it is one of our prettiest flowering trees in midsummer, it is not very handsome in winter when bare of foliage.

Carolina silverbell (*Halesia carolina*) is an interesting small tree whose branches are covered with small, pendant white flowers in spring. It does best in protected, partially shaded spots and prefers a rich moist loam that is reasonably well drained.

Smoke tree (*Cotinus coggygria*) is a low-growing tree which is hardy from Boston through Ohio, Missouri, Kansas, Oklahoma, and eastern Oregon. Its flowers, which hang in misty panicles about 8 inches long in June and July, give the tree an appearance of being enveloped in purplish smoke. It will grow in dry soils of low fertility.

The *Japanese snowbell* (*Styrax japonica*) is a small tree of flat-topped form with curved, horizontal branches, dense foliage, and pendulous, bell-shaped flowers appearing after the foliage, but not hidden by it. It does best in light, well-drained soil and in open situations. The fragrant snowbell (*S. obassia*) is a more or less columnar tree with large leaves. When in bloom in early summer it is exceptionally beautiful.

Sorrel tree or *sourwood* (*Oxydendrum arboreum*) is an eastern tree that grows well in moist, acid soils, and in partial shade. Its late fall flowers are very striking, as is the scarlet color of its leaves at the same season. It is relatively free from pests and diseases and hence needs little if any spraying.

Stewartia (*Stewartia koreana*), though little known, is an excellent, hardy tree with interesting bark resembling that of the sycamore, white flowers in summer, and orange-red leaves in autumn.

Less Common Species

Cucumber tree (*Magnolia acuminata*) is more suited for lawns than for streetside planting. It has large, interesting leaves and a cucumber-like fruit, which is inedible. Its winter silhouette is excellent, and it is very resistant to wind and ice storms.

The *Dawn redwood* (*Metasequoia glyptostroboides*) existed in North America about 15 million years ago. It was reintroduced to this hemisphere via seeds collected from living trees in central China in 1945 through the efforts of the late Dr. E. D. Merrill of the Arnold Arboretum.

Dawn redwood somewhat resembles the deciduous bald cypress in foliage character although its flat needles are nearer the size of a hemlock's. The trees are not particularly handsome.

Hundreds of Dawn redwood trees are now growing over the United States. Young trees are available from several nurseries, including Kelly Brothers, Dansville, N. Y. Some seedlings make as much as 4 feet of growth in a single season. Their rapid growth rate suggests they may eventually serve as a good source of lumber. They are not recommended for streetside planting and are best suited for lawns.

Kentucky coffee tree (*Gymnocladus dioicus*) grows best in rich, light soils in the northern United States. It is more useful for lawns than for streetsides. It is unusually resistant to storm damage and has an attractive silhouette in winter.

Japanese pagoda tree (*Sophora japonica*) grows well in the eastern United States except in the very cold regions. Its pea-like white flowers appear in late summer. It tolerates city conditions but is rather subject to branch breakage during ice storms and occasionally to winter injury.

Trees with Upright Growth Habit

In the last 10 years, the demand has increased for trees with upright (fastigiate) or columnar form for planting along narrow streets and on small city lots. The old familiar Lombardy poplar and the Bolleana poplar were the only kinds available for such planting thirty or so years ago. The high mortality among poplars affected by the Dothichiza canker disease (p. 360), however, makes the planting of these trees a gamble in most parts of the country.

Many nurserymen and most particularly, Edward H. Scanlon and Associates, are now offering upright forms of shade and ornamental trees. Following are a few of the better kinds to plant along narrow streets or where space is a limiting factor:

Two narrow upright forms of the sugar maple are Newton Sentry and Temple's Upright. The Newton Sentry has many ascending branches, no dominant central leader, and dark green leaves with leathery texture and wavy margins like those of black maple. The Temple's Upright has few ascending branches; a strong central leader, which produces stubby lateral branches, especially when secondary branches are removed; and yellow-green, nonleathery leaves. Both trees were originally introduced into the trade in the late nineteenth century.

The Scanlon red maple (*Acer rubrum conica* 'Scanlon') (Fig. 55), one of the newest forms of red maple, is an exquisitely handsome pyramidal tree with strong crotches and upsweeping branches. Its green leaves turn vivid red and orange in fall. The author planted one of these trees to good effect in a narrow corner of his property.

Another good selection of upright red maple is Armstrong, which resembles Lombardy poplar in growth habit.

FIG. 55. Edward H. Scanlon of Olmsted Falls, Ohio, is standing in front of a 10-year Scanlon Red Maple, a pyramidal-shaped, exquisitely handsome tree.

The Sentry ginkgo (*G. biloba fastigiata*) maintains its narrow columnar habit and hence is ideal for planting along very narrow city streets.

A fastigiate form of the black maple (*Acer nigrum*), which is taller-growing and with clearer-yellow fall color and more compact growth than the species, is also available.

Other kinds of fastigiate trees available are: Improved Columnar Norway maple (*Acer platanoides columnare*), Upright English oak (*Quercus robur fastigiata*), and Upright American linden (*Tilia Americana fastigiata*).

The Augustine ascending elm is an unusually fine columnar-shaped elm for use along narrow city streets and for landscaping that needs an accent. It is said to be unusually resistant to wind and ice storms and eventually reaches 100 feet in height.

Other superior upright trees are *Crataegus monogyna stricta, Acer pseudoplatanus pyramidalis, Sophora japonica pyramidalis, Fraxinus excelsior pyramidalis, Prunus Sargentii columnare,* and *Sorbus aucuparia columnare.*

Not only can the upright trees be used on narrow streets or small properties, but they have the following advantages over the parent species from which they were derived: first, maintenance costs are lower because they require no heavy pruning; second, their root system is more restricted and compact, and they can be planted, therefore, under crowded conditions near driveways, sewers, and underground conduits without causing so much trouble as the standard types; third, fastigiate trees are less susceptible to damage by sleet, snow, and wind because their branches offer less resistance to the elements; finally, they interfere less with street lighting than do the ordinary types.

Standard Trees for Streetside and Lawn Planting

Because many of the more desirable trees mentioned earlier in this chapter will not be widely available for some time, it is necessary to list the species of the tall-growing or standard trees which are still in plentiful supply. It is well to reiterate that while many of these trees have highly desirable characteristics, such as ease of transplanting and wide availability, they eventually grow to great size, making their maintenance costs very high and their removal expensive, if they die from one cause or another. Moreover, they conflict more frequently with power and telephone lines, often causing a complete loss of light, heat, and telephone services when they fall during hurricanes or ice storms.

These trees are better planted in parks, along very wide streets, or in backyards where their height does not interfere with public utilities and where their roots will not clog sewer pipes, lift sidewalks, or push curbstones out of line.

The most widely used groups—oaks, maples, elms, planes, and lindens—are discussed in greatest detail because they presently constitute the great majority of trees planted along eastern and midwestern streets. Lists of trees for these and other parts of the country are presented later in this chapter.

Ash (*Fraxinus*)

Ashes have several characteristics that make the standard species undesirable as street trees. They are subject to infestations by scale insects, are not particularly attractive at any season, drop their leaves early, produce large numbers of viable seed, and have wood that is inclined to be brittle.

White ash (*F. americana*) is probably the best of the standard ashes as a streetside tree where the soil is particularly favorable.

Red and black ashes thrive only in swampy soils and are, therefore, of little value as streetside trees.

Fig. 56. The cork tree, *Phellodendron Lavallei*, makes an excellent tree for streetside or lawn planting.

Cork Trees (*Phellodendron*)

Cork trees (*P. amurense, P. Lavallei, P. sachalinense,* and *P. chinense*) are slow-growing species with feathery leaves, which give as much shade as littleleaf linden and more than silk tree (Fig. 56). They grow well in dry soils, are resistant to smoke, and consequently thrive in large cities. They are rather free from insect attack, but some species produce fruit that litters the streets.

Elm (*Ulmus*)

The unfortunate introduction of the Dutch elm disease (p. 316) into the United States and the appearance of the virus disease phloem necrosis (p. 319) have somewhat dampened enthusiasm for planting elms (Fig. 57).

Fig. 57. The American elm, one of our most beautiful trees, is rapidly losing its popularity because of its great susceptibility to insects and diseases.

Despite its susceptibility to these highly destructive diseases, the American elm is still one of our most popular trees. It is easily transplanted and becomes quickly established. Because it requires considerable room, it should be planted only on wide streets where the two diseases have not made inroads. The American elm is susceptible also to several insect pests. These, however, can easily be controlled by periodic applications of the proper insecticide.*

Another elm that might be considered for planting in areas where the Dutch elm disease is prevalent is the Hansen Manchurian elm. This species was introduced into America by the late Dr. N. E. Hansen, who collected it in the rugged mountains near the Manchurian-Siberian border, where it withstands temperatures of 60 degrees below zero. It has strong, sturdy limbs which branch out with firm crotches that do not split easily like those of the American and Chinese elm. The leaf is similar to that of the Chinese elm but is larger and more glossy green. The Manchurian elm grows fast, as much as 8 to 12 feet in one season, and is decidedly resistant to the Dutch elm disease.

Golden elm (*U. carpinifolia wredei*) has a pyramidal shape and rich golden leaves that hold their color all season.

Moline elm (*U. americana moline*) is a rapidly growing strain of the American elm, with darker green foliage. It must be carefully pruned to eliminate the formation of weak crotches to which it is subject. It is more upright and less spreading than American elm. It does best in the Rocky Mountain area.

Chinese elm (*U. parvifolia*) has small leaves, which turn bright yellow in the autumn. Its flowers are formed in August and September. It is easily transplanted but is very subject to insects and breakage by wind and ice.

Siberian elm (*U. pumila*) is a small tree with sparse growth and brittle wood. It grows well in poor soil and is resistant to the Dutch elm disease. This elm should not be planted with the expectation that it will take the place of the American elm, which it does not resemble in shape or form.

English elm (*U. procera*) is a medium-large tree with a rounded head. It is subject to elm-leaf-beetle infestation and it suckers freely, but is more resistant to the Dutch elm and phloem necrosis diseases than the other elms, with the exception of the Siberian elm.

Buisman elm (*U. carpinifolia* Buisman), a strain of European elm, is resistant to Dutch elm disease.

* Where elms are planted close to a house, the owner must choose between insect control, with the resulting unsightly accumulation of spray residues on the roof, windows, and screens, and the possibility of a serious insect infestation on the foliage.

Hackberry (*Celtis*)

Hackberry (*C. occidentalis*) is tolerant of city conditions, but in many localities its use is not recommended because of its extreme susceptibility to witches'-broom (p. 329). The Chinese hackberry (*C. sinensis*) and the southern hackberry (*C. mississippiensis*) are less subject to this abnormality. Both species, especially the latter, should be used more extensively along city streets.

Horsechestnut (*Aesculus*)

Horsechestnut (*A. hippocastanum*) is valued for its shape and for its large white flower clusters. The nuts are objectionable because they litter the streets or are sought by boys, who often break branches in collecting them. Nut formation can be prevented, however, by spraying the trees at blooming time with a dilute solution of the plant hormone sodium 1-naphthaleneacetate sold under the name App-L-Set by the Dow Chemical Company, Midland, Mich.

The most serious objections to the horsechestnut are the continuous dropping of foliage during the summer and the susceptibility to leaf-blotch disease.

The double-flowering variety, *A. hippocastanum Baumanii*, does not produce nuts, but it is likely to scorch badly in summer and is also susceptible to leaf blotch.

The pink-flowering horsechestnut (*A. carnea*) is a very beautiful tree when in bloom. It is a cross between the common horsechestnut and the red-flowering buckeye.

The ruby-red horsechestnut (*A. carnea briotii*) is similar to the common horsechestnut but is smaller and produces deep red flowers, which are outstanding.

Linden (*Tilia*)

Lindens are commonly used as street and shade trees. They thrive best in fertile, moist soil. Most of the species are favorite hosts for Japanese beetles.

Silver linden (*T. tomentosa*), with leaves that are dark green above and silvery white beneath, produces an interesting effect in the wind. It grows into a compact specimen, especially desirable for lawns or parks. Among the lindens it withstands dry sandy soils best, it loses fewest leaves in summer, and is most resistant to red spider attack. A weeping form of this linden, the silver pendent linden, *T. petiolaris*, has more finely serrated leaves on its pendulous branches.

American linden (*T. americana*) thrives in deep, moist soil and is therefore not very satisfactory as a streetside tree. A valuable forest tree, it is surpassed by other varieties for landscape purposes.

Little-leaf European linden (*T. cordata*) is a small tree that develops a compact, pyramidal head to 30 feet. It is the best of the lindens for streetside plantings in most localities.

Crimean linden (*T. euchlora*) one of the finest of the lindens, with very glossy foliage, grows to about 40 feet. The big-leaf European linden (*T. vulgaris*), with very large leaves, though occasionally planted along city streets, is a less desirable kind.

LOCUST (*Robinia and Gleditsia*)

Black locust (*R. pseudoacacia*) is rapidly disappearing from the eastern landscape because of the ravages of the locust borer (p. 342).

Honey locust (*G. triacanthos*) is more commonly used as a lawn tree than along streets. Its finely divided foliage and its open spreading habit provide partial shade. It is very hardy, grows fairly rapidly, although at first its branches droop considerably and it is difficult to keep erect. Its falling leaflets in autumn curl and dry up, and consequently are not so much of a nuisance as most other deciduous tree leaves. The long fruit pods, which litter the ground in autumn, and the numerous branched thorns, however, are highly undesirable. In playgrounds where children are likely to climb trees, only the thornless variety (*G. triacanthos inermis*) or the Moraine locust, mentioned below, should be planted. Some shade-tree men consider thornless honey locust one of the best trees, especially for the larger eastern cities.

Honey locusts are particularly difficult to maintain in good condition in the Washington, D. C.-Philadelphia area because of their great susceptibility to the mimosa web-worm (p. 343).

The best known, most generally available of the new honey locusts is the Moraine locust (*G. triancanthos inermis*, 'Moraine') (Fig. 58), developed by the Siebenthaler Nursery Company of Dayton, Ohio. It is a thornless variety which is said to be sterile and hence forms no fruit pods.

A more recent introduction is the locust Sunburst developed by the Cole Nursery Company of Painesville, Ohio. It has bright golden bronze foliage, is said to be thornless, and to be more resistant to insect pests than the common honey locust. Its brilliant coloring is limited to 8 or 10 inches of branch tips early in the season. By fall the contrast is not sharp because all the foliage turns green. Another recent honey locust introduction by the same company is Imperial. It is said to grow straight and strong without staking and produces a dense, uniform crown with a minimum of trimming.

FIG. 58. A seven-year-old Moraine locust.

Princeton Nurseries at Princeton, N. J., recently introduced Shademaster honey locust, a fast, upright grower that frequently retains the leader branch, an unusual property of honey locusts.

Two other excellent locusts are the globe (*Robinia pseudoacacia umbraculifera*) and the Idaho (*Robinia pseudoacacia* × *hispida*). Although these must be sprayed occasionally, they are beautiful and functional.

Maidenhair Tree (*Ginkgo*)

Maidenhair tree (*G. biloba*), a native of Asia, is now used extensively as a street tree in large cities. It does well in all parts of the United States except in the extremely cold areas. Its greatest assets are ability to withstand city conditions, wide tolerance of soil types, ease of transplanting, resistance to ice and wind storms, and remarkable freedom from insect and fungus attack. Its ability to produce shade can be augmented by pruning to increase the branch spread. The leaves are distinctly fan-shaped.

It is best to plant male trees only, because female trees bear brown, olive-like fruits with a disagreeable odor. Fruit formation can be prevented, however, by spraying the trees at blossom time with a dilute solution of maleic hydrazide (sold as MH-30 and MH-40 by the Naugatuck Chemicals Division of the U. S. Rubber Company, Bethany 15, Conn.).

Maple (*Acer*)

The maples constitute a very important group of streetside trees. Some species have many excellent qualities, whereas others are virtually worthless.

Silver maple (*A. dasycarpum*) has been planted most extensively in the past, despite the fact that it is the least desirable maple. Because its brittle wood breaks readily even in mild windstorms, its rooting habit lifts pavements and clogs sewers, its extreme susceptibility to rapid wood-decay and to infestations by many species of insects, this species should never be considered for streetside plantings. Its use is prohibited in many cities and towns.

Norway maple (*A. platanoides*), the most widely planted maple at the present time, is a fairly satisfactory city tree. It is easily transplanted, grows rapidly but not too large, and is ideal for the average street.

Because of the dense foliage and shallow rooting habit of Norway maples, lawns cannot be maintained beneath them. Branches must be pruned frequently to allow sufficient penetration of light and water and fertilizers applied occasionally to compensate for the heavy withdrawal of these materials by the trees' roots. The shallow roots also break or lift pavements, walks, and street curbs. Although relatively free from insect pests, this species is rather susceptible to the Verticillium wilt disease (p. 247). During midsummer the leaves, and certain sucking insects on them, excrete a sticky material that disfigures automobiles and other objects left beneath the trees.

Many municipal tree men are now planting seedling Norway maples less frequently along city streets. The constant pruning required to remove low hanging branches and to keep them well shaped greatly increases maintenance costs. Specially selected strains of Norway maple do not have the many faults of the ordinary kinds grown from seeds. Hence these will be used more extensively in time. Among the selections in wide use now are: Charles F. Irish, Cleveland (Fig. 59), and Almira.

Though the Norway maple is considered hardy by some arborists, it is often injured by extremely low winter temperatures. Moreover, it is extremely susceptible to the fungus Ganoderma (p. 288), which appears as reddish conks, or fruiting organs, with coatings like varnish or lacquer, at the base of the trunk or on surface roots.

One of the newest strains of Norway maple is Crimson King, a seedling of the Schwedler maple, mentioned later. Unlike its parent, it holds its bright red color throughout the growing season. A strain closely resembling Crimson King, Faassen's Red-Leaf has dark red foliage.

Sycamore maple (*A. pseudoplatanus*) is more attractive than the Norway maple but in some areas it is not so satisfactory as a streetside tree. It grows fairly rapidly with a more spreading and open habit than the Norway. It has not attained widespread use because of its more exacting soil requirements. When planted, it should be given a large growing area.

Two varieties of sycamore maple are particularly outstanding: wine-leaved sycamore maple, *A. pseudoplatanus Spaethii,* and the variegated *A. pseudoplatanus Leopoldii.*

Sugar maple (*A. saccharum*) requires a relatively large area and a moist but well-drained soil for best development. It is subject to sunscorch, is sensitive to the heat reflected from city pavements, and consequently does better along roadsides and highways than in cities. It has fine symmetry, produces moderate shade, and its leaves are brilliantly colored in autumn.

Red maple (*A. rubrum*) thrives only in rich, moist soils and does not retain its vitality when planted in dry situations and under city conditions. It does best in suburban areas. Two excellent selections of this maple are Gerling and Tilford.

Black maple (*A. nigrum*) is sometimes used in highway plantings in the Midwest. It is a close relative of our eastern sugar maple.

Schwedler maple (*A. platanoides schwedleri*) looks much like the Norway maple except for its dark red foliage throughout the early growing period.

Although more subject to winter injury than Norway maple, it is being planted rather extensively in many cities.

Boxelder (*A. negundo*), the only common species of maple with compound leaves, is unsatisfactory as a streetside tree because of the extreme brittleness of its wood.

FIG. 59. The Cleveland Norway Maple is a globe-headed tree with a better growing habit than seedling Norway maples.

Many other maples including upright or fastigiate forms are discussed on page 123, and low-growing forms, some of which are suitable for streetside plantings, are discussed on page 118.

Oak (*Quercus*)

As a group, the oaks are probably one of the best of the older street trees. They are hardy, strong-wooded, long-lived, and relatively free of serious insect and fungus troubles. In addition, they add a great deal of color to the fall landscape.

Oaks have not been planted more extensively because they are thought to be slow growers and difficult to transplant. Although this is true of white and live oaks, most of the other species used as street trees grow fairly rapidly and can be transplanted rather easily. In fact, oak trees that are properly watered and fed will outgrow many other species, the trunk increasing from ½ to 1 inch in diameter a year.

Red oak (*Q. borealis maxima*), one of the fastest-growing and best oaks for street planting in most sections of the country, is one of the few species that will thrive along ocean fronts. It requires a relatively large area for proper development. Its growth habit is more open than that of pin oak, thus facilitating the passage of electric wires through it. The red oak becomes established rather slowly, and best results are obtained if it is moved with a soil ball.

Scarlet oak (*Q. coccinea*) grows best in dry, sandy soils, because its roots prefer good air drainage. It withstands droughts better than other oaks and requires less room than red oak. Its fall color is more striking than that of most other oaks. It is very difficult to move.

Pin oak (*Q. palustris*) is one of the most commonly planted species in some eastern cities (Fig. 60). It does not do well in some midwestern areas. It is easily transplanted, becomes established quickly, grows rapidly, thrives in very wet or extremely dry soils, and is resistant to most insect and fungus attacks. Its single, erect trunk with many small drooping branches on the lower part distinguishes it from the other commonly planted species. It is pyramidal when young and assumes a rounded crown at maturity.

The most objectionable characteristics of the pin oak are its drooping branches, which must be pruned to prevent interference with traffic, and its tendency to become chlorotic in limestone soils and in filled-in areas. The latter condition, however, usually can be remedied (p. 196). Other objectionable features are the tendency of the leaves of some trees to hang on during most of the winter, and the dense branch formation that necessitates frequent pruning.

FIG. 60. A row of fine pin oaks along a street in Mount Vernon, New York.

White oak (*Q. alba*) is probably one of the slowest-growing and longest-lived species in the eastern United States. Because it is difficult to transplant, it is rarely planted along streets. Its spreading branch habit allows electric wires to pass through without interference when it is used as a street tree. Although its leaves do not color so brilliantly in fall as do those of the scarlet oak, its broad, round head and spreading branches make it one of the most beautiful oaks at all other seasons (see Frontispiece).

Live oak (*Q. virginiana*) is planted commonly in the South and on the west coast as a street tree. Several other species of oaks are also used for street planting in these sections of the country.

Willow oak (*Q. phellos*) is one of the finer-textured oaks. Its light green, shiny, narrow, pointed leaves, 2½ to 5 inches long, resemble those of willow. It is easy to transplant, does best in fertile, acid soils but is not hardy in the colder regions of the country.

Shingle oak (*Q. imbricaria*) and Shumard oak (*Q. shumardii*) are also desirable species to plant in the more southerly parts of the country.

Plane (*Platanus*)

Planes constitute an important group in the tree population of eastern and midwestern cities and towns. For example, of the 158,000 trees planted along Philadelphia streets, 50,000 are plane trees, and all except two are London planes (*P. acerifolia*). Their odd method of shedding patches of olive and brown bark, exposing lighter colored bark beneath, is known to nearly everyone.

Some confusion persists regarding the names of several species used. The tree that is generally planted along city streets and that thrives well is the London plane (*P. acerifolia*). This vigorous tree is a hybrid between buttonwood (*P. occidentalis*), also commonly known as sycamore, and oriental plane (*P. orientalis*). The true oriental plane is rare in the United States, despite its general appearance in nursery catalogues. It has deep-lobed leaves and single fruit balls. The hybrid (London plane) is intermediate, with lobes starting nearly at the middle of the leaf and with fruit balls usually borne in pairs but at times singly or in threes. The seeds in the fruit balls of the London plane have sharp points; those of the buttonwood do not.

London plane, which is unquestionably the best of the planes, is better adapted than most other trees for planting in unfavorable city environments. It is occasionally subject to freezing injuries, is extremely susceptible to the destructive Cankerstain disease (p. 357) and Dothiorella canker (p. 359), and is difficult to maintain in an erect position for the first two or three years after it is planted along city streets.

An upright form, *P. acerifolia pyramidalis*, makes a fine tree for streetside planting.

Buttonwood is a native tree that grows to great size in the woodlands. It does not withstand city conditions so well as the London plane, and drops much of its foliage during the summer. It is repeatedly attacked by the blight disease, to which it is much more susceptible than the London plane.

Oriental plane (*P. orientalis*) can withstand a trying environment better than the buttonwood, but is rarely used as a street tree. For some unknown reason, it is virtually immune to the blight disease.

Poplar (*Populus*)

Poplars are not recommended for streetside plantings. Some of the species, in fact, have such outstanding faults that many communities prohibit their use along public right-of-ways.

Eugene poplar (*P. canadensis eugenei*) has very brittle wood, which breaks easily and is a constant menace to the safety of pedestrians. Though it tolerates city conditions and is a rapid grower, it is short-lived. Its roots clog sewers and break up pavements and walks.

Lombardy poplar (*P. nigra* var. *Italica*) is rapidly disappearing in the eastern United States because of its extreme susceptibility to the canker disease (p. 360). Its rapidity of growth and columnar habit have made it a favorite for windbreaks and crowded situations.

White poplar (*P. alba*) appears to thrive under city conditions but has many of the undesirable qualities of the Carolina poplar. The variety Bolleana, with an upright form resembling Lombardy poplar, is often used in parks and open areas. This variety is not much better than Lombardy poplar, however.

Sweetgum (*Liquidambar*)

Sweetgum (*L. styraciflua*), with its star-shaped leaves, excellent fall coloration, and deep corky bark, is one of our most beautiful trees. It requires a deep, friable soil and full sunlight. It grows well between the curb and walk in some cities. Drastic pruning of the top at transplanting time greatly increases its chances of survival. It should be used more extensively as a streetside tree where it can be given ample space. Its hard, spiny fruits, which fall on lawns and streets, are objectionable because they damage lawn mowers and even automobile tires that pass over them.

Tulip Tree (*Liriodendron*)

The tulip (*L. tulipifera*) is one of the tallest of our native eastern trees and is considered one of the most beautiful. It requires deep, rich soil and is very difficult to transplant. An excellent tree for private estates and other landscape plantings, it is not recommended as a streetside tree, because of its extremely brittle wood and its intolerance to city smoke.

Yellowwood (*Cladrastis*)

Yellowwood (*C. lutea*) is an excellent tree for cities and towns and should be planted more extensively. Some authorities believe it is more suitable than American elm. It has a graceful top, a relatively short trunk with yellow wood and smooth bark, and often reaches a height of 50 feet.

It should be planted in a sunny location. In late spring or early summer it produces hanging clusters of white, fragrant flowers resembling those of white wisteria. Unfortunately, multibranched specimens have a strong tendency to split at the crotches during heavy wind and ice storms.

Trees for Seashore Planting

Because of the unusually difficult conditions along seashores, relatively few trees do well. Some, however, are markedly tolerant of the salt-laden air, constant winds, and sandy soil.

Japanese black pine (*Pinus thunbergii*) is one of the most satisfactory evergreen trees for seashore planting along the east coast. It withstands salt- and sand-laden air and light sandy soils better than most trees. This tree, however, is not reliably winter-hardy in some parts of Canada. For example, the unseasonable frost which hit the Pacific coast of northwestern United States and southwestern Canada on November 11, 1955, killed Japanese black pine that had been growing there for 20 years. Austrian pine (*Pinus nigra austriaca*) suffered no ill effects from the same frost, and it is almost as resistant as Japanese black pine to salt spray on the west coast of Canada.

Red oak (*Quercus borealis maxima*), mentioned earlier, is one of the best oaks to plant along the eastern United States seashore.

Blackjack oak (*Q. marilandica*), sometimes referred to as 'scrub oak,' also makes an excellent tree for seashore plantings.

Russian olive (*Eleagnus angustifolia*), also mentioned earlier, is an unusually hardy tree that can withstand seashore conditions.

Other trees that can tolerate trying seashore conditions, provided they are given some slight protection or are planted a bit farther inland than the trees mentioned above, are mugho pine and pitch pine, American holly, post oak, honey locust, weeping willow, laurel willow, red maple, Norway maple, London plane, gray birch, hawthorns, shadbush, and crabapple.

Trees as Windbreaks

The Northern Great Plains area of the United States with its extremely low winter temperatures and very dry summers is one of the more difficult areas in which to grow trees, especially those used as windbreaks. In tests to determine hardiness for farm windbreaks conducted over a 38-year period, workers in the United States Department of Agriculture found that boxelder (*Acer negundo*), green ash (*Fraxinus pennsylvanica*), and the shrubby silver buffaloberry (*Shepherdia argentia*) were most satisfactory.*

Planting Streetside Trees

Since trees are more difficult to grow in the unnatural environment of cities, successful culture of streetside trees involves many problems not usually encountered in forest and country plantings.

Trees in large cities have a particularly difficult time to do their best. City conditions are so unfavorable for most trees that it is a wonder they grow at all. Smoke and other air pollutants, mechanical injuries, a disrupted water table, highly compacted soil, lack of organic matter, limited root space, reflected heat from buildings, visitations by dogs, lack of water, use of salt on sidewalks to melt ice—these are but a few of the external forces which make the lot of a city tree so difficult. Add to these, invasions by insects and infection of the roots by soil-inhabiting parasitic fungi, and one is almost ready to forego planting trees in cities. Despite these hazards and handicaps, however, millions of trees do grow in large cities, including Brooklyn.

In cities, the run-off of rainfall is nearly 100 per cent. Hence the small open soil area between the street curb and the sidewalk, where most city streets are planted, has little chance to absorb rain water.

Another important point to bear in mind is that a soil which may have had adequate aeration at the time a tree was planted, will deteriorate until the aeration is entirely inadequate. Trees growing along city streets are an excellent case in point. When the tree is first planted in the area between the street and the sidewalk, conditions may be satisfactory. As the roots extend into the areas covered by the street and sidewalk, conditions become less

* The most detailed publication on the use of trees as windbreaks *Trees Against the Wind* was published as P M W Bulletin No. 5, January 1953, by the State College of Washington Extension Service and the U. S. Department of Agriculture, co-operating, Pullman, Washington.

favorable. In addition, trampling of the open soil area by pedestrians also compacts the soil and makes it as nearly impervious to air and water as concrete.

Still another factor which makes the growing of trees along city streets difficult is the lack of adequate plant food. Dr. Curtis May of the United States Department of Agriculture once wrote, 'Street trees are even more likely to decline from lack of soil fertility than lawn trees, because the soil mass in which their roots can develop well is likely to be even more restricted than it is for lawn trees.'

Three factors largely govern success in planting trees along public right-of-ways—selection of the proper species, good growing conditions, and reasonable after-care and protection.

Selection of Species for City Streets

Tall-growing trees like the London plane, the ginkgo, and the honey locust are known to tolerate city conditions better than trees like the sugar maple. The usual explanation is that the foliage of the first three is more tolerant of air pollutants. The author is of the opinion that some deciduous trees are more tolerant because their roots can survive under the extremely poor soil conditions prevalent in most cities; that is, they are better able to tolerate excessively dry soils, poor soil aeration, high concentrations of soluble salts, and other adverse below-ground factors.

A great variety of trees suitable for planting along city streets, especially in the eastern and midwestern parts of the United States are described earlier in this chapter. Trees suitable for these areas and for other parts of the country are listed at the end of this chapter.

Persons entrusted with the selection and care of streetside trees should adopt programs for proper selection and maintenance. Many of the low-growing more desirable trees need more frequent spraying than the older tall-growing kinds. But in the long run, the former will be better and less expensive. One example will suffice to indicate how expensive the American elm plantings in the eastern and midwestern United States can be. Not only do such trees require frequent sprayings to control scale insects, elm leaf beetles, and European bark beetles, but once they become infected by a fatal disease such as the Dutch elm, or phloem necrosis, then great amounts of money are needed to remove them. The City of New York appropriated

$200,000 a few years ago just to remove diseased and dead elms. Such costs will continue each year until all susceptible individuals have been removed.

Size of Tree. The best size for street planting varies from ¾ to 1½ inches in caliper at the base. Larger trees are also used but become established more slowly and are more expensive.

Spacing. Most city trees are planted too closely. No trees should be planted closer than 40 feet, and large spreading types should be spaced 60 to 75 feet apart. Narrow streets are best planted in ginkgo, European linden, or some of the pyramidal varieties of maple, oak, elm, and hornbeam described earlier in this chapter.

Location. Street trees are customarily placed between the walk and street curb. Planting one of the tall-growing kinds in this area is bound, in time, to conflict with overhead wires, and with the sidewalks and curbs. This area can be properly planted with the kind of tree that will not lead to such difficulties. The decision on where to locate the tree will depend on site factors.

Where tall-growing trees are used, the area betweeen the walk and property line is more satisfactory (Fig. 61). This, of course, is feasible only where the house is set back some distance from the property line. The advantages of such an arrangement are many. It removes trees from the area where they are continually subject to injury by street traffic, the trees have more favorable soil conditions, they are less apt to tangle with overhead wires, the crowns do not form over the street area, and the street looks wider. Moreover, less damage to the trees will occur if the street is widened at a later date.*

Trees should be planted alternately along the street rather than opposite one another, to allow more space for development.

Time of Planting. Most street trees are planted in spring, fall, or winter, depending on local conditions. They can also be planted during the summer, provided special precautions are taken (p. 45).

Preparation of the Planting Area. The soil along many city streets is so impoverished that some good soil should be substituted. The usual practice is to remove about a cubic yard of soil and replace it with good topsoil to which a liberal amount of manure or peat moss has been added. The holes are approximately 4 feet square by 3 feet deep. If possible they should be dug beforehand to reduce delay at planting time.

* A disadvantage of locating trees near or on the property line is that the jurisdiction over such trees is taken away from the Shade-Tree Commissioner or Department of Parks. In Massachusetts, municipally managed trees can be planted as far as 20 feet in on private property.

FIG. 61. The ideal location for street trees is not along the curb but back on the property line. Trees are less subject to vehicular injuries and have more favorable growing conditions.

Treatment of Top. Most shade-tree men prune all lateral branches back to one or two buds. Young trees thus treated appear rather thin at first, but they soon regain their shape.

Supporting. An angle-iron fence post, 8 feet long, should be inserted about 1½ feet down on the street side; or wooden supports described in Chapter 7 can be used.

Protection. The benefits derived from protecting the trunk with burlap, kraft crepe paper, wax, or other materials, and from supporting the trunk with stakes have been discussed in the chapter on Transplanting.

After-Care. The first and second summers after planting are the most critical periods in the life of the young tree. Artificial watering must be practiced, especially if the rainfall is low. Installation of a Rex-Keyser system, as illustrated on page 43, will make the job easier and more effective. In the absence of such a system, a mulch of wood chips around the base of the newly planted tree is suggested. Such mulch will not only conserve moisture, but will also help control weeds and prevent soil compaction and deep freezing.

TREES FOR DIFFERENT AREAS OF THE UNITED STATES *

Shade Trees for the Northeast

Maples—Norway and varieties, Schwedler, sugar, red and varieties
Elms—American, rock, English
Oaks—pin, red, scarlet, white
Ginkgo
Tulip tree
Sweetgum
American linden
American sycamore
London plane tree
American yellowwood
European beech
Thornless honey locust
Cherries—Columnar Sargent Cherry, Sargents—Double Flowered Mazzard, Kwanzan, Akebono, Daikoku, Shirotae
Hackberry
Magnolia Loebneri and *M. Kobus*

Trees of Limited Use for the Northeast

American mountain ash
Black tupelo
Amur cork tree
Goldenrain tree
Japanese pagoda tree
Paper birch
Silver maple
Green ash
Kentucky coffee tree
Black walnut
Ohio buckeye and common horsechestnut
Weeping willow (around lakes or ponds—not on city lots)

* Compiled largely from *Trees, Yearbook of Agriculture 1949*, U. S. Government Printing Office, Washington, D. C.

Evergreen Trees for the Northeast

Eastern hemlock
Red pine
Eastern white pine
White fir
Nikko fir
Oriental spruce
Colorado blue spruce
White spruce
Eastern arborvitae

Shade Trees for Southeast

Live oak
Southern magnolia
Camphor tree
Willow oak
Red maple
Flowering dogwood
Sweetgum
American holly
American beech
Eastern redbud
Water oak
Mimosa or silk tree
Winged elm
American elm

Less-Used Trees for the Southeast

Sugarberry
Pecan
Eastern red cedar
Goldenrain tree
Tulip tree
Canary date
American sycamore
Carolina laurel-cherry
White oak
Scarlet oak
Pin oak
Laurel oak
Cabbage palmetto
Weeping willow

Southern Florida

Oxhorn bucida
Australian pine
Coconut
Royal poinciana
Benjamin fig
Mango
Cajeput tree
Cuban royal palm
African tulip tree
West Indian mahogany

Shade Trees for the Plains

Green ash
Cottonwood
American elm
Siberian elm
Hackberry
Bur oak

TREES FOR DIFFERENT AREAS OF THE UNITED STATES

Trees of Limited Use for the Plains

Ailanthus
Boxelder
Northern catalpa
Kentucky coffee tree
Chinese elm
Sugarberry
Black locust
Common honey locust
Silver maple
Russian olive
American sycamore
London plane tree
Eastern black walnut
Weeping willow

Evergreens for the Plains

Eastern red cedar
Ponderosa pine
Douglas fir
White fir
Austrian pine
Scotch pine
Colorado blue spruce
Black spruce
Western white spruce

Shade Trees for the Rockies

Ailanthus
Green ash
White ash
Boxelder
Northern catalpa
American elm
Common hackberry
Thornless honey locust
Linden
Black locust
Norway maple
Red mulberry
Poplars (lance-leaf, narrow-leaf, Lombardy, and Plains poplar)
Russian olive
Tamarisk

Evergreens for the Rockies

Arizona cypress
Eucalyptus
Rocky Mountain juniper
Aleppo pine
Austrian pine
Canary pine
Colorado pinyon pine
Ponderosa pine
Colorado spruce
Englemann spruce

Shade Trees for California

California live oak
Camphor tree
Red iron-bark
California pepper tree
Cape chestnut
Ginkgo
Norway maple
London plane tree
Sweetgum
Modesto ash
Pin oak
Southern red oak

Shade Trees for California (*Cont.*)

California black walnut
Chinese pistache
Panicled goldenrain tree
White mulberry
Canary pine
Coulter pine
Lawson cypress
California incense cedar
Deodar cedar
European white birch

Shade Trees for the North Pacific Area

Common hackberry
American yellowwood
Sweetgum
Tulip tree
Northern red oak
Oregon white oak
Pin oak
Big-leaf maple
Norway maple
American elm
Pacific madrone
California sycamore

Conifers for the North Pacific Area

Atlas cedar
Lawson cypress
California incense cedar
Himalayan pine
Common Douglas fir

Part II

SPECIFIC ABNORMALITIES OF TREES

9

DIAGNOSING TREE TROUBLES

During the last twenty years the author has been asked by commercial arborists, city park departments, landscape architects, public utility companies, and private home owners to diagnose the cause of decline or death of shade and ornamental trees. At the request of a public utility company, during the summer of 1956, for example, he personally examined and then diagnosed the cause of many abnormalities among more than 300 trees in New York City.

As a result of this experience, the author feels he is in a position to help others in diagnosing tree ailments. After all, successful treatment of any abnormality of trees, as of human beings, depends primarily on correct diagnosis. Some troubles occur so frequently on certain trees and have such specific symptoms that they are readily diagnosed by the competent tree man; others are so complex that even the most expert arborist cannot fathom them.

A reliable arborist will not hesitate to admit his inability to diagnose some abnormalities nor will he hesitate to recruit the aid of a specialist to help in the diagnosis, much as medical doctors often consult specialists in difficult cases. He will try corrective treatments of unknown value only as a last resort —after all attempts at a correct diagnosis have failed.

Qualifications of a Tree Diagnostician

The most important qualification of a good diagnostician is plain ordinary common sense; some refer to it as knack, intuition, or good judgment. He must also have a thorough understanding of a so-called normal tree, including the species he is examining. Not only is it necessary that he know the name of the tree, but he must know its characteristics as to winter-hardiness, tolerance to dry and wet soils, and reactions to other environmental factors.

The good diagnostician will not be afraid to ask questions. Some tree owners and park officials, for some mysterious reason, withhold information essential for a proper diagnosis. The diagnostician must know the history of the tree, such as when and how it was planted. He must also have some information on the past climatic history. This is acquired either from his own records or by asking the owner or person in charge of the tree about it. He must have information regarding drought periods, severity of previous winters, and prevalence of hurricanes or other unusual weather.

The tree diagnostician must also have a thorough understanding of the relation between the soil and the tree. Is the soil properly drained; is it well-aerated; does it hold enough moisture; is it fertile? Soil fertility can be determined only by a complete soil test. The expert must also have a working knowledge of entomology and plant pathology. Many of these phases are treated in some detail in various parts of this book. But additional reading and certainly extensive field experience are necessary before one can really qualify as an expert diagnostician.

Sometimes it is unwise to become so absorbed in detailed symptoms that important gross symptoms are completely overlooked. It is here that common sense is important.

Correct Diagnosis Important

An improper diagnosis at the start will quite naturally lead to improper treatment. For example, a few years ago the author was called to explain the branch die-back in several large oaks growing on an estate in Westchester County, New York. The oaks had been under the care of a professional arborist who had diagnosed the problem as due to lack of food. The trees were fed for several years, but branches continued to die back until the top of one of the trees was nearly bare.

A little digging at the base of the trees and probing into the bark just below the soil line showed that the shoestring root rot fungus (p. 243) had invaded and killed virtually all of the important tissues beneath the bark. In all probability, this fungus had been present in the roots for years and was the prime cause of the dying back of the top. All the plant food in the world applied to trees whose roots were diseased would never have done any good.

Following is another example of how an improper diagnosis can lead to an incorrect conclusion. During the summer of 1957, the author was asked by a public utility company to diagnose the cause of death of several dozen trees growing along streets of a New Jersey community.

One, a 16-inch DBH Norway maple, was said by a plant pathologist who examined the tree earlier, to have been killed by escaping natural gas from leaking mains. The only basis for his diagnosis was that a gas leak had been repaired in the general vicinity. The author removed the soil around the base of the tree and found that someone had girdled the tree completely below-ground at least a year or so before. The uneven cuts made by an instrument were still plainly visible! Here was a case, then, where a little digging by the first 'expert' might have resulted in an entirely different diagnosis.

One of the more difficult diagnoses involves troubles caused by nutrient deficiencies. One reason is that fewer studies have been made on shade and ornamental trees than on fruit trees and vegetables.

Generally speaking, foliage deficiencies of the various elements become apparent either at the base or at the tip of the current season's growth. A deficiency of nitrogen, phosphorus, potassium, magnesium, and zinc is apparent on the older leaves. A deficiency of iron, manganese, sulfur, calcium, boron, and copper is first apparent on the young leaves.*

A recently developed machine called the spectrograph enables the scientist to determine quickly whether a plant is suffering from a nutrient deficiency or from an overdose of some nutrient.

The procedure is to collect, in July or August, some leaves from a tree suspected of having a nutritional disorder, dry them, and then grind them to a fine powder and reduce them to ashes at a temperature of 500-600°. The ash is then dissolved and the solution placed on an electrode and burned with a high-voltage electric spark or arc. Each of the elements present will emit its own characteristic light, which is passed through a prism and recorded on a photographic plate as a series of dark lines or bands. These bands indicate the amounts of the various elements present. The spectrograph can determine how much calcium, magnesium, phosphorus, manganese, copper, boron, iron, and zinc are in leaves. Potassium is determined with a so-called flame photometer and nitrogen by chemical methods.

Several state colleges of agriculture now provide this type of leaf analysis

* Space limitations prevent a detailed discussion of nutrient deficiencies. More details can be obtained in Dr. L. C. Chadwick's article in the *Proc. 19th Natl. Shade Tree Conf.*: 5-18, 1943; from the books, *Hunger Signs in Crops*, Amer. Soc. Agron. & Amer. Fert. Assoc., Washington, D. C., 1941, and *The Care and Feeding of Garden Plants*, Amer. Soc. Hort. Sci. & Amer. Fert. Assoc., Washington, D. C., 1954.

service for professional nurserymen and arborists. In time, it is hoped that all such colleges will have facilities for making leaf analyses to aid tree men and owners of trees in calculating the fertilizer needs of plants and diagnosing ailments due to nutrient deficiencies.

Procedure in Diagnosing

Following are the standard procedures used by specialists in diagnosing tree troubles. They will vary somewhat with the individual. Some diagnosticians take more stock in symptoms above ground than below ground. Others, like the author, feel that the most serious tree troubles and those most difficult to diagnose are more frequently associated with below-ground symptoms and factors. More about this later.

Before examining any part of an ailing tree, one should study the general surroundings. Are other nearby trees healthy? Have any special treatments been given prior to the appearance of the abnormal condition? Is the tree under diagnosis so situated that a leaf bonfire beneath it may have played a part in its decline? After these and related questions have been answered, one should then proceed with the direct examination of the tree.

Examination of Leaves. The leaves constitute the best starting point because they are most accessible and are first to show outwardly the effects of any abnormal condition. Here, also, a complete understanding of a normal leaf is essential, because the size and the color of normal leaves vary greatly among the different tree species and even among trees of the same species.

Insect injuries to leaves are rather easily diagnosed, either by the presence of the pest or by the effect of its feeding. The leaves may be partly or completely chewed, or they may be yellowed as a result of sucking of the leaf sap, blotched from feeding between the leaf surfaces, or deformed from feeding and irritation.

Leaf injuries produced by parasitic fungi are not diagnosed so readily, however, because the causal organisms are usually not visible to the unaided eye. In some instances, tiny, black pin-point fungus bodies in the dead areas, visible without a hand lens, are indicative of the causal nature. Lesions resulting from fungus attack are more or less regular in outline with varying shades of color along the outer edges. They may range in size from tiny dots to spots more than half an inch in diameter. When several spots coalesce, the leaves may wilt and die.

Atmospheric conditions preceding the appearance of spots on leaves can often be used to advantage in determining the cause of the injury. For example, when leaf spots appear after a week or 10 days of continuous rains and cloudy weather, one can safely assume that some parasitic organism is responsible, because such conditions are favorable for its spread. When leaves are spotted or scorched following a week or more of extremely dry, hot weather, lack of water as discussed under Leaf Scorch, on page 194, may be responsible. Low temperatures in late spring may also result in much injury to the tender leaves (p. 167).

Changes in leaf structure, appearance, or function may result from such widely different causes as fume injury, deficient or excessive moisture, lack of available food, poor soil aeration, root injuries, or diseases. All but the last two of these can be disregarded if nearby trees of the same species as the one under study are perfectly healthy.

Examination of Trunk and Branches. A careful inspection of the branches and trunks should follow the leaf examination. Sunken areas in the bark indicate injury to tissues beneath. They may have been produced by fungus or bacterial infection or by nonparasitic agents, such as low and high temperatures. The presence of fungus bodies in such areas does not necessarily indicate that the fungus is the primary cause. Only a person with considerable mycological training can distinguish the pathogenic from the nonparasitic species. The diseased wood beneath the bark, if it is the direct result of fungus attack, shows a gradual change in color from diseased to healthy tissue. It is usually dark brown in the earlier, more severely infected portions, then changes to a deep green or a light brown, and finally to a lighter shade of either in the more recently affected parts. On the other hand, injuries resulting from low or high temperatures are usually well defined, an abrupt line of demarcation appearing between affected and unaffected tissues.

The bark of the trunk and branches should also be examined for small holes, sawdust frass, and scars or ridges. These indicate borer infestations in the inner bark, sapwood, or heartwood. As a rule, most borers become established in trees of poor vigor. Because of this, it is necessary to investigate the cause of the weakened condition rather than to assume that the borers are primarily responsible. Branches and smaller twigs should always be examined for infestations of scale insects. Although most scales are readily visible, a few so nearly resemble the color of the bark that they are sometimes overlooked.

Branches or twigs with no leaves or with wilted ones should be examined for discoloration of the sapwood, the usual symptom of wilt-producing fungi.

The service of a mycologist is needed for determining the species of fungus involved, because positive identification can be made only by laboratory isolations from the discolored tissue.

The appearance of suckers or watersprouts along the trunk and main branches may indicate a sudden change in environmental conditions, structural injuries, disease, or excessive, incorrect, and ill-timed pruning.

Although galls or overgrowths occasionally present on the main trunk may be caused by parasites and by disturbances in the water relations between the soil and the tree, many are produced by factors not clearly understood by scientists.

The general vigor of a tree usually can be ascertained from the color in the bark fissures and the rapidity of callus formation over wounds. In vigorously growing trees, the fissures are much lighter in color than the bark surface. A rapidly developing callus roll over the wound also indicates good vigor.

Examination of Roots. Because of their inaccessibility, roots are rarely inspected by many arborists. In diagnosing a general disorder in a tree, however, the possibility of root injury or disease must be carefully considered. More than half of the abnormalities in the hundreds of street and shade trees examined by the author during the last 20 years were found to be caused by injuries to or diseases of the root systems.

The sudden death of a tree usually results from the destruction of nearly all the roots or from the death of the tissues at the trunk base near the soil line. The factors most commonly involved in such cases are infection by fungi such as the shoestring fungus (p. 243), the wilt fungus (p. 247), and the *Ganoderma* fungus (p. 288), winter injury, rodent damage, heavy concentrations of manufactured illuminating gas, lightning strikes, and toxic chemicals like gasoline, oil, salt, and weed killers. Trees that become progressively weaker over a period of years may be affected by girdling roots, decay following sidewalk and curb installations or road improvement, poor soil type, poor drainage, lack of food, changes in grade, low concentrations of manufactured illuminating gas, and excessively deep planting from the start.

Diagnostic Tools

Certain implements are essential in diagnosing tree troubles (Fig. 62). Most important of these is a shovel or small spade for removal of soil around the base of the trunk to facilitate detection of girdling roots, rodent damage, winter injury, and infection by the shoestring fungus or other parasitic fungi.

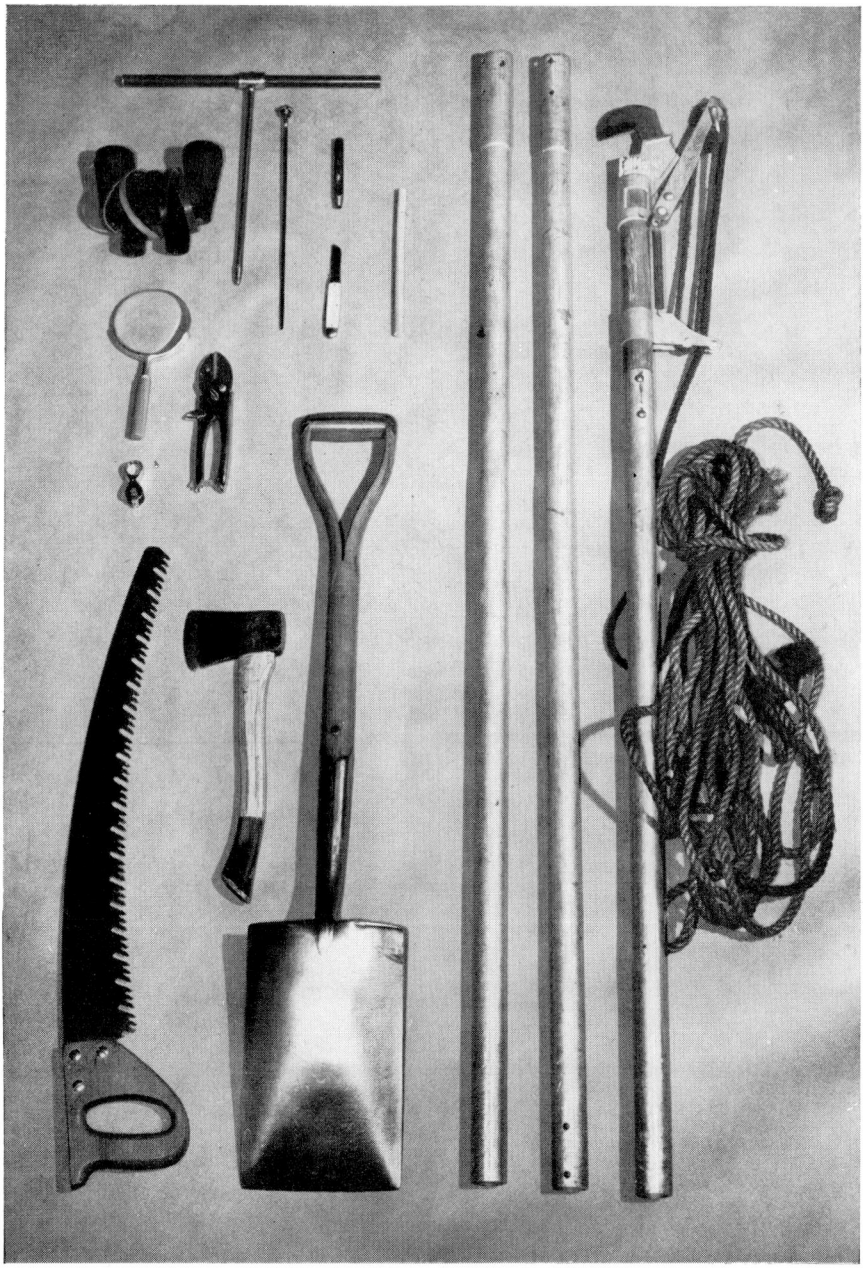

FIG. 62. Some of the implements used by a tree disease diagnostician.

In addition, it is used for digging soil in the vicinity of the smaller roots to be examined for the presence of decay. A soil auger is essential for collecting representative samples of soil for analytical tests and for determining the nature of the subsoil and the drainage. Pruning shears are necessary for cutting twigs and small branches to determine the presence of discoloration in the sapwood caused by wilt fungi and other injuries in the inner bark, cambium, and wood. A small saw is helpful where branches larger than can be handled with pruning shears must be cut. The author finds a sectional pole pruner, consisting of four 4-foot lengths, is excellent for cutting branch samples. This pruner is easily assembled and then disassembled and stored in the automobile luggage compartment.

A most important tool is a chisel or a curved gouge with a heavy handle for tapping the bark to locate dead areas and for making incisions into the bark and sapwood. Bark and sapwood specimens needed for further study are also collected with this instrument with the aid of strokes from a 2- or 3-pound composition mallet. A sharp instrument such as an ice pick is also useful to help determine the extent of soft, decayed wood.

An increment borer facilitates sampling for study of the annual growth rate of the tree, for determining the extent of heartwood decay, and for checking on the presence of vascular infections of fungi like Verticillium. One must be reasonably careful with this instrument on living trees, for it is known that discoloration and decay frequently develop around holes made by it. An increment hammer is also available for collecting more shallow samples of the bark and wood.

A hand lens is valuable for detecting the presence of red spider mites and tiny fungus bodies in leaf lesions.

Screw-top vials should be available for collecting insects to be submitted to a professional entomologist for identification. Glassine bags for holding leaf specimens and Kraft paper bags or other containers for holding bark and sapwood specimens, or soil samples, should also constitute part of the diagnostician's working kit.

A high-powered microscope for determining the presence of a foreign body in the plant tissues is helpful to persons familiar with its use and capable of recognizing what it reveals.

The following chapters treat problems that are rather readily diagnosed by the use of most of the diagnostic tools described above.

10

NONPARASITIC INJURIES

Damage Caused by Grade Changes around Trees

The addition or removal of soil may seriously disturb the delicate relationship between roots and soil and thus result in considerable damage or death of the trees. Yet such procedures may be necessary when new homes and highways are built, lawns or terraces graded, or street improvements made.

The author is often asked to diagnose the cause of the gradual weakening and even the death of beautiful oaks, maples, or beeches on properties in newly developed residential sections. He first examines the trunk base to see whether it flares out. Where no fill has been applied, the trunk is wider at the soil line than it is a foot or so above. If the trunk has no flare at its base and it enters the ground in a straight line, then a fill has probably been applied. Of course, some digging with a good spade will then reveal the depth of the fill.

Other visible symptoms are small yellow leaves, the presence of numerous suckers along the main trunk and branches, many dead twigs, and, in some instances, large dead branches. When the owner is questioned as to what precautions were taken to minimize the shock of placing the roots in a new environment, the usual answer is that none was thought necessary. The building contractor dug the subsoil from the cellar location and spread it promiscuously around the trees. He placed a foot of good topsoil over this, and his job was done!

Persons who plan to build a new home, or who have just moved into a new home surrounded by trees, or who plan construction changes near large trees, would be well advised to pay particular attention to the discussion that follows.*

Why Trees Suffer from Soil Fills. Before methods of reinvigorating trees suffering from existing grade changes and ways of preventing injury from fills and removals of soil are considered, it is well to stress again some of the basic principles involved. Air (primarily oxygen) and water are essential for normal functioning of roots. Soil micro-organisms, necessary to break down organic matter that serves as food for the roots, also need air. In addition, when air is lacking, certain gases and chemicals increase and become toxic to roots. The tree roots have become established within a certain area in the soil where the essential materials are present. The moment a blanket of soil is placed over the existing soil level, a marked disturbance occurs in the balance between the roots and these essential materials. As a result, the roots die, and the symptoms soon become visible in the above-ground parts. These symptoms may appear within a month, or they may not appear for several years.

Piling around trees, large heaps of sand or soil to be used for road and building construction, may also have a detrimental effect.

Factors Governing Extent of Injury from Fills. The extent of injury from fills varies with the species, age, and condition of the tree; the depth and type of fill; drainage; and several other minor factors. Sugar maple, beech, dogwood, oak, tulip tree, and evergreens are most severely injured; birch, hickory, and hemlock suffer less; and elm, poplar, willow, plane, pin oak, and locust are least affected. Trees in weak vegetative condition at the time the fill is made are more severely injured than trees in good vigor.

Obviously the deeper the fill, the more marked is the disturbance to the roots, and consequently the more serious are the effects. Clay soil fills cause most injury, because the fineness of the soil shuts out air and water most completely. The application of only an inch or two of clay soil may cause severe injury. Gravelly fills cause least trouble, because both air and water permeate them more readily. As a rule, the application of a layer several inches deep of gravelly soil, or even of the same kind of soil in which the tree has been growing, will do no harm. In such cases, trees eventually become accustomed to the new situation by producing additional roots near the surface.

* Additional information is presented in Farmers Bulletin 1967 *Reducing Damage to Trees from Construction Work*, available from Superintendent of Documents, Washington 25, D. C., for 10 cents.

Methods of Saving Trees in Existing Fills. Unfortunately, little can be done to save trees that have been suffering from grade fills over a long period. Much can be done, however, to restore trees surrounded by recent fills, or trees that do not yet appear to have suffered seriously from older fills. Corrective treatments, which should be undertaken as quickly as possible, must have as their objective the duplication, in so far as possible, of conditions existing before the fill was made. The following are the best methods:

Where a relatively shallow layer of soil has been added, the soil is periodically broken up or cultivated, compressed air is forced into the soil, and fertilizers and water are added.

Where the fill is moderately deep, the soil directly around the trunk is removed to the original level and a dry well is constructed. Holes are dug to the original level every few feet over the entire root area, and 6-inch bell tiles are placed in them. Fertilizers are then placed in the soil as near to the root zone as possible.

Where the fill is deep or where a particularly valuable tree is involved, more work and expense are required. The following procedure will help to overcome some of the harmful effects: Remove the soil to the original level around the trunk. Dig trenches radiating from this area to points below the outer spread of the branches. The level of the radial trenches should slope downward from the trunk to the outside in order to drain water away from the trunk. Connect the ends of the radial trenches by digging a circular trench to the same level. Add hydrated lime and fertilizer in the exposed area in the radial and circular trenches, but *not* around the trunk. Build a well of loosely joined stones around the trunk. Place lines of 6- or 8-inch split sewer pipe, 4-inch horseshoe drain tile, or 4-inch agricultural tile in the radial and circular trenches. The first two types are laid downward. Join the tile tightly enough to hold out soil, but loosely enough to allow air and water to enter. Cover the joints with cardboard, old tin, or some other material to prevent clogging by the soil. Place 6-inch bell tile upright over the junction of the tile in the circle and the radii, and pile stones around to hold them in position. Then place rocks followed by crushed stone and gravel over the ground tile, and fill the remaining space (about 12 inches) with good topsoil.

Preventing Injury from Soil Fills. Although the initial cost of proper precautionary measures is high, prevention of injury is much more effective than the curative measures just described. Before undertaking such a program, however, one should carefully consider the following points:

First, the age of the tree: An old or weak tree may not be worth the expense involved unless it has historical or aesthetic value. A young tree can be replaced at less cost than that necessary for precautionary measures. Second, vigor: If the tree is in low vigor or has an extensive cavity, its possible life span may be too short to justify the expense. Third, species: If the tree belongs to a short-lived species (Norway maple or tree-of-heaven), the added expense may not be justified, though it might be with long-lived trees, such as oaks or beeches. Fourth, the number of other trees in the immediate vicinity: obviously, if only one or two large trees are on the property, any attempt to save them will be worth while, since they greatly enhance the beauty of the home. Lastly, the susceptibility to some deadly disease: there might be some question, for instance, of the justification of precautionary measures around an elm if the tree is in an area infested by the Dutch elm disease (p. 316).

Preparation for Raising Grades. When all the aforementioned factors have been considered and the decision to undertake the necessary measures has been made, the procedure illustrated in Figure 63 should be followed.

The ground around the base and underneath the branch spread is cleared of all plants and sod, and the soil is then broken up without disturbing the roots. Commercial fertilizers are applied according to the method outlined on page 52. Four- or six-inch agricultural tile or split sewer pipes are laid in a wheel-and-spoke design with the tree as the hub. The radial lines of tile near the tree should be at least 1 foot higher than the ends joining the circle of tile. A few radial tiles should extend beyond the circle and should slope sharply downward to ensure good drainage. An open-jointed stone or brick well is then constructed around the trunk up to the level of the new fill. The inner circumference of the stone well should be about 2 feet from the circumference of the trunk. Six-inch bell tiles are placed above the junction of the two tile systems, the bell end reaching the planned grade level, and stones are placed around the bell tiles to hold them erect. All ground tiles are then covered with small rocks and cobblestones to a depth of 18 inches. Next, the large rocks are covered with a layer of crushed stone, and these in turn with gravel, to a level of about 12 inches from the final grade. Cinders should never be used as a part of the fill. A thin layer of straw or hay should be placed over the gravel to prevent soil from sealing the air spaces. Then good topsoil should be spread over the entire area except in the tree well and the bell tile.

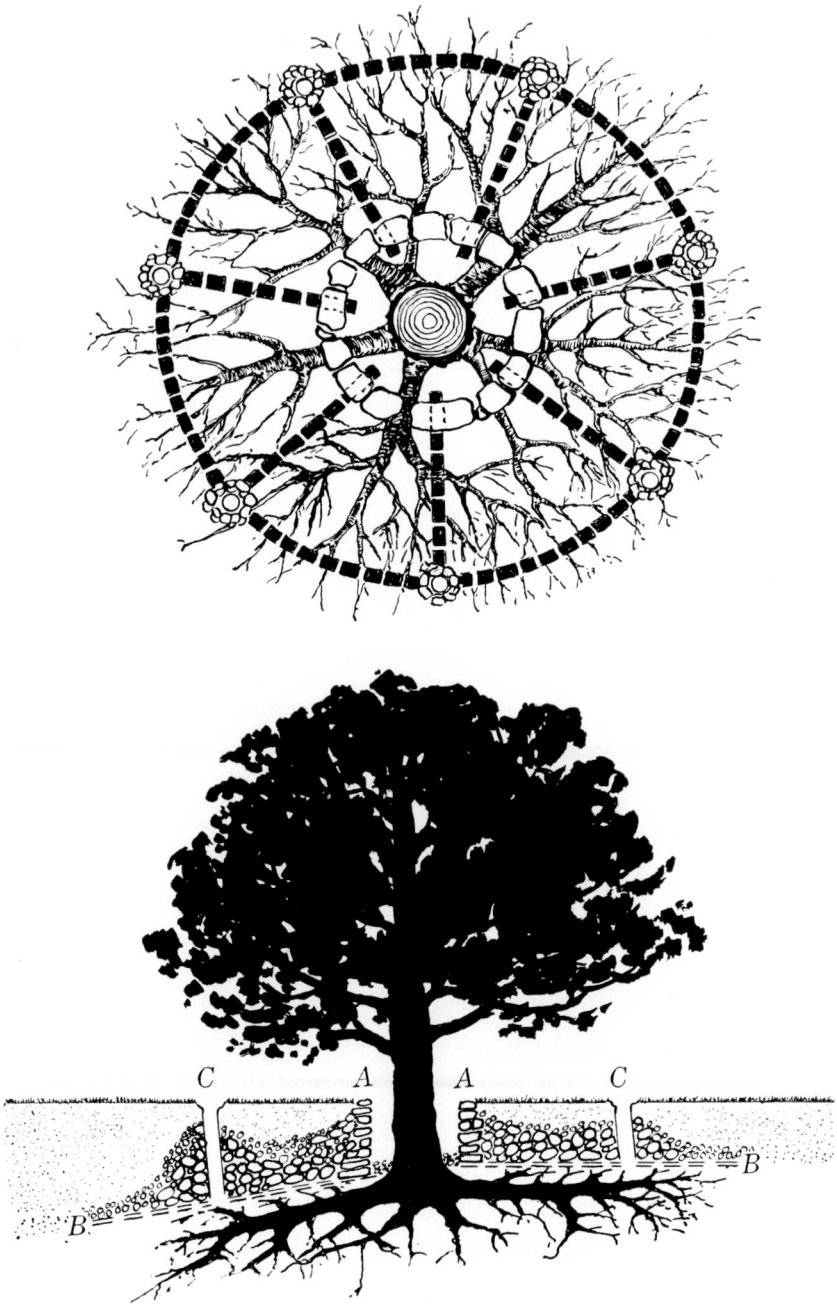

FIG. 63. Preventing injury by soil fills. UPPER: Bird's-eye view. LOWER: Side view showing *A*, the dry well; *B*, ground tile; and *C*, upright bell tile.

To prevent clogging, crushed stone should be placed inside the dry well over the openings of the radial tile.

The procedure outlined can be followed where a grade is to be raised around several trees in a group. In such cases the tile can run from one tree to another, increasing the air circulation.

To prevent children from falling into the hole, the area between the trunk and the stone well should be either covered by an iron grate or filled with a 50-50 mixture of crushed charcoal and sand. The latter practice will prevent not only filling with leaves and consequent impairment of the efficiency of air and water drainage, but also rodent infestation and mosquito breeding.

Children are often tempted to fill bell tiles with soil. To prevent this, the tiles should be filled with crushed rock and covered with a screen.

Other Methods. The method outlined constitutes an ideal one to ensure the least disturbance by fills, but a number of other methods can be substituted. Where the water drainage is not serious, coarse gravel in the fill can be substituted for the tile. This material has sufficient porosity to ensure air drainage. Instead of bell tile in the system, stones, crushed rock, and gravel can be added so that the upper level of these porous materials slants toward the surface in the vicinity below the outer branch spread.

Lowering Grades. Although removing soil around trees is not so disastrous as increasing the soil level, four changes take place: valuable topsoil is removed, feeding roots are exposed to drying out and low temperature injury, roots below the removed soil may be damaged, and the water table is changed. Most trees, however, will not suffer appreciably when only a few inches of topsoil are removed, since the roots soon become adjusted to the new conditions.

Injury to the tree as a result of lowering the grade can be reduced to a minimum by avoiding damage to roots. Exposed roots, if cut or broken, should be treated and covered with peat moss to prevent drying. Leaf mold, peat moss, or rotted manure should be incorporated in the remaining soil to increase the water-holding capacity. If large roots are cut or damaged, a few branches should be pruned to maintain the proper balance between these parts. Fertilizers should be applied, especially on the upper or undisturbed area.

Under some conditions injury can be reduced by terracing the new slope around the tree, rather than cutting the soil away in a uniform gradient.

Slime Flux

The foul-smelling and unsightly seepage from wounds in the bark or wood of various shade trees is known as slime flux (Fig. 64). It occurs most commonly on profusely bleeding trees, such as birch, elm, and maple, especially on those in declining vigor.

There are two types of slime flux, brown and alcoholic, distinguished on the basis of origin, color, and mode of development. Brown slime flux originates from heartwood sap. Before reaching the wounded surface, the sap is a clear watery fluid containing several nutrients. On the surface it soon changes to a brown, slimy ooze, as a result of the feeding by fungi, bacteria, and insects. Alcoholic slime flux, on the other hand, develops from the sap of bark and sapwood. It is white and frothy and usually forms near the base of the trunk. The sap is rich in starches, sugars, and proteins and thus makes an excellent growing medium for various bacteria and yeasts. These organisms produce fermentation, and the resulting odor is attractive to many species of insects.

Slime flux contains acids and aldehydes that are toxic to tree tissues. When the exudations persist for some time, much of the bark and cambium over which they flow is killed.

Treatment of Slime Flux. Thus far no effective preventive or curative measure is known. Alcoholic flux is usually of short duration and results in less damage than the more persistent brown flux. Bark tracing of the wounded area, shellacking, and dressing with a good paint to stimulate rapid healing are suggested for the former. Brown flux is more difficult to check, since the sap from which it originates is forced out of the heartwood by positive pressure. The sap may continue to flow despite cauterization of the wound with a blow torch followed by a covering of thick tar, the treatment most commonly recommended. Some arborists find that considerable reduction in the flow can be obtained by boring holes into the heartwood below the wound and inserting pipes to drain off the sap. This relieves the positive pressure and allows the wound to heal. After the wound heals, the drain pipes are removed and the holes are plugged with a filler. Trees with slime flux should be fertilized to promote vegetative growth, which in turn hastens callus development over the wound.

Occasionally a report appears in professional aborists' and nurserymen's

FIG. 64. Slime flux oozing from a pruned branch stub of American elm.

magazines on the use of sulfa drugs and antibiotics for preventing slime flux. Such reports, however, have not been substantiated by scientific tree workers. It is to be hoped that an effective chemical will be devised in the near future to control slime flux.

Injuries Caused by Low Temperatures

Low temperatures may injure trees in either vegetative or dormant condition. Injury is most common in regions where seasonal variations are greatest. For purposes of discussion, low-temperature injuries may be divided according to the season of the year in which they occur, that is, spring, fall, and winter.

Spring Frosts. Sudden drops to freezing temperature or below result in wilting, blackening, and death of tender twigs and leaves of deciduous trees, or reddening of the needles and defoliation of coniferous trees. Injury is most severe when the low temperatures occur in very late spring, when new growth is well advanced. Because of frost pockets, trees growing in hollows and valleys are damaged more often than those on higher ground.

Warm days in early spring stimulate premature growth, which is readily injured by the cold days and nights that follow. Among plants susceptible to this type of injury are: boxwood, Japanese dogwood, Pyracantha, red and white oaks, hackberry, and white ash.

Autumn Frosts. A cool summer followed by a warm autumn prolongs the growing season. Under such conditions twigs, buds, and branches fail to mature properly and therefore are more subject to injury by early autumn frosts. Unseasonable cold waves in the fall may also result in much damage. The disastrous frost which hit the Pacific Northwest on November 11, 1955, severely injured and killed thousands of trees and shrubs.

Winter Injury. Despite their dormant condition, trees frequently suffer severely during the winter. Their winter-hardiness is influenced by drainage, location, natural protection, species of tree, and character of the root system, as well as by a combination of unfavorable weather conditions.

The roots of trees are more likely to freeze in poorly drained than in well-drained soils, maple, ash, elm, and pine roots being most susceptible. The effects of frozen roots are seldom noted until the following summer, when the above-ground parts wilt and die. Injury to roots occurs most commonly during winters of little snowfall or in soils bare of small plants and other vegetation.

The ability of the tree to withstand low temperatures is also governed by the maturity of the wood. Trees fed with excessive amounts of high nitrogenous fertilizers, or those making considerable growth in late fall, are most commonly injured.

Aside from extensive damage to branches and roots, unfavorable temperatures during the winter produce more or less localized injuries on the trunks and branches. These are known as frost cracks, cup-shakes, and frost cankers.

FROST CRACKS: Longitudinal separations of the bark and wood are known as frost cracks. These openings may be large enough to permit the insertion of a hand, and may extend in a radial direction to the center of the tree or beyond.

Frost cracks are most likely to form in periods of wide temperature fluctuations. The water in the wood cells near the outer surface of the trunk moves out of the cells and freezes during sudden drops in temperature. The loss of water from these cells results in the drying of the wood in much the same way as green lumber drys and cracks when exposed to the sun. At the same time, the temperature of the cells in the center of the trunk remains much higher and little drying of these cells or shrinkage of the wood takes place. The unequal shrinkage between the outer and the inner layers of wood sets up great strain, which is released only by the separation of the layers. The break occurs suddenly along the grain of the wood and is usually accompanied by a loud report.

Cracks formed in this manner appear principally on the south and west sides of the trunk, since these are heated by the sun's rays and a greater gradient prevails when the temperatures drop suddenly during the night. If all exposures of the trunk were heated evenly, tension and frost cracks would not develop, as all tissues would shrink at the same rate.

Deciduous trees appear to be more subject to cracking than do evergreens. Among the former, Norway maple, Schwedler maple, elm, horsechestnut, linden, oak, and plane suffer most. Isolated trees are more susceptible than those growing in woodland areas, and trees at their most vigorous age (6 to 18 inches in diameter) are more subject than very young or old ones. Probably because of the higher moisture content of their tissues, trees growing in poorly drained sites are more subject to cracking than are those growing in drier, better drained soils.

When warmer temperatures arrive, the frozen tissues thaw, absorb more water, and the crack closes. The cracked zone in the heartwood never closes completely even though the surface may be sealed by callus formation. Because of insufficient support, however, the same area again splits open the following winter. The repeated splitting and callusing eventually results in the formation of a considerable mass of tissue over the affected area.

The most serious aspect of frost cracks is that they provide conditions favorable for the entrance of wood-decay fungi. The open surfaces should be treated and dressed as are open cavities. The installation of bolts as described under Bracing the Cavity (p. 92) may prevent the repeated annual splitting and thus enable adequate bark and sapwood to develop without disturbance.

CUP-SHAKES: Wide temperature fluctuations within a relatively short period produce another type of injury known as cup-shakes. This results from conditions that are the reverse of those responsible for frost cracks. A sudden heating by sunshine of the outer tissues of the trunk, following low temperatures, will cause these tissues to expand more rapidly than the inner tissues, resulting in a cleavage or separation along an annual ring.

FROST CANKERS: Low-temperature injuries may be confined to small localized areas on the trunk, branches, or in crotches. The lesions or cankers resulting from this limited injury are common on maple and London plane, and are confined principally to the southern and western exposures of the trunk. Scalding by the sun's rays on these exposures may penetrate as deeply as the cambium, resulting in a well-defined lesion.

The inner bark and cambium at or near the soil line can also be damaged by unseasonably low temperatures. The live tissues in these areas mature later than the twigs and branches. When low temperatures occur before the tissues are sufficiently hardened, they are apt to be damaged. The area affected may be confined to one side, or it may extend completely around the trunk and cause girdling. Old Baldwin apple trees appear to be especially susceptible to this type of damage. The practice known as bridge grafting can be adopted to save trees so injured.

Winter Drying of Evergreens. Winter injury to such plants as rhododendron, laurel, holly, pines, spruces, and firs is rarely caused by excessive cold during the winter. The damage is caused, rather, by excessive and rapid fluctuations in temperature or by late spring freezes after the plants have resumed activity.

Evergreen plants lose water continuously. Although water loss is less during the winter months, it may be considerably increased when the plants are subjected to drying winds or are growing in warm sunny spots. Unless the water lost during this period is replaced by absorption through the roots, the leaves wilt, turn brown, and many die.

In the earlier stages, winter injury on broad-leaved evergreens is evident as a scorching of the leaves at the tips and along the outer margins (Fig. 65). The color of the affected parts tends to be brown rather than yellow. On narrow-leaved evergreens, the needles are browned entirely or from the tips downward along part of their length. The terminal buds and twigs are brittle and snap readily when bent.

FIG. 65. Winter injury to rhododendrons and other broad-leaved evergreens appears as a browning along the edges of the leaves.

Prevention of Winter Injury. From the foregoing discussion it is obvious that the ability of a plant to withstand low temperatures depends on many factors beyond man's control. In some instances, however, several precautions can be taken to reduce the possibility of winter damage.

The selection of well-drained soils as sites for trees cannot be overemphasized. Trees considered susceptible always withstand low temperatures better when planted in soil with good aeration and water drainage.

Fertilization and soil aeration not only increase root growth but also tend to encourage deeper rooting, both of which help to reduce the possibility of winter injury. Better aeration can be obtained by spading the open soil area around the tree and by improving the water drainage. Digging should be shallow, so as not to disturb the feeding roots growing near the surface.

To protect broad-leaved evergreens, such as rhododendron and laurel, windbreaks of coniferous evergreens, which as a rule suffer less severely, are to be encouraged. Heavy mulches of oak-leaf mold or acid peat moss to prevent deep freezing and thus to facilitate water absorption by the roots are of great assistance in avoiding winter injury. To ensure an ample water supply during the winter months, soils around evergreens should be thoroughly soaked before freezing weather sets in.

Winter-browning can be prevented, or at least greatly reduced, by spraying the leaves with the liquid plastic Wilt-Pruf (p. 45). This material should be applied with a pressure sprayer on a mild day in late November or early December in the latitude of New York City. The author has seen many kinds of evergreens including rhododendrons, boxwood, and mountain laurel which came through the winter undamaged following an application of Wilt-Pruf in late fall or early winter.

Care of Winter-Injured Plants. In attempting to aid a winter-injured plant to regain its vigor, several precautions should be noted.

Drastic pruning of such plants is not advisable, since additional harm is apt to occur. Moreover, pruning should be deferred until the buds open in the spring, when dead wood can easily be distinguished from the live. A moderately pruned tree will recover more quickly than one severely pruned or one not pruned. It is impractical to prune winter-injured needles or leaves of evergreens, but all dead wood should, of course, be removed.

Some disagreement exists among tree men as to the specific fertilizer practices to be followed on severely winter-injured trees. Those who favor omission of fertilizer for the entire season following injury claim that heavy spring applications may either further injure already damaged roots, or result in the production of much foliage. A reduction in the root mass would lower the water intake, whereas an overabundance of foliage would require more water than could be transported by the winter-injured conducting tissues. These advocates say that a moderate application of fertilizer might be justified about the first of July, since some new water-conducting tissue will have been formed by that time.

Advocates of medium to heavy applications of fertilizers in the spring following winter injury claim that such treatments encourage the formation of new tissue, thus making it possible for the tree better to withstand summer conditions.

Lightning Injury

Every year hundreds of shade and ornamental trees are struck by lightning throughout the country. The amount and the type of injury are extremely variable and appear to be governed by the voltage of the charge, the moisture content of the part struck, and the species of tree involved.

The woody parts of the tree may be completely shattered (Fig. 66) and may then burn, a thin strip of bark parallel to the wood fibers down the entire length of the trunk may be burned or stripped off, the internal tissues may be severely burned without external evidence, or part or all of the roots may be killed. The upper trunk and branches of evergreens, especially spruces, may be killed outright, while the lower portions remain unaffected. In crowded groves, trees close to the one directly hit will also die. In many cases, grass and other vegetation growing near the stricken trunk will be killed.

Tall trees or those growing alone in open areas and trees with roots in moist soils or those growing along bodies of water are most likely to be struck. Though no species of tree is totally immune, some are definitely more resistant to lightning bolts than others. Beech, birch, and horsechestnut, for example, are rarely struck, whereas elm, maple, oak, pine, poplar, spruce, and tulip are commonly hit. The reason for the wide variation in susceptibility is not clear. Some authorities attribute it to the difference in composition of the trees: for example, trees high in oils (beech and birch) are poor conductors of electricity, whereas trees high in starch content (ash, maple, oak, etc.) are good conductors. In addition, deep-rooted or decaying trees appear to be more subject to attack than are shallow-rooted or healthy trees.

It is commonly believed that lightning never strikes twice in the same place. This is not so, because some trees have been struck by lightning as many as seven times, judging from the scars on their trunks.

Repairing Lightning-Injured Trees. The injured tree should be inspected carefully before any attempt is made to repair the damage. Many trees are severely injured internally or below ground, despite the absence of external symptoms, and will soon die regardless of treatment. Consequently, expensive treatments should not be undertaken until the tree appears to have a good chance of recovery.

Where external damage is not great or where the tree is particularly valuable, several immediate measures are justifiable. Some benefit is derived by tacking on and covering with burlap any long thin pieces of bark that have

FIG. 66. Shattering of the trunk on a lightning-struck tulip tree.

been split or lifted from the sapwood. Shattered limbs and torn bark should be removed carefully and all open surfaces dressed. In addition, some rapidly available fertilizer should be placed around the root area to stimulate vigorous growth.

Lightning Protection. Trees growing in locations favorable for strikes, and large, rare, or otherwise valuable ones can be adequately protected at comparatively small cost by the installation of proper equipment.*

The principles governing the protection of buildings and other structures from lightning bolts are employed also in safeguarding trees. The major difference is in the type of materials used and their method of installation. Tree protection systems must have flexible cables to allow for the swaying of the trunk and branches, and adjustable units to allow for the tree's growth.

A vertical conductor is fastened along the trunk with copper nails, from the highest point in the tree to the ground. The ground end is fastened to several radial conductors, which are buried underground and extend beyond the root area. These, in turn, are attached to ground rods 8 feet in length, which are driven vertically into the ground.

The system should be inspected every few years, the air terminals extended to the new growth, and other adjustments necessitated by expansion of the tree should be made. Ground terminals should also be checked periodically, and worn-out parts replaced.

Two well-known companies that handle lightning protection equipment are Electra Protection Co., Inc., Albany 6, New York, and Independent Protection Co., Inc., Goshen, Indiana.

Other Electrical Injuries

Wires carrying electric current may also cause some injury to trees. Suitable nonconductors should be placed over wires passing in the vicinity of the trunk and branches. Much of the injury by wires can be avoided by judicious pruning, as discussed under Line Clearance (p. 75), so that branches and wires do not interfere with each other. Outdoor evergreens bedecked with lights at Christmas time can be damaged if too many poorly placed bulbs or worn

* The theories and details of installing lightning-protection equipment are too involved to permit discussion in this book. This information is available in *Tree Preservation Bulletin No. 5*, which may be obtained from the Superintendent of Documents, Washington 25, D. C., for 20 cents. Installation must be entrusted to a specialist in tree care and should never be attempted by the layman.

equipment is used. The bulbs should be placed so that they do not come into direct contact with the needles or twig tips.

Gas Injury

Much has been written about the effect of escaping gas on trees. Almost all of this information is based on the illuminating gas manufactured from coal and oil, the principal gas used in this country during the first 40 years of the present century. The manufactured gas contains a number of ingredients known to be toxic to the roots of plants. Among these are hydrogen cyanide, which forms hydrocyanic gas in solution; carbon monoxide; and a number of chemicals known as unsaturated hydrocarbons. In the last group are such chemicals as ethylene, acetylene, benzene, butylene, and propylene. Such famous research scientists as Doctors P. W. Zimmerman, A. E. Hitchcock, and William Crocker of the Boyce Thompson Institute, Yonkers, New York, have shown all these chemicals to be harmful to vegetation.

In areas where gas for heating and cooking purposes is still being manufactured from coal and oil, the danger of damage to roots of trees and other plants exists whenever such gas escapes from leaking mains and bathes the roots (Fig. 67).

Fig. 67. Manufactured illuminating gas escaping from broken pipes below ground killed the London plane trees in the foreground.

The toxic effect of escaping manufactured gas is most noticeable when the trees are in leaf. The symptoms produced vary greatly with the species of tree, the amount and constituents of the gas, the kind of soil, and the character of the surface cover.

The leaves of trees exposed to large amounts of manufactured gas suddenly wilt and turn brown, then twigs and branches die. These symptoms appear gradually over a period of one or two years where the flow of manufactured gas is slow.

Some trees exhibit very striking responses when their roots are subjected to escaping manufactured gas. For example, long narrow cankers appear near the trunk base of London planes. The inner bark tissues in the cankered area become water-soaked, swollen, and, most striking of all, distinctly watermelon-pink. The inner bark of Carolina poplars exposed to manufactured gas also increases in thickness and appears water-soaked, but undergoes no striking change in color.

The woody portion of the smaller roots in some trees may turn bluish-black after long exposures to manufactured gas. This, however, is not a reliable symptom because the roots of some trees may turn the same color from other factors where no free manufactured gas exists in the soil.

Determination of Manufactured-Gas Leaks. Where gas is escaping in considerable quantity, the characteristic odor can usually be detected in the air near the injured tree. Lesser amounts can often be detected by smelling the air in holes dug to a depth of 12 to 24 inches, or by using special gas-detecting apparatus such as the Davis Gas Tester, manufactured by Davis Emergency Equipment Co., Newark 4, N. J.

In many instances, however, manufactured gas is present in concentrations too low to be detected by man or machines, but still high enough to cause injury. A tomato plant, which is about 200 times as sensitive as the human nose in detecting ethylene in manufactured gas, can be used to detect extremely minute quantities of gas in the soil. The potted tomato is placed overnight in a hole near the suspected leak. The ethylene fraction in manufactured gas will cause the tomato stems to bend sharply downward (Fig. 68) instead of maintaining their normal erectness (Fig. 69). The downward bending does not result from wilting, but is a growth response to the ethylene fraction in the gas.

If young tomato plants are not readily available, sweet peas or carnations,

FIG. 68. Test plants indicate the presence of manufactured illuminating gas in the soil. The branches of this tomato plant responded to the ethylene fraction in the gas by bending sharply downward.

FIG. 69. LEFT: This tomato plant shows normal branch habit.

FIG. 70. RIGHT: Tomato plant subjected to natural gas keeps its normal shape. There is no sharp downward bending of the branches as in Figure 68.

obtainable from most florists, may be substituted. The ethylene fraction causes prompt closing of open flowers and prevents any buds from opening.

Natural Gas Is Nontoxic. The widespread conversion to natural gas in the last 20 years has completely changed the gas-damage picture. Most communities in this country are now using natural gas, which is piped largely from the southwestern parts of the country, particularly Texas. In some areas, gas companies are still using a mixture of the older manufactured gas and natural gas. Here the symptoms of the combination somewhat resemble those produced by straight manufactured gas, though they are less pronounced.

Communities using straight natural gas should no longer be troubled with damage to trees and other vegetation. The reason is that pure natural gas does not contain any of the unsaturated hydrocarbons, hydrogen cyanide, or carbon monoxide present in manufactured gas (Table IV), all of which are toxic to vegetation.

Table IV

APPROXIMATE PERCENTAGES OF SUBSTANCES COMPRISING MANUFACTURED AND NATURAL GAS

Substance	Manufactured	Natural *
Unsaturated hydrocarbons— ethylene, acetylene, benzene, butylene, propylene	8.0	None
Carbon monoxide	33.0	None
Cyanogen compounds and hydrocyanic acid	5.0	None
Hydrogen	33.0	
Carbon dioxide	1.5	1.0
Oxygen	1.0	0.4
Nitrogen	6.0	0.6
Saturated hydrocarbons— methane, ethane, propane, etc.	12.5	98.0

* Another natural gas—liquid petroleum—is usually propane or butane, also saturated hydrocarbons which are stored in a liquid state and vaporized for consumers.

During 1956 and 1957, the author subjected the roots of young Norway maples, pin oaks, and London planes to natural gas and was never able to detect any toxic effect from such a treatment. Trees of the same age subjected to a mixture of manufactured gas, however, were killed within a few weeks. Tomato plants exposed to natural gas did not show the responses so typical of those produced by manufactured gas (Fig. 70).

There is virtually no reliable information in the printed literature on the effect of natural gas on trees. As mentioned earlier, most of the printed material is based on experiences with manufactured gas. Nevertheless, because of ignorance, lack of adequate information to the contrary, or so-called 'good public relations,' many public utilities continue to pay for trees alleged to have been killed by natural gas.

The author examined nearly 500 dead and dying streetside trees during 1956 and 1957. Almost every tree was alleged, by arborists and other persons called in to diagnose the trouble, to have been damaged by gas. The external symptoms visible on these trees were variable and the diagnosis in most instances was based on the fact that a gas leak had been repaired in the vicinity. The author found, however, that other more plausible factors were responsible for the decline and death of many of the trees. Among these were malicious girdling below ground (p. 153), girdling roots (p. 180), gasoline and oil in the soil, and fungi infecting the roots and main trunk (Chapter 14).

One important characteristic of manufactured gas is that it kills not only trees but also grass, hedges, and other vegetation in the immediate vicinity of the dead trees. This was never observed by the author in his investigations of trees alleged to have been killed by natural gas.

Conceivably, natural gas can cause injury to vegetation when a leak occurs along a transmission line, where the gas is under 200 or more pounds pressure. In such a situation the flow of gas is so intense that the soil dries out rapidly, which in turn results in damage to the vegetation. Even ordinary air forced through soil at high pressures would dry out the soil and thus damage vegetation. Strong leaks will also change the structure of the soil—a change which is discernible for some distance. Leaks in high pressure gas transmission lines in the Southwest can actually be detected by the change in color and texture of the soil.

But gas leaks that occur along most city streets do not come in the same category because the gas is usually under only 2 or 3 pounds pressure, far too low to cause drying out of the soil.

Treatment of Gas-Injured Trees. Trees injured by escaping manufactured gas can usually be saved if the leak is detected early or if the flow of gas has been rather slow. First, the leak must be repaired by the gas company as quickly as possible. A trench is then dug on the side of the tree nearest the leak and the soil aerated as thoroughly as possible. Because the water-soluble constituents are usually most toxic, large quantities of water should then be applied. Severely injured roots should be pruned and painted to prevent rapid decay of the wood. Following such treatments, the soil should be replaced and the entire root area fertilized to stimulate new top and root growth. The author has witnessed several recoveries due to this procedure.

Where trees have been killed by high concentrations of manufactured gas, the best procedure is to repair the leak, remove the trees, and aerate the soil for several months. If new trees are to be replanted immediately, probably the safest procedure is to remove the soil and replace it with fresh soil. Where this is not feasible, air may be forced into the soil under compression, followed by heavy watering.

Injury by Girdling Roots

Many trees are weakened and some are killed by the growth habits of certain of their roots. Such roots grow closely appressed to the main trunk or large laterals in such a way as to choke the members surrounded, much as does a wire left around a branch for a number of years. This choking action restricts the movement of nutrients in the trunk or in the strangled area of the large roots. When a large lateral root is severely girdled, the branches that depend upon it for nutrients commonly show weak vegetative growth and may eventually die of starvation. If the taproot of a tree is severely girdled, the main branch leader may die back. As a rule, trees affected by girdling roots do not die suddenly, but become progressively weaker over a 5- or 10-year period despite good pruning and feeding practices.

Girdling roots which develop below ground level can often be detected by examining the trunk base. If the trunk ascends straight up from the ground, as it does when a soil fill has been made (p. 159), or is slightly concave on one side, instead of showing a normal flare (swelling) or buttress at the soil line, then one can suspect a girdling root. Of course digging the soil alongside the trunk should reveal whether the strangling member is present.

Trees growing along paved streets always suffer more from girdling roots than do those in open areas, and middle-aged or old trees more than younger ones.

How Girdling Roots Develop. Roots are deflected from their normal spreading course by unfavorable environmental conditions. They grow most rapidly in areas where air, moisture, or food supply is best. A good case in point is that of a tree growing along a city street. The roots of such a tree, when newly planted, are usually spread out in all directions by the planter. For a number of years, the roots spread in the general direction in which they were originally placed. Because of unfavorable conditions under the pavement, the larger roots and especially the younger ones will eventually grow away from the street toward the open areas between the curb and sidewalk and toward lawns. In the course of bending, the younger roots cross over older roots and become closely appressed to the base of the main trunk. Such roots grow and expand rapidly because of the more favorable conditions and eventually press against or strangle the roots they cross or the trunk base they touch (Fig. 71).

Root girdling frequently occurs where a street tree is planted in a hole filled with good soil and surrounded by hard, impervious subsoil. Inability to penetrate the more unfavorable soil forces the roots to develop within the filled area, thus favoring girdling.

Careless placement or overcrowding of the roots at transplanting time may also lead to root girdling and the rotting of the crowns. This has been observed not only with shade trees along city streets, but also with red, white, and Scotch pines planted in reforestation projects. In the latter case, a direct correlation exists between careless placement of roots and the amount of subsequent infection in the root crown by fungi and bacteria. As twisted roots develop, air and water pockets are formed. These make ideal breeding places for organisms that produce root and crown decay.

In a few instances, fertilization of the soil in the immediate vicinity of young street trees has stimulated new root production, produced overcrowding, and eventually resulted in the formation of girdling roots.

Girdling roots are always more numerous in compacted than in loose soils, and on soft-wooded than on hard-wooded trees.

How To Diagnose Girdling Roots. The best time to determine the presence of girdling roots is in late fall just before frost. At that time, the side of the tree over the girdled area will show lighter green leaves, which tend to absciss earlier than normally. One must always bear in mind, however, that other factors may produce similar symptoms and that the most positive sign is the discovery of girdling roots. Such roots can be uncovered by carefully removing the soil around the trunk on the side showing weak top growth, poor bark

FIG. 71. Treatment of girdling roots. TOP: A girdling root. CENTER: Removing the root with a chisel and mallet. BOTTOM: The girdling root has been removed. The cut should be dressed with tree paint.

development, or a pronounced swelling at the base of the trunk. Usually the girdling root is to be found at or a few feet below the soil surface. Occasionally is may be deeper if the taproot is being strangled.

Treatment for Girdling Roots. If the girdling root has not yet seriously impaired the tree's chance of recovery, it should be severed with a chisel and mallet (Fig. 71) at its point of attachment to the trunk or to the large lateral root. A few inches should be cut from the severed end to prevent its reuniting with the member from which it was severed. Then the cut surfaces, on the trunk or main root and on the severed root itself, should be painted to hinder insect and fungus invasion. Finally the soil is replaced in its original position. If the tree has been considerably weakened, judicious pruning should be practiced and some good fertilizer applied.

Smoke and Soot Injury

Smoke emanating from chimneys of manufacturing plants, apartment house incinerators, and other instruments of combustion, including automobiles, contain ingredients which are harmful to the leaves of trees and other plants.

Smoke from factories frequently contains one to several chemicals which do the actual damage. The three major pollutants are sulfur dioxide, fluorine compounds, and the so-called smog typical of the Los Angeles area. The first two are so-called acid-forming gases, whereas smog is a peculiar combination of many different chemicals.

Sulfur Dioxide and Fluorine Pollutants. Sulfur dioxide, perhaps the most common ingredient in the smoke from factories, can injure vegetation in concentrations as low as one or two parts in a million parts of air. Although highly toxic to plants, it is relatively nontoxic to humans at these low concentrations. At concentrations high enough to injure plants, it may irritate the mucous membranes of individuals who are exposed to it for long periods.

Both sulfur dioxide and fluorine produce leaf markings that are quite distinctive. When the gas concentration is high, the leaf cells become water-soaked and then collapse and dry out rapidly, leaving blanched or scorched areas. Mild attacks will merely cause the leaves to turn yellow.

Maple and other broad-leaved trees exposed to sufficiently high concentrations of sulfur dioxide show ivory-white markings, mostly between the main veins. Sometimes the markings occur along the margins, with branches extending into the interveinal areas. With some broad-leaved trees the markings are darker brown in color.

Sulfur dioxide causes a reddish discoloration of the needles on Douglas fir and Ponderosa pine, followed by shriveling of the affected tissues. The discoloration may occur at the tips only or as a band over the entire length of the needles.

When the concentration is too low or the time of exposure to sulfur dioxide is too short to cause any visible lesions or yellowing of the leaves, the growth of the plant will not be affected. In other words, sulfur dioxide is not a systemic poison and its action is not cumulative. As a matter of fact, at low concentrations of about three tenths of a part per million parts of air, it is absorbed by the leaves and changed to the sulfate form.

When relatively high concentrations of one or more parts per million are absorbed for an hour or more, sulfurous acid or sulfite forms in the interior cells. These compounds cause visible damage to the leaf cells and prevent them from functioning normally.

Middle-aged leaves are most susceptible to sulfur dioxide, while the young or growing leaves are markedly resistant. Brownish-red or yellow areas between the leaf veins are a rather characteristic symptom. The margins or tips of some leaves may also be discolored or blanched. On coniferous trees, the tips, bases, or centers of the needles turn red and then, especially when the gas concentration is high or the gas has been present for a long period, turn completely brown and drop off.

Often, sulfur dioxide and fluorine damage cannot be told apart in the field. Only a chemical analysis of the leaves will reveal which gas is responsible.

Fluorine compounds are much more toxic to vegetation than is sulfur dioxide. Industries which produce aluminum, steel, ceramics, and phosphates are the principal contributors of fluorine compounds to the atmosphere.

Plants vary widely in their tolerance to fluorine. Cotton, for example, can absorb as many as 5,000 parts per million parts of air without showing injury, whereas gladioli may be injured when they absorb as little as 20 to 30 parts per million. Peach, prune, and conifers also are very sensitive to fluorine.

Smog—Unusual Pollutant. Smog is a term used to designate a mixture of smoke and fog. In the Los Angeles area, however, this term refers to a complex of liquids, solids, and gases, comprising more than 50 chemical elements and compounds.

The harmful ingredients are believed to be products resulting from a combination of unsaturated hydrocarbons and the ozone in the atmosphere. The hydrocarbons come from oil refineries and related industries as well as from automobile exhausts.

Silvering, glazing, streaking, and speckling of the leaves are among the more common symptoms of the Los Angeles type of smog. The symptoms vary according to the type of vegetation and the structure of the leaf. Leafy vegetables are more susceptible than trees. They will show a metallic sheen or glazed appearance on their lower leaf surfaces from 1 to 3 days after exposure to smog.

A great deal of research is now being conducted in California to determine ways of eliminating smog or of reducing its damage. One investigator has already observed that smog damage to certain crops can be prevented by applying dusts or sprays containing dithiocarbamates, a widely used group of fungicides. Others have reported that a light spraying with ascorbic acid (Vitamin C) of the leaves of some plants prevented the burning and bronzing symptoms of smog.

Many factories, particularly in the Los Angeles area, which formerly released large quantities of sulfur dioxide into the atmosphere, have installed special equipment to trap the harmful gases. Others have built extremely tall chimneys so that the smoke is well diluted before it reaches susceptible crops.

Chlorine and Hydrogen Sulfide Pollutants. Two other gases may also cause damage to vegetation—chlorine and hydrogen sulfide. The former produces most unusual markings. Dead areas develop between the veins and generally toward the center of the leaf, resulting in a netted appearance. Hydrogen sulfide (which has the odor of spoiled eggs) differs from most other pollutants in its effects. The youngest leaves are most sensitive and hence are marked first. The markings are usually light tan to white.

Smoke. Though not so toxic as sulfur dioxide or fluorine, soot, the solid residue of smoke, may also damage plants. This material slowly settles on the leaves and eventually may coat them completely. Here, the damage does not result from clogging of the breathing pores (stomata), as some believe, but from the screening out of sunlight. The sharp reduction in light reduces the capacity of the leaf to manufacture food, and consequently results in a serious impairment of the general vigor of the plant. Because the leaves of evergreens remain attached throughout the year, and in some species for several years, a considerable deposit may accumulate and thus greatly reduce the chances of survival of these trees. Soot is one factor that makes it difficult to grow evergreens in large cities.

Trees vary considerably in their tolerance to smoky atmospheres. Tree-of-heaven, ginkgo, and Carolina poplar seem to withstand such conditions best, and London plane, pin oak, Norway maple, elm, linden, and tulip show decreasing tolerance in the order mentioned. As has been pointed out, other factors, including tolerance of roots to harmful chemicals, also enable some trees to tolerate adverse city conditions better.

Toxic gases in the atmosphere may harm vegetation indirectly as well as directly. The acidity of a soil may increase as a result of the interaction between acid gases and the lime particles in the soil. Such a change not only reduces the population of certain beneficial soil bacteria, but also decreases the activity of others. Excessive soil acidity, whether speeded up by air pollutants or from natural leaching by rainfall or irrigation, can be counteracted by the application of lime.

Obviously one way to reduce injury from smoke is to eliminate the production of the most harmful constituents at their source. Where this is not feasible, trees most tolerant to city conditions should be planted. Injury to small evergreens by soot deposits can be minimized by spraying with a soapy solution followed by syringing with clear water. A material known at Calgon (sodium hexametaphosphate) can also be used to remove soot from broad- and narrow-leaved evergreens.

Arborists and others interested in the care of trees should not be too hasty in concluding that leaf abnormalities are always caused by air pollutants. They should bear in mind that unfavorable weather, insects, fungi, bacteria, spray materials, growth regulators, and viruses also produce symptoms that might be confused with symptoms of air pollution.

Over the last 20 years the author has examined literally hundreds of cases of alleged air-pollution injury to vegetation and has found that only a small percentage of the damage was actually caused by gas. And this experience was gained in areas that were highly industrialized! Before blaming poor growth of vegetation on air pollutants, therefore, one would do well to call in a competent diagnostician. Plant pathologists at some of the state agricultural colleges are equipped to render this service.

Lichen Injury

Lichens often appear as a green coating on the trunks (Fig. 72) and branches of trees. They are actually two plants in one, being composed of a fungus body and an algal body, which live together in complete harmony. The alga supplies

FIG. 72. Lichens growing over a tree trunk.

elaborated food to the fungus and, in turn, receives protection and some food from the fungus.

Lichens do not parasitize trees, but merely use the bark as a medium on which to grow. They are more unsightly than injurious, although when extensive they may interfere with the gaseous exchange of the parts they cover. They are common on trees throughout the southern states and appear less frequently in the northern states. Because of their extreme sensitivity to sulfur dioxide released from factory chimneys, lichens seldom appear on trees in the larger cities. They rarely develop on rapidly growing trees, because the bark is shed before the lichens have an opportunity to grow over much of the surface.

As a rule, lichens can be eradicated by spraying the infested parts with bordeaux mixture 2-2-50 or any ready-made copper spray.

Chemical Injuries

Weed Killers. The various chemicals used as weed killers severely damage or even kill trees in the immediate vicinity. The author has noted many cases of toxicity to trees caused by such materials. If the weed killer contains arsenic, diagnosis may be confirmed by a chemical analysis of the wood. Some state agricultural experiment stations are equipped to make such an analysis.

When carelessly used, the new selective weed killers containing 2,4-D (2,4-dichlorophenoxyacetic acid) have caused widespread damage to trees (Fig. 73).

Persons treating lawns for weed control or using weed killers for controlling poison ivy and other undesirable plants along highways and right-of-ways, should exercise extreme caution to avoid drifting of the 2,4-D and similar materials to nearby trees.

Symptoms Vary Greatly. Symptoms depend on the concentration of material which reaches the tree, the kind of tree, temperature, and other factors. The most general one is a distortion of leaves and twigs. Leaves roll upward or downward, at the midrib or along the edges, so as to be cupped. Leaf stems may curl downward, in some cases forming loops resembling a curled pigtail. From a distance, severely affected trees appear wilted because of this leaf and stem distortion.

The author has noted 2,4-D injury on many trees including elm, dogwood, box elder, hawthorn, Norway maple, tuliptree, pin, red, white, and black oaks, and sassafras. Unless the trees received an unusually heavy dose or repeated lighter doses, they usually recovered.

To avoid possible damage to trees, it is suggested that one of the less volatile forms of 2,4-D and 2,4,5-T be used, and that it be applied on windless days, and in a coarse spray (at low pressure), rather than as a fine mist such as that used for spraying trees with fungicides and insecticides.

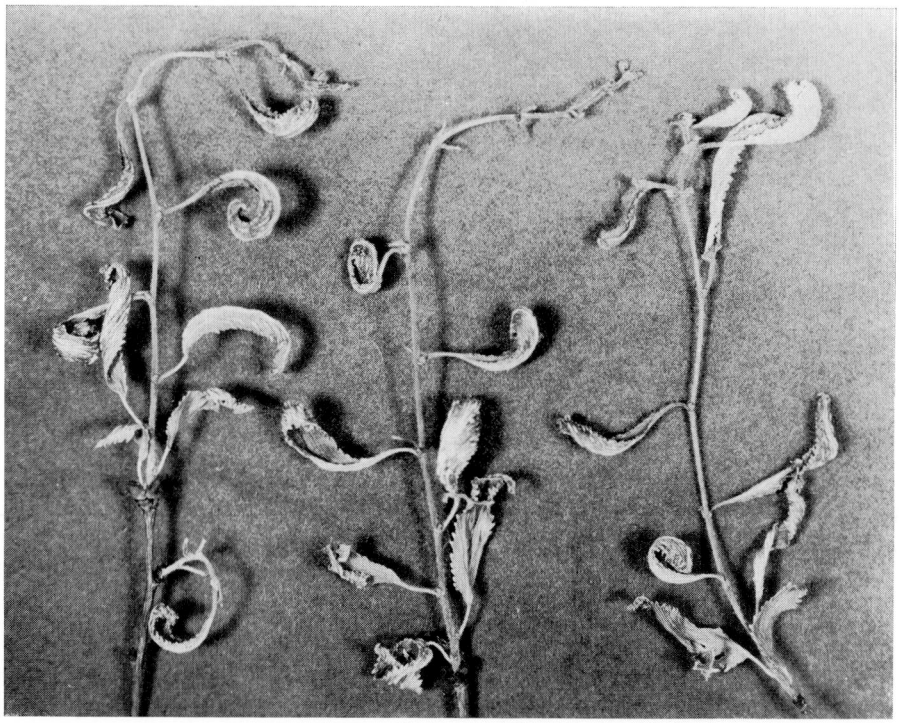

FIG. 73. 2,4-D injury to English elm leaves.

The author was among the first to draw attention to the possibility that trees such as pin oaks growing in lawns treated in the fall with 2,4-D can absorb enough of the hormone in their dormant buds to produce distorted leaves the following spring. Arborists should bear in mind that wooden spray tanks cannot be made free of 2,4-D once this hormone has been used in them. Such a sprayer cannot safely be used later for normal pest control operations.

Where a sprayer with a metal tank has been used, the 2,4-D can be removed as follows: Fill the tank with warm water and household ammonia at the rate of one quart of ammonia for each 25 gallons of water. Then pump out a few gallons to wash pump parts, hose, and nozzles, and allow the remainder to stay in the tank for 24 hours, or preferably 48. Then drain the tank and rinse several times with clear water.

It is well to stress the fact that even minute traces of 2,4-D in such sprayers will cause severe distortion of leaves and twigs. Injury to susceptible plants has been observed with dilutions as low as ½ part 2,4-D in a million parts of water.

Calcium Chloride. Trees growing along country roadsides may be damaged by the calcium chloride used to keep down dust on dirt roads. Heavy rains wash the chemical off the roads and carry it down to the roots where it is absorbed, then transported to the leaves where it causes injury. Correct grading of the roadside so as to facilitate removal of flood waters and prevent their being soaked into the soil around the roots will greatly reduce the chances of damage.

Tree species vary in their tolerance to calcium chloride. Red oak, white oak, and American elm are most tolerant, whereas balsam fir, white spruce, beech, sugar maple, cottonwood, and aspen are least tolerant.

Damage to the leaves appears as a leaf scorch which is difficult to distinguish from physiological leaf scorch caused by lack of water in the soil.

Salt. Sand treated with salt and piled along roadsides for winter use may also damage trees. The salt is leached down into the roots and causes burning. Sand piles thus treated should, obviously, be placed at some distance from the root areas of the trees.

Salt is frequently scattered over sidewalks in winter to melt ice or prevent water from freezing. When the salty solution finds its way to open soil near roots it can be very toxic. As a mater of fact, sodium chloride (ordinary table salt) is five to ten times more toxic to some trees than calcium chloride. Sand or sawdust should be used to prevent slipping in areas near a tree and other plant roots.

Cinders. Hard and soft coal cinders, especially those that have been unexposed to weathering, should not be placed over the soil beneath trees. Toxic materials in such cinders will be washed down in rain water and harm the roots.

Salt Spray. Trees growing along the Atlantic coast are often injured by salt spray blown from the ocean. During hurricanes the spray actually damaged leaves 50 miles from salt water. Most of the damage, however, is usually confined to within a few miles of the coast.

Among the broad-leaved trees, elm, magnolia, Norway maple, sugar maple, tulip, and sourgum appear most susceptible to salt-water spray, whereas red oak and horsechestnut are very resistant. Among the evergreens, white pine, hemlock, junipers, and Scotch pine are most harmed, whereas Japanese black pine, Colorado blue spruce, Austrian pine, holly, spruce, and yew suffer least.

Removal of badly damaged trees and pruning of mildly affected ones are all that can be suggested.

Spray Injuries. The various sprays recommended in the latter part of this book for disease and insect control may occasionally injure trees. In general, damage from fungicides and insecticides is more severe on undernourished trees and on those growing in poorly drained soils than on vigorous specimens. Insect and frost damage also predispose leaves to spray injuries. Cool, damp weather favors the chances of injury from bordeaux mixture and other copper fungicides. High temperatures and humidities increase the chances of injury from lime sulfur and other sulfur sprays.

Copper in Rainwater. The author is of the opinion that rainwater that passes through copper leaders and drainpipes of homes may injure trees. He has noted the gradual weakening and death of a number of trees the roots of which were continuously wetted with such water.

To prevent damage, the water passing through copper leaders should be drained directly into the sewer or the street gutter.

Copper Sulfate. This chemical is frequently mentioned as a control for algae (green scum) in ponds and lakes. The usual concentration is 1 part of the copper sulfate in a million parts of water, by weight. When the concentration is in excess of this figure, there is danger of harming trees growing around the edge of the pond or lake.

Tree Bands. Banding trees with a sticky substance can be dangerous unless means are taken to prevent the inner bark and cambium from absorbing the material. If the banding material is applied directly to the bark, or if the thicker outer bark is shaved or cut away and the material then applied, severe damage may follow.

Sugar maples appear to be more susceptible than other trees to banding substances applied directly to the bark.

Hail Injury

At times, hailstorms cause considerable damage. They may completely defoliate the trees or shred the leaves sufficiently to check growth sharply. They are most injurious to young trees or to those with undeveloped foliage in early spring. Areas in the bark and cambium may be severely bruised or even killed from the impact of the hailstones. When the bark injuries do not callus over rapidly, they often serve as entrance points for wood decay fungi.

Mechanical Injuries

Damage to trees, especially the trunks, by motor vehicles, by children, and even by thoughtless adults occurs too commonly to warrant much discussion. Injury by automobiles and trucks is often unavoidable.

Most mechanical injuries are confined to the outer portions of the trunk. These should receive immediate treatment. Cut the bark along the edge of the wound in the form of a pointed ellipse (Fig. 74), shellack the cut edges, and then dress the wound with a tree paint.

Fig. 74. Two examples of correct treatment of bark injuries. Note the pointed cuts and the excellent callus development.

Where constant danger of mechanical injury exists, young trees should be protected by a tree guard.

Wire, unless covered with a piece of garden hose or some other suitable material, should never be used to aid in the support of trees. As the trunk expands, the bare wire will tend to girdle the trunk (Fig. 75) and cause severe weakening or even the death of the tree.

Accumulations of snow, especially on evergreens such as yews, hemlocks, and junipers, can cause severe branch breakage. Such accumulations should be gently shaken off as soon as possible with a broom or bamboo rake. Ice-covered trees should be handled even more carefully. If the temperature is on the rise, rinsing the ice-coated trees with water from a hose will thaw the ice.

Some deciduous trees are prone to cracking and splitting of branches in ice and wind storms. Among those most subject are the ashes, catalpas, hickories, horsechestnuts, red maple, Empress tree, Siberian elm, silk tree, tulip tree, and yellowwood.

FIG. 75. Wire placed around a tree will eventually girdle the trunk.

The lawnmower is a more destructive instrument for debarking trees than are automobiles and malicious children. Careless use of this instrument results in severe damage to the inner bark and cambium near the soil line. Parasitic fungi like Ceratocystis, which infects plane trees (p. 357), and wood-destroying fungi can get started in wounds made by lawnmowers.

If no grass is planted close to the base of a tree, there is less chance of causing damage. Installation of a so-called No-Trim Tree Guard eliminates the need for edging and trimming grass around lawn trees.

Leaf Scorch

Although the trouble known as leaf scorch is most prevalent on Japanese red, sugar, silver, Norway, and sycamore maples, it also occurs on horsechestnut, ash, elm, and beeches and to a lesser extent on oaks and other deciduous trees.

The Katsura tree (p. 121) is particularly subject to leaf scorch when planted near hard-surfaced pavements. The tree apparently cannot tolerate a paved surface over its roots.

Symptoms. Scattered areas in the leaf, first between the veins or along the margin, turn light or dark brown. The edges of the discolored areas are very irregular. All the leaves on a given branch appear to be affected more or less uniformly. When a considerable area of the leaf surface is discolored, the canopy of the tree assumes a dry, scorched appearance (Fig. 76). In severe cases, the leaves dry up completely, then fall prematurely. When such defoliation occurs before midsummer, new leaves are formed before fall. In mild cases, the leaves remain on the tree, and little damage results.

The position of the most severely affected leaves can be used, in many instances, as a means of distinguishing leaf scorch from leaf spot and blight damage caused by fungi. Leaves affected by scorch are usually most abundant on the side of the tree exposed to the prevailing winds or to the most intense rays of the sun, whereas spotted and blighted leaves resulting from fungus attack are usually scattered throughout the treetop, the greatest numbers being on the lower, more densely shaded parts.

Cause. Leaves are scorched when the roots fail to supply sufficient water to compensate for that lost at critical periods. They are most severely affected during periods of high temperatures and drying winds. The inability to supply the necessary water is influenced by the moisture content of the soil and by the location and condition of the root system. Obviously, even the most extensive root system cannot supply enough water to compensate for the tremendous amounts lost through the leaves if the soil moisture content is low because of a prolonged drought.

Though of less common occurrence, scorch may appear in trees growing in excessively wet soils. Under such conditions the water-absorbing capabilities of the roots may be greatly reduced because of lack of air.

Trees with diseased roots, those whose root systems have been reduced as a result of curb installation or building operations, and those whose roots are restricted or covered with impervious materials are most subject to leaf scorch. For example, sugar maples adjacent to concrete-paved driveways and near street intersections are commonly observed to have more scorch than those

FIG. 76. The tree in the center shows severe leaf scorch.

more favorably situated along the same street or those growing on lawns or along country roads.

Where injury to a part of the root system is severe and permanent, or where soil conditions are continuously unfavorable, leaf scorch may occur regularly.

In fact, some trees show leaf scorch every year regardless of whether the season is cool and moist or hot and dry.

Heavy infestations of aphids and other sucking insects usually contribute to the severity of leaf scorch. Some investigators feel that the trouble is associated with a deficiency of potash in the leaves. They believe that potash-deficient leaves tend to lose water more rapidly and consequently become scorched more readily.

Control. Any practice that increases root development and improves the tree generally will help to reduce leaf scorch. The application of chemical fertilizers, especially those containing large quantities of potash, is suggested. In addition, insects and fungus diseases should be kept under control.

The tops of trees with restricted or injured root systems should be reduced by judicious pruning to increase the supply of water to the remaining leaves.

Where soils are likely to become dry, water should be applied, especially during seasons of low rainfall. Where the soil is heavy and compacted, some improvement will result from breaking up the surface layers or from inserting upright tile to facilitate penetration of water. In the rarer cases where scorch results from the inability of the roots to absorb water because of excessively wet soils, the installation of drain tiles in the manner suggested under Injuries Caused by Grade Changes (p. 159) will help.

Leaf scorching in particularly valuable trees can be prevented, or at least reduced, by spraying the leaves with Wilt-Pruf before the regular 'scorching' period.

Chlorosis

The uniform yellowing of leaves resulting from a reduction in the normal amount of chlorophyll is termed chlorosis. The loss of the green coloring matter reduces the efficiency of the leaf in manufacturing food.

Chlorotic leaves may result from fungus, virus, or insect attack; low temperatures; toxic materials in the air or soil; excessive soil moisture; surpluses of soil minerals; lack or nonavailability of nutrients.

Chlorosis of many shade trees, especially of pin oak, appears to be associated with nonavailability of iron rather than with a lack of this element. This is especially true in soils containing limestone, ashes, or other alkaline materials, where the pH of the soil varies from 6.7 to 8.5. Iron may be present in such soils, but in a form that cannot be absorbed by the plant. Whether the chlorosis of boxelder, catalpa, cottonwood, bur and white oaks, sweetgum, and red, silver and sugar maples, like that of pin oak, is caused primarily by nonavailability of iron is not known definitely. Chlorosis in many other trees, especially fruit, may result also from deficiencies or nonavailability of zinc, magnesium, boron, or nitrogen, elements apparently rarely involved in chlorosis of shade trees growing along city streets.

Symptoms. The leaves of affected trees first turn uniformly yellowish-green, or they may remain green along the veins but turn yellow in the intravenal areas. The terminal growth of twigs is small, and the tree is generally stunted. The tissue between the leaf veins or along the leaf edge may die on trees affected for several years with chlorosis. Eventually whole branches or the entire tree may die prematurely unless the condition responsible is corrected.

Control. Chlorosis due to a deficiency or nonavailability of iron can often be corrected by special treatments. The newest treatment involves the use of so-called iron chelates. These are generally more effective than such chemicals as iron sulfate and iron ammonium citrate used formerly.

The iron chelate is strongly held in soluble form against the action of phosphates and other chemicals in the soil. Because it remains in solution rather than in dissociated form, the iron is readily absorbed by the plants. Moreover, the iron in chelated form is needed in much smaller amounts than the chemicals formerly used. For example, a gram of chelated iron applied to the soil around an iron-deficient tree is as effective as a pound of iron sulfate.

Iron chelates are applied in either of two ways: in powdered form directly to the soil, or dissolved in water and sprayed on the leaves.

For best results with soil applications the iron chelate should be distributed uniformly over the soil surface beneath the tree and thoroughly watered to a depth of 6 inches. To assure an even distribution, the proper amount of chelated iron powder should be mixed with fertilizer, or with some inert material such as sand, or the required amount should be dissolved in water and sprayed or sprinkled over the soil. Pin oaks from 3 to 6 inches DBH will require about one pound of iron chelate; those over 6 inches DBH about 2 pounds.

Leaf sprays should be applied as thorough cover sprays, that is, sufficient liquid should be applied to wet thoroughly all leaf surfaces. The sprays are best applied when the leaves are fully expanded and should not be applied to shade trees in blossom or to fruit trees in blossom or in fruit. Pin oaks can be sprayed with a solution containing 8 ounces of iron chelate in 25 gallons of water.

Sequestrene of iron, made by the Geigy Company, New York, N. Y., was the first of several iron chelates to appear on the market. Presently this company markets its products as Sequestrene 330 Fe Iron Chelate for use on alkaline or slightly acid soils, and its Sequestrene NaFe Iron Chelate for more acid soils.

Dow Chemical Company, Midland, Mich., markets its own chelates under the names of Versenol F and Versene F.

Iron chelates are also sold under trade names like Perma Green Iron 135 by Refined Products Corp., Lyndhurst, N. J., and Iron Tetrine by Glyco Products Co., Brooklyn, N. Y.

Perhaps it might be well to restress what has been said earlier. One should not conclude that all yellowing of leaves is caused by a deficiency of iron. Chlorotic leaves may also result from fungus, virus, insect, or mite attack; low temperatures; lack of nitrogen; toxic materials like 2,4-D in the air, or other chemicals in the soil; excessive soil moisture; surpluses of such metals as copper, manganese, or zinc. Correct diagnosis (see Chapter 9) is therefore important to obtain preventive or curative results with iron chelates.

Other metallic chelates are available for curing deficiencies of elements other than iron.

Other Methods for Curing Iron Deficiency. Until the development of iron chelates, iron deficiency was cured by spraying leaves with a solution containing 5 pounds of ferrous sulfate and 2 pounds of soybean flour in 100 gallons of water. A few ounces of a wetting agent added to the water, before the ferrous sulfate and flour were mixed in, helped to stick the spray on the leaves and make it more effective.

Iron salts such as iron ammonium phosphate injected into holes bored into the trunk also was used to cure trees affected with chlorosis. This method has several drawbacks, including the danger of initiating infections by wood-decay fungi in the vicinity of the bore holes.

Finally, applications of a 50-50 mixture of ferrous sulfate and powdered sulfur to the soil were used. One pound of this mixture for each inch of trunk diameter is evenly distributed early in spring throughout the root area in the manner outlined for inserting dry chemical fertilizers (p. 54).

Fire Damage

Most home owners assume that it is safe to rake up and burn leaves under trees. Unfortunately, most trees are easily damaged by the heat produced. The cambium layer beneath the bark may be killed and the bark above become susceptible to invasion by borers.

Some do not realize that branches high up in a tree may also be damaged by a leaf fire in the street below. Heat from the fire rises and is sufficiently intense to damage the branches severely.

About six years ago, the author and Mr. Edwin C. Rundlett, arboriculturist of the New York Department of Parks, studied many streetside trees which apparently were first injured by leaf fires and then became infected by the *Botryosphaeria* (*Dothiorella*) fungus (p. 359) and infested with American plum borers (*Euzophera semifuneralis*).

Rodent Damage

Young trees with thin, succulent bark are subject to damage by rodents such as field mice, pocket gophers, and rabbits during winters of heavy snowfall, or where the trees have been mulched with some strawy material. Placing a collar of a quarter-inch wire mesh or even ordinary window screening around the base of such trees will prevent much damage. Use of strawy mulches in which mice are apt to nest should be avoided. The trunk bases of susceptible trees can also be painted with a specially developed repellent. Ringwood Repellent, made by Panogen, Inc., Woodstock, Ill., is effective against rabbits.

11

INSECT CONTROL

A working knowledge of insect pests, of how they injure plants, and of methods of combating them is essential to any tree preservationist. The author does not intend to present a detailed discussion on methods of identifying insects. This is best left to texts written specifically for that purpose. Moreover, accurate identification of any insect can be obtained by submitting the specimen to the Bureau of Entomology and Plant Quarantine of the United States Department of Agriculture at Washington, D. C., or to the state entomologist usually located in the state capitol or in the state agricultural experiment station.

A brief discussion of the life history and structure of insects, however, will help to clarify many of the terms used in the latter part of this book.

Life History of Insects

One of the most interesting and unfathomable mysteries of nature is found in a study of the life cycle of the insect as it passes through various stages from egg to adult. The process that embodies all these changes in form is known as metamorphosis.

The life cycle of butterflies, beetles, sawflies, true flies, and moths comprises four separate stages: egg, larva, pupa, and adult. Insects with this life cycle are said to have complete metamorphosis.

After mating, the female deposits her eggs in a safe place that will prove a favorable environment for the worm-like animals, larvae, when they emerge; or she produces them in such large numbers that there is a sufficient survival to keep the insects abundant. The names of the larval forms vary according to the groups to which they belong. The larvae of butterflies and moths are called caterpillars; those of beetles, grubs (Fig. 77A); those of sawflies, false caterpillars; and the legless larvae of flies, maggots.

As a larva grows it sheds its skin, or molts, several times, each successive skin being larger and better able to accommodate the animal's ever-increasing body. Finally the larva becomes full grown and is ready to enter the third stage of development, the pupa. Larvae of many insects prepare a protective covering for this stage. The cocoon of the moth and of the sawfly, the pupal case of true flies, and the pupal cell of the beetle represent this phase of the transformation.

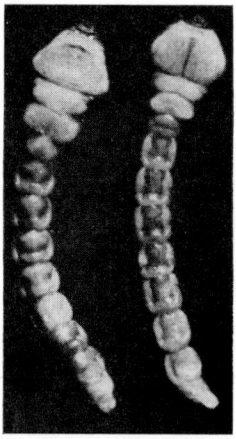

FIG. 77A. LEFT: Flat-headed borers.

FIG. 77B. RIGHT: Adult beetle of one of the flat-headed borers. The wing covers are hard and horny.

Although the pupal stage is considered a resting period, it is marked by considerable metabolic activity and many important changes, which culminate in the emergence of the adult insect.

A different form of development characterizes some insects. Grasshoppers, plant lice, plant bugs, lace bugs, and leafhoppers belong to a group having incomplete metamorphosis. The egg develops into a small wingless creature, called a nymph. Resembling the adult except for its ungainly head and small body, the nymph passes through several periods of molting. Wings become more prominent with each successive molt, and gradually the entire body assumes the appearance of the adult insect. Thus, the last stage is attained directly, no pupa being formed in the life cycle.

Structure of Insects

The body of an adult insect is composed of 3 principal parts: head, thorax, and abdomen. The head consists of 1 section to which are attached the feelers or antennae; the mouth parts; and the eyes. The thorax, just behind the head, has 3 segments from which emanate 6 true legs and usually 4 wings. Segmentation of the abdomen, the posterior body part, varies with different insects. The spiracles, small breathing pores, are located in the segments of the abdomen and thorax. In females a small, sharp-pointed organ for depositing eggs, the ovipositer, is found at the posterior end of the abdomen. Small appendages, called 'prolegs,' are attached to the abdomen of the larvae of many moths, butterflies, and sawflies.

Sawflies, moths, beetles, and countless other insects have 2 pairs of wings. Often the front pair is extremely hard and serves as a protective covering for both the hind wings and the greater part of the abdomen. Such hardened wings, called 'elytra,' may be observed in the beetle (Fig. 77B).

Control Methods

Prior to the development of DDT and the many other insecticides which followed it, the materials employed to combat insects were governed largely by the feeding habits of the insects and to a lesser degree by their life histories. For example, insects that chewed leaves were controlled with a stomach poison such as lead arsenate, and those that sucked out plant juices by means of specialized mouthparts were controlled by contact insecticides such as nicotine sulfate. Because many of the newer insecticides are capable of acting as stomach poisons, contact insecticides, and some also as fumigants, making a distinction in the feeding habits of the insects is no longer necessary.

Despite the introduction of the more highly effective materials, however, it is still important to time the spray applications properly so as to achieve maximum control. For example, DDT spray is effective against the boxwood leaf miner if it is applied when the adults are flying about before they lay their eggs on the leaves. To apply this material after the eggs have hatched and the tiny larvae have already entered the leaf is largely wasted effort.

Exact timing is also necessary to combat the European pine shoot moth (p. 414). The worms bore into the pine tip needles as soon as they hatch out of their eggs. Hence a spray must be present on the needles before the tiny worms bore in. A second spray should be applied 10 days after the first to catch any latecomers.

Correct timing is not quite so important when combating leaf-chewing insects such as canker worms and bagworms, which feed externally. But even these are controlled more readily if the spray is applied when the worms are still very young.

Insecticides and Miticides

Following is an alphabetical list of most of the newer insecticides and miticides together with a few of the older kinds. The latter are still generally available and still capable of controlling many kinds of insects.

Aldrin is one of the newer of the so-called chlorinated hydrocarbons now being used primarily for controlling certain soil-infesting insects and for grasshoppers. It will control some leaf-mining insects, sucking insects, and caterpillars.

Aramite is an excellent mite killer for use on shade and ornamental trees and shrubs. Its use on food crops is prohibited. Marketed as a 15 per cent wettable powder, as a 25 per cent emulsion, and as a 2 to 4 per cent dust, it is especially effective against the two-spotted spider mite, the boxwood mite, and the spruce mite.

Chlordane (also known as Octachlor and Velsicol 1068) is a complex chlorinated hydrocarbon which is a persistent stomach and contact poison to most species of insects. It gives good control of soil-infesting pests such as beetle grubs, ants, and termites. It is available as a 50 per cent wettable powder and in emulsifiable form for use as a spray and as a 5 per cent dust for use in combating soil-infesting insects. Chlordane is not compatible with alkaline materials such as lime and lime sulfur. It should be used with great care, and contact with the skin should be avoided.

Chlorobenzilate, another of the more recently developed chemicals, is effective in controlling most species of spider mites. It has good residual properties and is compatible with most insecticides and fungicides, but not those that are alkaline or strongly acid. It will not kill insects, including beneficial ones like bees, and it has a low order of toxicity to warm-blooded animals.

DDT (dichloro diphenyl trichloroethane) was perhaps the most widely used insecticide for the 10-year period following World War II. Among the reasons for its popularity during that period were its ability to control a wide variety of insects and its long-lasting residual effect.

At the present, except for some highly specific uses, such as control of elm bark beetles which spread the Dutch elm disease fungus (p. 317), of borers in trees and shrubs, and of gypsy moth, it is being rapidly supplanted by other insecticides.

Among the reasons for its loss of popularity are the following: First, many insects which at first quickly succumbed to it are no longer affected by it. Second, it is ineffective against mites and certain species of aphids; in fact, promiscuous use of DDT frequently results in a great increase of mites. Third, widespread publicity was given to its toxicity to birds, fish, and perhaps human beings.

Despite these drawbacks, however, there are many situations where DDT can be used to advantage, especially on shade and ornamental trees.

DDT comes in a wide variety of formulations. For the commercial arborist with standard hydraulic sprayers, concentrates containing 50 per cent DDT in the form of wettable powders and 25 or 30 per cent DDT in emulsion or miscible form are perhaps best.

The 50 per cent wettable powders usually are used at the rate of 2 pounds in 100 gallons of water. This dosage contains one pound of actual DDT and is generally effective against most shade tree pests. The emulsion or miscible forms containing 25 or 30 per cent DDT are used at the rate of 2 to 6 quarts in 100 gallons of water. The arborist who plans to use the miscible type is warned to purchase an emulsion prepared specifically for spraying trees.

For use in ultramodern mist blowers (p. 218), a 25 per cent emulsifiable DDT concentrate should be used at the rate of 25 gallons of the concentrate in 25 gallons of water.

DDT is compatible with most of the newer insecticides and fungicides and with the high analysis, all-soluble foliar fertilizers. Where there is a possibility of a build-up of mites, it is wise to include a miticide such as Aramite in the DDT spray.

DDT will not control all pests. It does not give good control of bagworm (p. 383), for example.

Delnav is one of the newest organic phosphates to appear on the market. It is effective against thrips, leaf miners, and various species of mites. It destroys mite eggs in addition to the mites themselves.

Derris extracts are obtained from the roots of certain South American plants. The principal effective ingredient is a material called 'rotenone,' which is nontoxic to humans and animals. Derris extracts are not so effective as pyrethrum extracts in controlling certain pests.

Some commercial arborists are using the rotenoid type of sprays, such as derris or cubè powder and rosin-residue emulsion, to avoid the unsightly white spotting and poisonous residue of lead arsenate. Satisfactory rotenoid sprays, such as Japellent and Japrocide, contain finely powdered derris or cubè with a suitable wetting and sticking agent. These sprays have considerable contact action against insects, but for chewing insects must be applied carefully as soon as the young insects appear. Rotenone is not compatible with alkaline materials such as lime sulfur, lime, calcium arsenate, and bordeaux mixture.

Dieldrin, a highly effective contact and stomach poison, is similar to aldrin. It controls a wide variety of insects on many kinds of plants. Its use on ornamental and shade tree pests is expanding rapidly. At present its major use is for soil-infesting insects, including Japanese beetle grubs, ants, and termites. As a foliage spray for leaf-infesting insects, its use must be under the immediate supervision of a qualified arborist or other persons acquainted with its highly poisonous properties.

Dimite (also known as DMC and DCPC) is an effective mite killer. It is compatible with most insecticides and fungicides except those with an acid reaction.

EPN is an insect and mite killer used on fruit trees and vegetables and for mosquito control. It is not recommended for shade trees because it is similar to parathion in toxic effect.

Guthion is yet another of the new organic phosphates, whose use at this time is limited to fiber crops such as cotton and to trees such as apples, crabapples, peaches, pears, and quinces.

Kelthane, one of the newer mite killers, is highly effective against a wide variety of mites, has a long residual action, is harmless to insects including bees, and is nonhazardous to humans and pets when used as directed.

Lead arsenate is the most widely used poison for chewing insects. It is effective, inexpensive, readily available, and safe for application on most plants. The standard dilution is 2 to 3 pounds in 50 gallons of water. This mixture will adhere more readily to the leaves and provide protection for a longer period if special adhesives are added. The use of adhesives and similar materials is discussed in detail under Spreading, Wetting, and Sticking Agents, in a later part of this chapter.

Lead arsenate is poisonous to man and to animals when taken internally. A few cases of poisoning of birds have been reported by commercial arborists, but recent tests conducted at an experiment station showed that there was little harm to birds fed insects poisoned by this material.

Calcium, magnesium, and *zinc arsenates* and *sodium fluosilicate* are used occasionally as stomach poisons in instances where these materials fill some particular need, or where lead arsenate is not entirely satisfactory.

Lime sulfur can be used either as a dormant or a summer contact insecticide. It also has good fungicidal properties. The dormant spray dilution is 1 part of concentrated lime sulfur (sold in horticultural supply houses) to 8 or 10 parts of water; and the summer dilution is 1 part in 49 parts of water. Although lime sulfur was at one time the principal dormant insecticide, it has now been largely replaced by oils, which are more effective, less costly, less irritating to the operator, and less destructive to spraying machine parts and accessories. Furthermore, use of lime sulfur near houses is inadvisable because it reacts with the paint.

Lindane, which is a purified form of the gamma isomer of benzene hexachloride, is widely used as a control for pests such as aphids, lace bugs, and leaf miners on trees and shrubs. It is available as a dust, wettable powders, and emulsions, and is compatible with most other insecticides and with most miticides and fungicides.

Malathion is one of the few widely available organic phosphate insecticides which has a comparatively low order of toxicity to warm-blooded animals. It is said to be less toxic than rotenone to warm-blooded animals.

Malathion is presently used in enormous quantities because it is very effective against a wide variety of insects, including 'hard-to-kill' kinds like scales and mealy bugs, and pests like mites. It is not effective against mature, armored scales but gives excellent control of the crawling and half-grown stages of these pests. It is especially useful for scale, aphid, and mite control where a dormant spray was not applied. For latitudes similar to that of New York City the first application is made in mid-May and a second about 2 weeks later.

Malathion has a disagreeable odor when first mixed and applied. Moreover, the emulsifiable concentrate applied with a mist blower (p. 218) has been known to injure the leaves of sugar maple when the air temperature is high.

Methoxychlor is an analogue of DDT and controls similar kinds of insects. It is supposed to be less toxic than DDT to warm-blooded animals and for that reason is used on edible crops more than is DDT.

Methyl parathion (also known as *Nitrox 80*) is similar to parathion and has similar insect-controlling properties with slightly less toxicity to warm-blooded animals.

Nicotine sulfate, sold under the trade name Black Leaf-40, was at one time the most widely used contact insecticide. It is still available, but newer materials like malathion and lindane are rapidly replacing it.

Aphids, young scales, and mealy bugs can be controlled with it. The standard dilution is ½ pint in 50 gallons of water (1-800), although dilutions may vary from 1 part in 400 to 1 part in 1,000 parts of water, depending on the type of insect to be controlled. Nicotine sulfate is more toxic to insects when used at relatively high temperatures, above 80° F. It is relatively ineffective when used at cool temperatures, 60° or below. It also is more effective when a wetting agent is used.

Oil sprays are now widely used by most arborists to control scale insects, aphids and other sucking insects, and red spiders. Depending on the type, these sprays may be used on trees either during the dormant period or in summer. Stronger oils or higher concentrations can be used with greater safety on dormant trees than on those in leaf.

Dormant oil sprays are applied when the trees are not in active growth. In the latitude of New York City this would be in late March and early April. Farther south, they should be applied earlier in the year.

Dormant oil sprays should never be used when the air temperature is below 40° F. or when it is likely to drop to freezing or below before the spray has dried on the tree. Moreover they should never be used on beech, black walnut, butternut, Japanese maple, hickory, sugar maple, and walnut. Certain evergreens, such as Cryptomeria, Douglas fir, the true firs (Abies), larch, hemlock, junipers, retinospora, Japanese umbrella pine, and yews, may also be damaged in some seasons. Oil sprays will temporarily remove the so-called bloom from Colorado and Koster blue spruces. The bloom gives these trees their attractive appearance. It will reappear, however, within a month or so after the oils are applied.

Dormant oil sprays are of two types: miscible oils and oil emulsions.

Miscible oils contain soap-like substances that form a milky emulsion when dispersed in water. They are much used for shade tree work because they can be easily handled and mixed, are stable, and form very small uniform oil droplets when diluted. Miscible oils are used at dilutions of 1 part in 15 to 25 parts of water. A few of the miscible oils available are sold under such trade names as Dendrol, Scalecide, and Sunoco.

Oil emulsions contain oils that have been emulsified with a small amount of water and with caseinates, gums, or certain soaps, to form a thick creamy preparation. They are not so stable as the miscible oils, and the oils tend to separate from the emulsifying agents. Consequently they must be shaken vigorously before being used. Though more difficult to handle and more likely to cause injury, they are less expensive than the miscible oils and they deposit the oil more readily on the sprayed object. Some of the oil emulsions on the market are known as Emulso, Free-mulsion, Kleenup, and Standard Dormant Emulsion.

Another group of dormant spray materials are the so-called dinitro compounds sold either as dry powders or as slurries under several brand names such as DN powder, DN Dry Mix, Elgetol, and Krenite. These types are largely used by commercial fruit growers, nurserymen, and some arborists against aphid eggs, the oyster shell scale, and other armored scales on fruit trees, flowering trees, and lilacs. Dinitro compounds are more disagreeable to use than are dormant oils and because of their staining properties, the spray operator must wear old clothes and rubber gloves. Dinitro compounds should be used only when the trees are fully dormant.

When applying dormant sprays, in addition to taking the precautions already listed, one would do well also to observe the following suggestions: Use a good sprayer that will deliver a fine uniform spray; cover all sides of the trunks, branches, and twigs thoroughly; start with a clean tank free of arsenicals, lime sulfur, or 2,4-D; never apply a dormant spray more than once on each tree each spring; do not allow the spray to puddle about the base of the tree.

Summer oil sprays, which are used when the trees are in leaf, are, of necessity, weaker than dormant oil sprays. They are very effective, nevertheless, because they are applied when the more susceptible stages of scales, red spider, and mealy bugs are about. Summer oil sprays of low concentrations appear to be safe on most evergreens except Norway, white, and blue spruces and upright junipers with cup-shaped leaves. Spray injury to such plants is always greater when the temperature and humidity are high.

Summer oil sprays are of two principal types: white oil emulsions and emulsive white oils.

White oil emulsions differ slightly from dormant oil emulsions in that they contain more highly refined oils and do not have soap as the emulsifying agent. Some summer oil emulsions available are Ortho-K, Summer-mulsion, Verdol, and Volck.

Summer oils are commonly mixed with nicotine sulfate to combat sucking insects. The mixture spreads over and adheres well to the sprayed objects. The two materials are much more toxic to insects when used in combination than when used separately. Summer oils may also be combined with arsenate of lead or with wetting agents.

Ovex (formerly known as Ovotran and also known as Chlorfenson) is a good mite killer, particularly of partly grown mites and the egg stage. Directions for use must be followed carefully to avoid damaging young leaves of holly, dogwood, and other trees.

Parathion, an organic phosphate and one of the most effective insecticides ever developed, is unfortunately too dangerous to handle because of its toxicity to humans and pets. Its use on shade and ornamental trees is not recommended because of its toxic properties. Nurserymen, commercial fruit growers, and florists acquainted with its toxic properties use it to control a wide variety of insects.

Pyrethrum extracts are obtained from the flowers of several European and Asiatic species of chrysanthemum. Pyrethrum or derris, at 2 pounds in 50 gallons of water plus a quart of summer oil, are commonly substituted for arsenate of lead when a spray that is nonpoisonous to humans and animals is required.

Rotenone, an extract of certain leguminous plants, is a stomach poison as well as a contact insecticide. It is harmless to man and animals, but extremely toxic to fish, and should never be sprayed on trees in the vicinity of pools.

Sevin is a carbamate insecticide developed by the Union Carbide Chemicals Company. It has been tested on apple trees where it controls some of the more-difficult-to-control pests, including codling moth, red-banded leafhopper, apple maggot, and plum curculio. It looks promising as a control for many insects that attack shade trees.

Soaps of various kinds, from mild laundry to the more efficient but unpleasant fish or whale oil, have long been used as components of insect sprays. Aphids infesting small trees can be controlled fairly well by spraying with a mixture of 1 pound of ordinary soap dissolved in 5 gallons of water. Though most soaps possess some contact insecticidal properties, they are much more effective when mixed with nicotine, pyrethrum, or derris extracts.

Sulfur Dusts. The sulfur dusts and sprays discussed in the chapter on fungicides have a limited use as contact insecticides. They are effective in controlling spider mites on many evergreens, especially those on which the use of oil sprays is not feasible.

Systemics. Systemic insecticides are chemicals which can be absorbed either through the roots or leaves of actively growing plants in sufficient amounts to kill insects and related pests feeding on or breeding in them. They have been found to be more effective on herbaceous plants than on woody plants like trees. Many of them contain organic phosphates and hence are very dangerous to handle. In the hands of a skilled arborist, nurseryman, or florist, however, they do an excellent job of insect and mite control. Directions for their use should be followed to the letter. Among the insecticides which act as systemics when applied as leaf sprays are Systox, OMPA, Sytam, Thimet, Isotox 200, and Isolan.

Tedion is a new mite killer which is especially effective against the egg and immature stages. It is said to be extremely safe to warm-blooded animals and is compatible with other pest control materials.

Tetram is an organic phosphate developed by British scientists. It is especially effective against mites but is not recommended for use on shade trees.

Trithion is still another organic phosphate similar to Guthion and methyl parathion. Like the others, it is not recommended for use on shade trees.

Perhaps it would be well to reiterate that the toxicity of most of the organic phosphates at present is such that these materials cannot be recommended for shade tree spraying. They can be used on fruit trees and smaller ornamental trees where the spray operator has good control of the spray drift, and where there are no persons or pets in the immediate vicinity.

Spreading and Sticking Agents

Any material that helps spread the spray in an even, unbroken film over the surface of leaves, bark, or insects' bodies is known as a spreader; this action is usually accomplished by lowering the interfacial tension of the spray. Materials that help the spray to adhere firmly to the leaves are known as stickers.

Soap, household detergents like Dreft, and some oils are good spreaders.

Some well-known spreaders and stickers are casein, calcium caseinate, ordinary wheat flour, powdered skim milk, and soybean flour. Calcium caseinate is prepared by mixing 1 part of casein with 3 parts of lime, 1 pound of the mixture being added to 100 gallons of spray. The flours are usually used at the rate of 2 to 4 pounds to 100 gallons of spray. Powdered skim milk is used at the rate of 1 pound to 50 gallons of spray.

Emulsified mineral oils used at the rate of 0.5 per cent oil in the diluted spray are also used as spreaders and stickers. A 0.25 per cent mixture of fish oil and linseed oil, 1 pint in 100 gallons of arsenate of lead spray, helps the lead to adhere firmly to the foliage.

Commercial spreaders and stickers widely available include DuPont Spreader-Sticker (SS_3), Triton B 1956, NMO, Orthex, and Spred-O-Stick. These should be diluted according to the directions on the package.

Borers

Any insect that feeds inside the roots, trunk, branches, or twigs of a tree is known as a borer. The borer is usually in the larval or worm stage, although some beetles, which are the adult stage of certain insects, also may bore into the tree.

Eggs are deposited in bark crevices and scar tissue or, in a few instances, below the bark surface, by the adult during spring, summer, or fall, depending on the species. Tiny larvae hatch from the eggs and penetrate the bark. Once they are below the surface, they cannot be controlled by the ordinary poisons used for leaf-chewing insects.

Factors Favoring Borer Infestations. Although a few borers can attack vigorously growing trees, most species become established in trees low in vigor. Consequently any factor that tends to lower the vitality of a tree predisposes it to borer attack. Among the more important of these factors are prolonged dry spells, changes in the environment unfavorable to the growth of the tree, loss of roots in transplanting, repeated defoliation by insect or fungus parasites, and bark injuries caused by frost, heat, or mechanical agents.

Drought periods weaken trees by checking normal growth and by causing premature defoliation. Rapid environmental changes, including changes in the height of the water table caused by digging drainage ditches, damming, or altering the grade, result in general weakening of the trees. Thinning out a grove makes the remaining trees more subject to sun scald and root injury, because the direct sunshine not only heats the bark but also greatly increases the soil temperatures. The effects of moving a tree to a new location have already been discussed in the chapter on Transplanting. Repeated defoliation by insects and fungi undermine the general health of the tree. Bark injuries, aside from weakening trees, result in the production of scar tissue, a preferred site for oviposition.

Prevention of Borer Attack. Many borer infestations can be prevented by improving, when feasible, any unfavorable situation around the trees. Proper fertilization, adequate watering during dry spells, periodic spraying for leaf-chewing insects and blighting fungi, and pruning infested or weakened branches are recommended. All bark wounds should receive immediate treatment to facilitate rapid healing and thus reduce the amount of scar tissue.

Borer infestations, common in recently transplanted trees, can be materially lessened by placing a barrier over the trunk and larger limbs immediately after the trees are set in their permanent location. Newspaper, wrapping paper, burlap, or specially prepared Kraft crepe paper is used. The last, also known as Tree Wrap, is now employed most extensively because it provides greater protection, is applied more easily, and is neater in appearance than the other materials. The crepe paper consists of two layers cemented together by asphaltum and is sold in 4- and 6-inch widths, wound in bolts of 25 or more yards. It is wrapped around the tree in much the same way as surgical bandage is applied, at an angle that permits sufficient overlap to make a double thickness. Binder twine is wound in the opposite direction to hold the paper in place. Both the twine and the paper remain in place for about two years before disintegrating. This is long enough to provide protection during the most susceptible period.

Asphaltum impregnated Kraft paper should not be treated with an oil base insecticide because the oil will dissolve the asphaltum.

Polyethylene wrap has not proved satisfactory because it holds too much moisture, which causes the tree to sweat and thus favors the development of fungus cankers.

Aluminum foil has been tried with considerable success. Not only is it easy to apply and attractive, but it does not need to be tied down with binder twine as do some of the other wrapping materials.

Control. The method used to control borers already established depends on the part of the tree infested and the species of borer involved. Twig and root borers are controlled by pruning and burning infested parts or by spraying or injecting certain chemicals into the unpruned infested parts.

Application of insecticides to the bark during the adult's egg-laying period or before the eggs hatch and the borers enter the bark is standard procedure for controlling trunk and branch borers. Among the most generally used materials at this time are malathion, DDT, or dieldrin. A sticking and wetting agent (p. 210) added to the spray will increase the efficiency of the material.

The bark on the main trunk and larger branches can be sprayed or painted with the solution. Unless one is also trying to control leaf-infesting insects, it is not necessary to spray the leaves with borer-controlling materials.

Three or four applications at weekly to 10-day intervals may be necessary with some borers to provide protection over the entire egg-laying period. Details on these applications are given later under the different trees susceptible to borers.

Borers that make galleries which lead to the bark surface (sawdust-like frass is a telltale sign) can be controlled by injecting a toxic paste into the holes and then sealing them with a small wad of chewing gum, grafting wax, or putty. These pastes are sold under trade names like Bor-Tox, Borer-Kil, and Borer-Sol.

Of course, some borers inside the tree can be killed by crushing them with a flexible wire inserted into the opening, or after probing for them with a sharp knife.

Termites and Ants in Trees

Termites are usually associated with damage to wood in buildings. But they are occasionally found in decaying wood of live trees. Once they become established in a tree, their feeding is not confined to the decaying wood; they also feed on sound heartwood and sapwood. Several different kinds of termites can infest trees. The so-called dry-wood kinds can cause damage, but the so-called subterranean species are more destructive. The latter nest underground and may enter the tree through injured or dead roots, or they may build so-called shelter tubes of mud along the outer bark which connect the underground galleries with the breeding and feeding place up on the trunk.

Ants of several kinds also are capable of damaging live trees. Some feed on the bark at the base of trees; others eat out the heartwood. Ants are built differently from termites. The principal difference is that ants have a noticeable constriction of the body between the thorax (where the legs and wings are attached) and the abdomen, whereas termites have a broad connection between these two sections. The wings of termites have finer venation and protrude farther beyond the abdomen than do those of ants.

Both termites and ants are controlled with either chlordane or dieldrin. A hand duster can be used to blow chlordane dust into the colony, or a small hand sprayer can be used to wet the infested areas with either a chlordane or a dieldrin solution. Cracks and crevices leading to the insects' nests and the surfaces over which the insects walk should be dusted or sprayed with the insecticide.

Chlordane and dieldrin give off toxic fumes which may harm persons unless precautions for their use are followed.

Certain carriers or solvents for some termite-proofing materials are extremely toxic to trees and shrubs. Hence any extensive use of such materials for termite-proofing homes should be used with due regard to their toxic properties, especially where tree and other plant roots are situated nearby.

Biological Control of Insects

Insects are subject to diseases caused by bacteria, fungi, and viruses. One interesting approach to controlling some insects, therefore, involves the use of certain parasitic agents. The best known example is that of bacterial spores to control Japanese beetle grubs.

In a few instances, man-manipulated parasites have been successfully used to control insects. For example, a virus has been used to wipe out an infestation of the great basin tent caterpillar (*Malacosoma fragilis*). The virus, taken from dead caterpillars which had succumbed naturally from the virus infection, was sprayed on healthy caterpillars in the laboratory. After these contracted the disease and died, their bodies were ground to make a suspension of the virus. A pint of ground, virus-killed caterpillars was sufficient to spray 30 to 50 acres of trees infested with tent caterpillars. Tests are now under way to determine whether the virus spreads from the areas where it is applied by man. If the artificially applied virus spreads a considerable distance, the treatment would be economically feasible.

Another example of biological control on a large scale was reported a few years ago. Pines infested with the European pine sawfly larvae were sprayed by airplane in May with an infectious virus suspension. Twenty-three days later, 89 per cent of the worms were dead. There was a 100 per cent kill on 60 per cent of the trees and only a slight defoliation. Trees not sprayed with the virus suspension were nearly 100 per cent defoliated.

12

SPRAYING EQUIPMENT AND PRACTICES

Successful control of insect pests is obtained not only by proper selection and preparation of the insecticide but also by correct placement of the material on the infested or susceptible parts of the tree. The latter is best accomplished with good spraying equipment.

Equipment

Small trees and shrubs can be sprayed with barrel or bucket type sprayers or a small sprayer carried on the operator's back, or with small power sprayers, (Fig. 78). Large trees can be properly sprayed only with large spraying machines (Figs. 79 and 80). Such machines are expensive and are owned principally by commercial arborists and park and shade-tree departments that do considerable shade-tree work.

Hydraulic Sprayers. The modern spray machine consists of a large wooden or steel tank, ranging in capacity from 100 to 1,000 gallons, for holding the spray material; an agitator inside the tank, which revolves to keep the spray material uniformly dispersed; and a gasoline-driven pump, which forces the material from the pump through the spray hose. These parts are usually mounted as a unit on skids to facilitate moving onto a truck or some other carrier, or they may be permanently mounted on a chassis and drawn by a truck, tractor, or other vehicle. Certain accessories, such as adequate hose and spray guns with special nozzles, complete the equipment.

Spray machines are available with pumping capacities ranging from 5½ gallons a minute at 300 pounds pressure, to 55 gallons at 800 pounds. Specially made hose of ¾- or 1-inch size must be used to withstand some of the higher pressures.

The following are among the companies that manufacture large spraying machines: John Bean, Div. of Food Machinery and Chem. Co., Lansing 4, Mich.; Friend Mfg. Co., Gasport, N. Y.; Hardie Mfg. Co., Div. of Vulcan-Iron Works, Wilkes-Barre, Pa.; and F. E. Myers and Bros. Co., Ashland, Ohio.

FIG. 78. A small power sprayer is excellent for spraying low-growing trees.

The stream of spray material and the height to which it will reach are governed by the size and construction of the nozzle, the amount of material discharged, and the pressure behind it. There are five principal types of spray nozzles. These are the disk nozzle, which throws a hollow cone-shaped spray; the Vermorel nozzle, which throws a misty, more or less cone-shaped spray; the bordeaux nozzle, which throws a flat, fan-shaped driving spray; the spray gun, which throws a cone-type to solid-stream spray; and the park nozzle, which throws a solid-stream spray. Each type has its own particular advantages and uses.

The pressure of the spray material at the nozzle opening, as well as the type of nozzle, governs the success of a spraying operation. Other factors re-

FIG. 79. A hydraulic sprayer applying a dormant spray.

maining the same—, for example, the size of the nozzle opening—any increase in pressure breaks up the spray material into finer particles, forces them to a greater distance from the nozzle, and results in the application of a greater volume of the insecticide or fungicide.

To spray a tree approximately 20 feet high, it is necessary to have a nozzle with a ⅛-inch opening in the disk, which will deliver 6 gallons of spray a minute at 180 pounds pressure. A tree 90 feet high can be adequately sprayed with a nozzle having a 5/16-inch opening, releasing 44 gallons a minute at a pressure of 300 to 400 pounds. Because of friction as the spray moves through the hose, the pressure at the nozzle is much lower than that developed by the pump at the spray tank. For each 100-foot increase in length of 1-inch hose from the tank to the nozzle in the example cited above, the pressure in the tank must be increased by 33 pounds in order to provide sufficient pressure at the nozzle to enable coverage of the 90-foot tree.

Sprays are applied to tall trees in a so-called solid stream, that is, the material leaves the nozzle much as water issues from a fire hose. This stream, forced out under great pressure, soon reaches a height, however, at which it breaks into a mist, which drifts onto the leaves and stems.

In contrast to solid-stream spraying, so-called mist spraying is used on trees less than 20 feet high. In the latter case the spray breaks into a fine mist as soon as it leaves the nozzle, giving rapid and complete coverage.

Hydraulic sprayers must be handled properly to keep them at peak efficiency. Following are some suggestions to achieve this objective: lubricate daily, particularly all moving parts; always pour chemical mixtures through a good strainer before they enter the tank; rinse out the tank, hose, and nozzles with clean water at the end of the day; carry, rather than drag, the spray hose over rough terrain.

Mist Blowers. The discovery of more potent insecticides, such as DDT, has speeded the development of new types of pest-control applicators. Water serves as the vehicle for the insecticides used in the conventional hydraulic sprayers, whereas the so-called mist blowers, or air sprayers, use blasts of air to propel the insecticide in liquid form. With such machines it is possible to cover more trees in shorter time and at far less cost. The use of highly concentrated materials not only speeds up the refilling time but sharply reduces run-off or drip waste which causes heavy loss of more dilute materials applied with the hydraulic sprayer.

Some of the mist sprayers can disperse spray particles for relatively long

FIG. 80. A 600-gallon capacity hydraulic sprayer in action on a tall tree.

distances, often up to 100 feet, against a mild wind. One should bear in mind, however, that if the liquid phase disappears before the insecticide reaches its target, the dry insecticide will not adhere to the leaf surface. In other words, to deposit a coating of insecticide properly, the liquid-insecticide mixture must reach the leaves before the liquid phase evaporates.

There are several sizes and types of mist blowers on the market. Some are more suitable for use on large shade trees, and others are better adapted for watershed, park, and forest plantings.*

Among the manufacturers of mist blowers are: Cooley Spray Equipment Works, Somers, Conn.; The Elgin Corp., 501 Fifth Avenue, New York 17, N. Y.; The Fog-Air Co., 415 Lexington Avenue, New York 17, N. Y.; and Todd Shipyards Corp., Brooklyn 31, N. Y.

Spraying Practices

Although correct spraying technique can be perfected only by actual practice, a number of suggestions to help in its acquisition are in order.

The intensity and direction of the wind must be considered before spraying operations begin. Unless the operator has had considerable experience, poor coverage and considerable waste of material are bound to occur on windy days. Maximum coverage on such days is obtained when the spray is directed toward the leeward sides of the leaves and twigs. As the spray is caught by the wind, it is blown back in the opposite direction, thus ensuring complete coverage. The wind factor is of even greater importance when using mist blowers.

Another factor to be considered is the proximity of the trees to buildings and other structures. The operator should determine beforehand how best to cover the tree with the least disfigurement to these objects. Where mist blowers are used near parked automobiles, the operator must take into account the possibility of coating the automobiles with the spray. The problem can be solved by enforcing regulations which prevent parking of automobiles under trees that are to be sprayed.

The spray gun should be adjustable to meet the situation in hand. It must

* Arborists and park superintendents interested in more details on mist blowers and the concentrated sprays used in them will find several excellent articles in the *Twenty-third Proceedings of the National Shade Tree Conference*, obtainable for a moderate price from Dr. L. C. Chadwick, Ohio State University, Columbus, Ohio. Even more helpful, is the recently published book *Concentrated Spray Equipment* by S. F. Potts, published by Dorland Books, Caldwell, N. J.

be opened to the solid-stream discharge rate when the tops of tall trees are being sprayed, and cut down to a mist spray where lower branches are involved. There is no ironclad rule as to the amount of spray that should be applied to any particular tree. The object of spraying is to cover every leaf, twig, and branch that is infested or that might later become infested by some insect. Once complete coverage is obtained, the operator should move on to another part of the tree or to another tree. It is not necessary to spray until the material drips from the leaves.

Safety Precautions

Following are some additional safety precautions:

Always read the label before using spray materials. Respect warnings and precautions listed.

Keep sprays out of reach of children, pets, and irresponsible adults. Do not store them near the pantry.

Store spray materials in their original containers.

Do not smoke while spraying.

Avoid inhaling sprays.

Do not spill spray materials on skin or clothing. Should some be accidentally spilled, remove contaminated clothing as quickly as possible and wash the skin thoroughly.

Wash hands and face thoroughly with soap and water and change to clean clothing after spraying.

Dispose of empty containers so they pose no hazard to humans, pets, or plants.

If symptoms of illness occur, call a physician or call the poison control center in your community.

Spray formulations, including insecticides, fungicides, miticides, and combination sprays, are presented toward the end of the next chapter.

13

TREE DISEASES AND THEIR CONTROL

Although disease control unquestionably constitutes one of the major phases of successful tree maintenance, it is the least clearly understood by the tree worker. This lack of understanding is due partly to the difficulty in recognizing disease symptoms and partly to the rather complex nature of the parasitic agents. Insufficient knowledge of the various chemicals used to combat diseases and of their proper applications is also a contributing factor.

Broadly speaking, a tree is considered diseased when its structure or functions deviate from the normal as discussed in the first chapter. Such a definition includes abnormalities caused by drought, excessive moisture, lack of nutrients, and chemical and electrical injuries. The present discussion, however, is limited to those diseases caused by infective agents such as fungi, bacteria, and viruses. The symptoms they produce, their nature, and the chemicals and arboricultural practices used to combat them are also included in the discussion.

Symptoms

A diseased tree exhibits one or several characteristic symptoms, depending on the species of tree, its environment, and the causal agent. Three general groups of symptoms occur: dying of the tissues, or necrosis; dwarfing or underdevelopment, hypoplasia; and overgrowth, or hyperplasia.

Necrotic symptoms include dead twigs or leaves—blights; sunken decayed areas on the trunk, branches, or roots—cankers (Fig. 81); small, well-defined dead areas on the leaves—leaf spots (Fig. 81); and dead or wilted tops owing to the killing or obstructing of the conducting tissue—wilts.

Chlorosis, or yellowing of the leaves resulting from underdevelopment of chlorophyll, is a hypoplastic symptom often associated with virus diseases. Stunting of twig growth and the production of undersized leaves are other hypoplastic symptoms.

FIG. 81. Symptoms of Fungus Diseases
UPPER LEFT: Cankers on hickory stem.
UPPER RIGHT: Circular spots caused by the fungus Phyllosticta.
LOWER LEFT: Pin-point fruiting bodies of the Nectria canker fungus.
LOWER RIGHT: Fruiting bodies of the fungus *Polyporus sulphureus*.

The various types of galls produced by both fungi and bacteria are hyperplastic symptoms. Probably the most striking example is the overgrowth produced on trees and shrubs affected by the crown gall disease discussed on page 240. Leaves affected by the so-called curl diseases, wherein small areas or the entire leaf grows to twice the normal size, exhibit another type of hyperplastic symptom.

In addition to symptoms produced on the plant attacked, certain signs, such as fruiting and vegetative structures of the causal organisms, help in diagnosing the disease. For example, many fungi produce on the affected areas fruiting bodies that are easily recognized and facilitate diagnosis. Such bodies vary in size from pin-point dots (Fig. 81), which are barely visible to the unaided eye, to bodies a foot or more in diameter (Fig. 81). The latter commonly occur in wood-decay diseases and appear as shelves or brackets adhering to the dead bark.

Additional signs may also occur that aid in determining the cause of a disease. For example, droplets of colorless or milky ooze appear on the bark of trees affected by the fire blight disease, and spruces affected by the cytospora canker disease nearly always have grayish-white resins exuding along the affected branches.

Nature of Parasitic Agents

Most diseases of shade and ornamental trees are caused by one of three agents: bacteria, fungi, or viruses.

Bacteria and fungi are plants that differ from those most familiar to everyone in that they do not contain the green coloring matter, chlorophyll, which is necessary for the manufacture of food. As a result, they are dependent on the higher plants for their sustenance. Bacteria and fungi that depend entirely upon decaying organic matter for their food source are known as saprophytes. These forms are frequently beneficial, in that they hasten decay and the liberation of minerals for use in future plant growth. Forms that are able to obtain their nourishment by direct attack upon living plant cells are known as parasites or pathogens. It is this group that is concerned in disease production.

Bacteria are microscopic, single-celled plants which multiply by simple fission, a single individual dividing to form two. Some types of bacteria possess flagella or hair-like appendages which effect a limited amount of locomotion in liquids. Bacteria have several forms, but those that parasitize plants are usually rod-shaped.

Fungi, as a rule, are much more complex than bacteria. They may possess two distinct stages—vegetative and reproductive. The vegetative stage consists of a weft or mass of microscopic, thread-like strands called 'hyphae.' Unlike the bacteria, a single fungus plant may be composed of innumerable cells, joined together. Certain tissues in this complex are capable of performing highly specific functions. The reproductive stage may be simple or very complex, either microscopic or visible to the naked eye, often reaching 12 or more inches in diameter. In all cases, however, the reproductive bodies produce microscopic spores of various shapes and colors, which function much as do the seeds of higher plants (Fig. 82). The spores may be ejected, drop out, or ooze (Fig. 83) from the reproductive bodies and are carried by some means to susceptible plant parts. In a moist environment, these spores germinate by pushing forth a tiny strand known as a germ tube. Once inside the host plant, the tube branches into numerous strands, which constitute a new fungus plant. Eventually this plant produces a new crop of spores directly, or fruit bodies capable of producing spores.

Viruses are ultramicroscopic principles generally carried from diseased to healthy plants by insects. The infective principle usually occurs in all parts of the plant and survives the winter in plant tissues. The best known virus disease of shade trees is phloem necrosis of elm.

Dissemination of Bacteria and Fungi

Bacterial plant pathogens and many fungus pathogens gain entrance into the plant tissue through the natural openings, such as the stomata, or breathing pores, and the hydathodes in the leaf tissues, or the lenticels in the stem. Wounds caused by insects or by mechanical means also afford convenient openings. Many fungi, however, are also able to penetrate directly the unbroken epidermis of the leaves. By growth or multiplication within the host tissue, fungi and bacteria absorb the cell contents and cause death, or by their presence cause overgrowth or dwarfing of the invaded tissues.

Fungus parasites are disseminated primarily in the spore stage. They may be splashed by rain or carried by air currents, by insects, by animals, on the hands of the tree worker, or on tools, from tree to tree or from place to place.

Wind-blown rain is probably the most important agent in the local dissemination of spores, particularly of those fungi that cause leaf spots, leaf blights, and twig cankers. Spores produced on the surface of infected leaves or in cankers are splashed or washed to nearby healthy leaves and twigs, on which new infections occur.

Air currents are also important agents of dissemination. For example, the

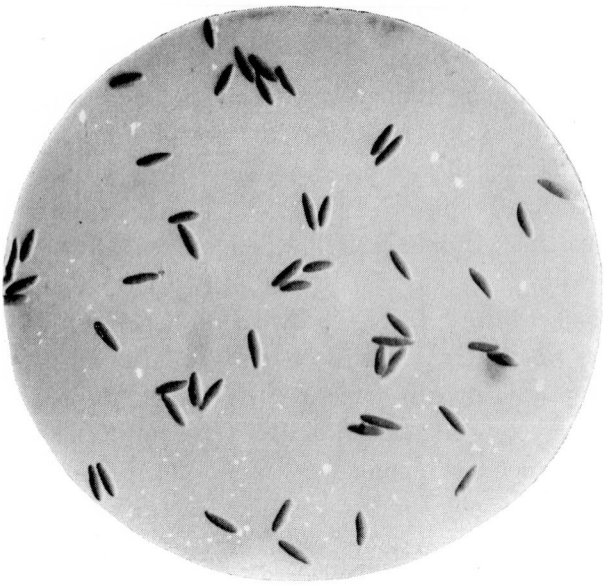

FIG. 82. Spores of the Dothiorella fungus (highly magnified), an important tree parasite.

FIG. 83. Spore tendrils of the fungus *Cytospora valsa* oozing from infected pine bark.

spores of the chestnut blight fungus are blown by the wind for many miles from the tissues where they are formed.

Insects are well-known agents of dissemination. Bees and flies are known to spread the bacteria that cause the fire blight disease. Several species of bark beetle are believed to be chiefly responsible for the spread of spores of the fungus causing the Dutch elm disease.

The chestnut blight fungus is known to be spread by birds and squirrels lighting on or traveling over extensive bark cankers, on the surface of which spores are produced in large numbers. These spores adhere to the talons or feet and may be deposited again on some distant tree.

Spores may be spread on tree workers' hands and clothing, as is the case with the Cytospora canker disease of spruce, especially when diseased branches are pruned during wet weather. The fungus responsible for plane canker may be disseminated on infested pruning saws, and even in many types of wound dressings (p. 357).

Pruning saws, shears, and other metal tools can be disinfested by dipping of wiping them with 70 per cent denatured alcohol. Tree-climbing ropes contaminated with fungi can be disinfested by confining the rope for 3 hours in a 10-gallon closed receptacle containing 4 ounces of liquid formaldehyde.

Soil-inhabiting organisms capable of producing diseases such as wilts and root and collar rots are carried largely by the transfer of infested soil on tools, on shoes, or by washing during heavy rains. Winds, lifting soil particles and carrying them from place to place, may also be a means of dissemination of this group of plant parasites.

Many soil-inhabiting organisms live as saprophytes in the soil until they come in contact with susceptible tree roots. They enter such roots, even unwounded ones, and become parasites. Wilt fungi grow directly into the conducting tissues, where they cause a partial clogging or secrete poisons that are carried in the water stream to the leaves. Here they accumulate and cause a collapse of the leaf tissues.

Many fungi that normally are weak parasites, which do little damage to trees growing under proper conditions, can readily destroy trees growing under adverse conditions.

To avoid damage by so-called weak parasites, one should provide the best possible growing conditions for trees. The trees should be well fed, the soil provided with organic matter, drainage improved if necessary, and the trees watered during extended periods of drought.

Overwintering of Bacteria and Fungi

Leaf-spotting and leaf-blighting bacteria and fungi overwinter in fallen infected leaves, on dormant buds, and in small twig cankers; canker-producing bacteria and fungi overwinter in the margins of the cankers; and wilt fungi pass the winter in the vascular tissues of infected trees and in the soil.

Principles of Tree Disease Control

The control of shade and ornamental tree diseases is based largely on the principles of eradication and protection, and to a lesser degree on exclusion and immunization.

Eradication. The removal and destruction of the fungus or bacterial parasites either at their source or at the point where infection occurs constitutes one phase of eradication. For example, the collection and burning of fallen diseased leaves eliminates an important source of inoculum. In the same manner, the pruning of branches infected by pathogenic fungi or bacteria also removes an important source of inoculum for subsequent infections.

Another commonly employed eradicatory practice is the application of dormant or delayed dormant sprays at the point where infection occurs. For example, lime sulfur solution is often recommended for the control of diseases such as leaf blisters and anthracnoses, in which spores pass the winter in the bud scales and on the twigs. Such a spray actually disinfests the buds and twigs before the spores have an opportunity to germinate or to penetrate the tree's tissues in the early spring.

The removal of decayed wood in trunks and branches may also be considered an eradicatory measure. As has already been pointed out in the chapter on Cavity Treatments, however, such a practice rarely, if ever, accomplishes its objective, inasmuch as the causal fungi extend well beyond the zone of visible decay.

Protection. Protection of trees from attacks by pathogenic organisms includes those measures involving the use of sprays and manipulation of the environment in order to make conditions unfavorable for the pathogen, favorable for the tree, or both.

Exclusion and Immunization. The principles of exclusion and immunization are less commonly used to control tree diseases. Exclusion involves the prevention of importation of trees known to be susceptible to certain highly contagious diseases into an area that is still uninfested, such as the movement of elms from the Dutch elm disease quarantine area to uninfested zones. Immunization involves the use of strains of trees that are resistant to some particular disease. This field is not so highly developed as it is with smaller plants, though some work is now being conducted in various parts of the country on breeding trees for disease resistance.

Manipulation of the Environment

Certain arboricultural practices can be executed to change the environment and effect disease control. For example, judicious pruning of many deciduous trees reduces the density of the crown and results in a reduction of the severity of the anthracnose disease. Most fungus spores must be exposed to high humidity or free water for at least several hours in order to germinate and infect. Thinning out the tree's crown enables more rapid drying after rains or heavy dews, thus materially reducing the number of infections.

Improving soil drainage by the installation of drain tiles increases the ability of the tree to withstand attacks by many soil-borne organisms. The addition of fertilizers high in nitrogenous materials will aid many trees to outgrow certain diseases. Probably the best example is in the treatment of maples mildly infected with the Verticillium fungus. The increased growth resulting from the fertilization enables the formation of a thick layer of sapwood, which seals the invading fungus in the heartwood, where it is harmless.

It is well to remember, however, that certain practices may change the environment so as to make conditions more favorable for the pathogen. Excessive applications of nitrogenous fertilizers, for example, will increase the susceptibility of such trees as apple, hawthorn, and mountain ash to the fire blight disease. Consequently, the application of fertilizers low in nitrogen and relatively high in phosphorus and potash is recommended in such cases.

It is now well recognized that trees growing in good, well-drained, and well-aerated soils usually show less disease than those in poor soils. Supplying such conditions, whenever feasible, will naturally result in healthier trees.

Spraying for Disease Control

Three points must be kept foremost in mind to ensure success with sprays: they must be applied at the proper time; they must be applied thoroughly; and they must contain the proper ingredients.

All protective sprays must be applied before rainy periods, because fungi and bacteria penetrate plant tissues when the plants are wet. Most of the failures to control leaf diseases that are reported by tree men are due to lack of appreciation of the importance of timely applications.

Complete coverage with the spray is also essential, for the most efficient material will do little good unless it is applied as a fine mist, which will deposit an even protective film. The lower surfaces of the leaves are the most common entrance points for leaf-infecting fungi. Consequently, these should be thoroughly coated. The various types of spray machines discussed under Spraying Practices are satisfactory for the application of disease-preventing sprays.

Fungicides and Bactericides

The two chemical elements still being used in protective sprays, copper and sulfur, are inexpensive and plentiful. They form relatively insoluble compounds, a property that permits their use on plants without injury to the tender leaves.

Copper Sprays. Bordeaux mixture is the oldest, best-known, and probably the most effective material for preventing most of the fungus and bacterial diseases of leaves. It is prepared by mixing copper sulfate, hydrated lime, and water in varying proportions, the ingredients being expressed in the formulae 2-2-50, 4-4-50, etc., the first number representing the pounds of copper sulfate; the second, the pounds of lime; and the third, the gallons of water. For example, the 4-4-50 formula contains:

Copper Sulfate Crystals	4 pounds
High-grade Chemical Hydrated Lime	4 pounds
Water	50 gallons

Fifty gallons of 4-4-50 bordeaux are prepared as follows: Dissolve 4 pounds of copper sulfate in 4 gallons of water by suspending the crystals overnight in a porous bag over a wooden or noncorroding container. Pour this solution into the tank and dilute it to 25 gallons. In a separate container, mix the 4 pounds of hydrated lime with enough water to form a thin paste. Then start the agitator in the tank, and pour in the lime mixture through the sieve until the total mixture in the tank equals 50 gallons. Powdered copper sulfate may be substituted for the crystal form. It may be poured directly into the tank and will dissolve rather quickly if the agitator is used. This eliminates the overnight delay necessitated by the use of crystals.

Bordeaux mixture is a complex gelatinous material that adheres well once it has dried on the foliage. Although often termed an insoluble material, the spray residue actually releases minute amounts of soluble copper, which are toxic to the germ tubes of fungus spores that endeavor to penetrate the sprayed portions. Surplus spray deteriorates rapidly and should not be stored for future use.

Bordeaux mixture adheres well to most shade-tree leaves and thus does not require the addition of a spreading and sticking agent. When used on the waxy leaves of pines and other conifers, however, casein-soap or fish-oil-soap should be added at the rate of 2 pounds in 50 gallons of spray.

The objectionable features of bordeaux mixture are that considerable time is needed in its preparation, it forms an unsightly residue on the leaves, and it may burn foliage during extended periods of cool, wet weather. The last two features hold for other copper sprays as well, although some appear to be less toxic.

A large number of copper substitutes for bordeaux mixture is readily available. These proprietary compounds are slightly more expensive, but most tree men feel that the time saved in preparing the spray for application more than compensates for the difference in price. Among the ready-made copper compounds that can be substituted for bordeaux mixture are: Basi-cop, Bordo, Coposil, Copper A, Copper Hydro Bordo, Cuprocide, and Spray Cop. These materials should be used at the concentrations recommended by the manufacturers.

Most copper sprays are corrosive to metals and should be thoroughly removed from spraying machines and accessories by rinsing with water after spraying operations are completed.

Sulfur Sprays. Many sprays containing sulfur as the active ingredient are also used for disease control.

Lime sulfur solution, mentioned as a disinfesting spray and as a contact insecticide, is also used as a protective spray. It is sold in concentrated liquid form and as a powder at most horticultural and farm supply stores and is diluted according to the need. As a dormant disinfesting spray, it can be safely used at a strength of 1 part in 10 parts of water. If the powdered form is used, it should be mixed at the rate of 20 pounds in 50 gallons of water. As a protective spray on leaves, it is used at the rate of 1 gallon of the concentrated liquid or 4 pounds of the powder in 50 gallons of water. Lime sulfur sprays react strongly with house paint and therefore should never be used on trees growing near homes.

A number of ready-made sulfur compounds, known as wettable flotation sulfurs, are also sold. They are used principally as protective sprays, and are much more agreeable to handle than is lime sulfur. These sulfur sprays are sold under such trade names as Flotox Wettable Sulfur, Kolofog, Magnetic Spray Sulfur, Micronized Wettable Sulfur, Microsulfur, and Mike Sulfur.

Sulfur sprays are used principally to combat mildew and rust diseases and, as has been mentioned under Insect Control, in combating red spider. In addition, they are used to combat most diseases on trees in the Rose family, such as apple, because such trees cannot safely tolerate copper sprays. It must be pointed out, however, that any spray containing sulfur may injure trees under conditions of extremely high temperatures; in fact, no spray of any type should be applied when the air temperature is above 90° F.

Newer Fungicides. Captan (N trichloromethyl thio tetrahydrophthalimide) is one of the newest fungicides to appear on the market. Sold under trade names Orthocide, Orthocide 406, Captan 50 W, it is effective against many leaf-spotting diseases of fruit and shade trees and certain leaf-curling diseases.

Dichlone (2,3-Dichloro-1,4 Naphthoquinone), sold under the trade name *Phygon XL*, is effective against many diseases of fruit and flowering trees including scab on apples, brown rot of peaches, and leaf spot of cherries.

Dinitro octyl phenyl crotonate, sold under trade names of *Karathane* and *Mildex*, gives excellent control of powdery mildews.

Dithiocarbamates are relatively new organic fungicides derived from dithiocarbamic acid and which possess many properties far superior to the older fungicides. Reasonably inexpensive at the effective dilutions, they do numerous jobs of disease control better than the older coppers and sulfurs. One of their best qualities, however, is that under all sorts of weather conditions they are far less toxic to plants than either of the long-established fungicides.

The best known and most widely used is *ferbam* (ferric dimethyl dithiocarbamate), a black, finely ground, wettable powder packaged under the trade names *Fermate, Ferradow, Coromate*, and *Karbam Black*. It is specific against the cedar rust disease of apples, and it controls scab, blotch, and other diseases of apples, as well as many leaf spots of other fruit and shade trees. Ferbam is compatible with many of the insecticides and fungicides commonly used in combination sprays but is incompatible with lime and monohydrated copper sulfate.

Ziram (zinc dimethyl dithiocarbamate) is similar to ferbam, except that it contains zinc instead of iron. This white, finely ground, wettable powder, sold under the trade names *Zerlate, Zincate, Karbam White, Corozate,* and *Methasan,* is not used on shade and ornamental trees so frequently as is ferbam.

Zineb, another dithiocarbamate fungicide, is sold under the names *Dithane Z-78* and *Parzate Zineb Fungicide.*

Maneb, a manganese-containing relative of ferbam, is excellent for controlling leaf spots of hawthorns and certain fruit trees. It is sold under trade names like *Manzate* and *Dithane M 22.* An aqueous suspension of manganous dithiocarbamate is sold under the name *Tennam.*

Nabam (disodium ethylene bisdithiocarbamate) when combined with zinc sulfate forms zineb. It is sold under the trade name *Dithane D-14, Chem Bam,* and *Parzate.*

The dithiocarbamates are generally ineffective against the powdery mildew fungi and bacteria.

The *omadines* are among the latest fungicides to appear on the market. They are metallic salts of 2-pyridinethione-1-oxide and have given excellent control of such fungus diseases as peach brown rot, apple scab, and fire blight of pear.

Organic mercury compounds are now being used to control several important diseases of shade and ornamental trees.

Puratized Agricultural Spray (phenyl mercury triethanol-lactate) has proved effective against anthracnose diseases. A single delayed dormant spray often gives as good control as 2 or 3 standard sprays of bordeaux mixture applied during the growing season.

Other organic mercury compounds are sold under the trade names *Agrox, Gallotox, Mema, Mer-Kote, Phix,* and *Setrete.*

Combination Sprays and Compatibility. Some materials may be safely combined with others to control fungus or bacterial diseases and insects in a single operation. Because improper mixing of these materials may cause injury to the sprayed plants or reduce the efficiency of one or several of the materials in the mixture, the author has compiled the following list to aid in avoiding trouble.

The materials are given numerical designations, and indicate other pesticides with which the material may be combined. For example, bordeaux mixture (3) may be combined with aldrin, dieldrin, endrin, heptachlor (2), dimite (7), dormant oil (8), lead arsenate (13), nicotine sulfate (17), Ovex (19), or summer oil (23). This chart does not indicate mixtures which are incompatible.

Spray Compatibilities

(1) Aramite: 2, 5, 6, 9, 13, 17, 18, 19, 20, 21, 22, 23.
(2) Aldrin, dieldrin, endrin, and heptachlor: 1, 3, 4, 6, 7, 8, 9, 10, 11, 14, 16, 17, 18, 19, 20, 21, 22, 23.
(3) Bordeaux mixture: 2, 7, 8, 13, 17, 19, 23.
(4) Captan: 2, 5, 6, 7, 9, 13, 16, 17, 18, 19, 20, 21.
(5) Chlordane: 1, 4, 6, 7, 8, 9, 10, 13, 16, 18, 19, 20, 21, 22.
(6) DDT, TDE, and methoxychlor: 1, 2, 4, 5, 7, 8, 9, 10, 13, 16, 18, 19, 20, 21, 22, 23.
(7) Dimite: 2, 3, 4, 5, 6, 9, 11, 13, 16, 17, 20, 22.
(8) Dormant oil: 2, 3, 5, 6, 17.
(9) Ferbam, maneb, ziram, zineb: 1, 2, 4, 5, 6, 7, 13, 16, 17, 19, 20, 21, 23.
(10) Fixed coppers: 2, 5, 6, 11, 13, 17, 19, 20, 21, 22, 23.
(11) Hydrated lime: 2, 7, 10, 13, 14, 17, 19, 22, 23.
(12) Karathane, Mildex: 2, 4, 5, 9, 13, 15, 21, 22.
(13) Lead arsenate: 1, 3, 4, 5, 6, 7, 9, 10, 11, 14, 16, 17, 18, 19, 20, 21, 22, 23.
(14) Lime sulfur: 2, 11, 13, 19, 22.
(15) Lindane, BHC (use wettable powders): 1, 4, 5, 8, 9, 10, 12, 13, 16, 18, 19, 20, 21, 22, 23.
(16) Malathion, demeton, and parathion: 2, 4, 5, 6, 7, 9, 13, 19, 20, 21, 22.
(17) Nicotine sulfate: 1, 2, 3, 4, 7, 9, 10, 11, 13, 19, 21, 22, 23.
(18) Organic mercury: 1, 2, 4, 5, 6, 13, 17, 19, 21.
(19) Ovex: 1, 2, 3, 4, 5, 6, 9, 10, 11, 13, 19, 21.
(20) Quinones (chloranil, dichlone) (use wettable powders): 1, 2, 4, 5, 6, 7, 9, 10, 13, 16, 19, 21, 22.
(21) Rotenone and pyrethrum: 1, 2, 4, 5, 6, 9, 10, 13, 16, 17, 18, 19, 20, 22, 23.
(22) Sulfur: 1, 2, 5, 6, 7, 10, 11, 13, 14, 16, 17, 19, 20, 21.
(23) Summer oil: 1, 2, 3, 6, 7, 9, 10, 11, 13, 17, 19, 21.

Spray and Dust Formulas *

Actidione—100 parts per million—one level teaspoon of 0.5 per cent Actidione, plus 5 level tablespoons of spreader-corrective, in 3 gallons water.

Aramite 15 per cent wettable powder, 2 pounds in 100 gallons water, or 2 tablespoons in 1 gallon.

Aramite 25 per cent emulsion, 1 pint in 100 gallons, or 1 teaspoon in 1 gallon.

Aramite 2 or 4 per cent dust.

Aramite 15 per cent wettable powder, 2 pounds plus Schradan (OMPA) ½ pound in 100 gallons of water.

* These are some of the more commonly used insecticides, miticides, fungicides, bactericides, and combination sprays presently in common use.

Azobenzene 70 per cent wettable powder, 1½ pounds plus Aramite, 15 per cent wettable powder, ¾ pound in 100 gallons water.

Azobenzene 70 per cent wettable powder, 1½ pounds, plus chlorobenzilate, 25 per cent wettable powder, ¾ pound in 100 gallons water.

Bordeaux mixture 4-4-50, copper sulfate 4 pounds, hydrated lime 4 pounds in 50 gallons water. To make 3 gallons, dissolve 2 ounces copper sulfate in 1 gallon water and 2 ounces hydrated lime in 2 gallons water. Then pour copper sulfate solution into limewater and use immediately.

Captan 50 per cent wettable powder, 2 pounds in 100 gallons water, or 2 tablespoons in 1 gallon.

Captan 5 per cent dust.

Chlordane 40 per cent wettable powder, 5 pounds in 100 gallons water, or 5 level tablespoons in 1 gallon.

Chlorobenzilate 25 per cent emulsifiable solution, 1 pint in 100 gallons, or 1 teaspoon per gallon.

Chlorobenzilate 25 per cent wettable powder, 1 pound in 100 gallons, or 1 level tablespoon per gallon.

Chlorobenzilate 25 per cent wettable powder, 1 pound plus Schradan (OMPA) ½ pound in 100 gallons water.

Copper fungicides (fixed). As directed by manufacturer.

Copper sulfate (snow), 8 pounds plus hydrated lime, 12 pounds plus linseed oil, 1 pint in 100 gallons water.

DDT 25 per cent emulsion, 2 quarts in 100 gallons water, or 4 teaspoons in 1 gallon.

DDT 50 per cent wettable powder, 2 pounds in 100 gallons water, or 2 tablespoons in 1 gallon.

DDT 50 per cent wettable powder, 2 pounds plus malathion 25 per cent wettable powder, 2 pounds in 100 gallons water, or 2 tablespoons of each in 1 gallon water.

DDT 5 per cent dust.

Dieldrin 25 per cent wettable powder, 2 pounds in 100 gallons, or 1 level tablespoon per gallon.

Dinitro powder (DN Dry Mix), 4 pounds in 100 gallons water, or ½ cup in 3 gallons.

Dinitro Slurry (Elgetol or Krenite), 1 gallon in 100 gallons water.

Dormant sprays, (Dendrol, Scalecide, Sunoco, etc.) as directed by the manufacturer.

Elgetol, 1 gallon in 100 gallons water, or 2½ tablespoons in 1 gallon.

Ferbam 76 per cent, 2 pounds in 100 gallons water, or 2 tablespoons in 1 gallon.

Ferbam 76 per cent, 1½ pounds, plus summer spray oil emulsion, 1 quart in 100 gallons water.

Ferbam 76 per cent, ½ pound plus wettable sulfur 3 pounds in 100 gallons.

Ferbam, 7 per cent dust.

Fermate—see ferbam.

SPRAY AND DUST FORMULAS (*Cont.*)

Ferradow—see ferbam.

Karathane, as directed by the manufacturer.

Kelthane 18.5 per cent emulsifiable solution, 1 pint in 100 gallons or 1 teaspoon per gallon.

Kelthane 25 per cent wettable powder, 1 to 1½ pounds in 100 gallons or 1 tablespoon per gallon.

Lead arsenate 6 pounds plus wheat flour 4 pounds in 100 gallons water, or 3 level tablespoons of lead arsenate and 2 of wheat flour in one gallon. Mix flour thoroughly with a little cold water to make a thin paste. Wash the dry lead arsenate into the sprayer and fill the tank half full. Add the flour paste with constant stirring. Then add the rest of the water.

Lime sulfur, 10 gallons concentrated in 100 gallons water, or one pint in one gallon.

Lindane 25 per cent wettable powder, 1 pound in 100 gallons water, or 1 tablespoon in 1 gallon.

Lindane 25 per cent emulsion, 1 pint in 100 gallons water, or 1 teaspoon per gallon.

Lindane, 1 per cent dust.

Malathion 25 per cent wettable powder, 3 pounds in 100 gallons water, or 3 tablespoons in 1 gallon.

Malathion 50 per cent emulsifiable liquid, 1½ pints in 100 gallons water, or 1 tablespoon per gallon.

Malathion, 4 per cent dust.

Malathion 25 per cent wettable powder, 4 pounds plus DDT 50 per cent wettable powder, 2 pounds plus ferbam 76 per cent 2 pounds in 100 gallons water, or 4 level tablespoons malathion, 2 level tablespoons DDT plus 2 teaspoons ferbam in 1 gallon.

Malathion 50 per cent emulsifiable liquid, 1½ pints, plus DDT 25 per cent emulsifiable liquid, 1 quart in 100 gallons water, or 1½ teaspoons malathion and 2 teaspoons DDT in 1 gallon.

Methoxychlor 25 per cent emulsion, 3 quarts in 100 gallons water, or 2 tablespoons per gallon.

Methoxychlor 50 per cent wettable powder, 3 pounds in 100 gallons water, or 3 tablespoons per gallon.

Methoxychlor 5 per cent dust.

Mildex (Karathane) as directed by manufacturer.

Nabam, 1⅓ quarts plus zinc sulfate, 1 pound plus hydrated lime ½ pound plus Triton X-100 one ounce in 100 gallons water.

Niagara Carbamate (see ferbam).

Nicotine sulfate 40 per cent, one pint plus soap flakes or soap powder 5 pounds in 100 gallons water, or one tablespoon nicotine sulfate and 6 tablespoons of soap in one gallon.

Parathion 15 per cent wettable powder, 1½ pounds in 100 gallons, or 1 tablespoon per gallon.

Phenyl mercury 5 per cent solution, 1 pint in 100 gallons water or 1 teaspoon per gallon.

Phenyl mercury 10 per cent solution, ½ pint in 100 gallons, or ½ teaspoon per gallon.

Phygon XL, one pound in 100 gallons or 1 teaspoon per gallon.

Puratized Agricultural Spray, 1 pint in 100 gallons, or 1 teaspoon per gallon.

Pyrethrum, 2 pounds plus summer oil, one quart in 50 gallons water.

Rotenone 1 per cent emulsifiable solution, 1 quart in 100 gallons, or 1 tablespoon per gallon.

Streptomycin, as directed by the manufacturer.

Sulfur (wettable), 3 pounds in 100 gallons, or 1 tablespoon per gallon.

Sulfur sprays, as directed by the manufacturer.

Superior dormant miscible oil (Orthol D, Neutrona, Superior), as directed by the manufacturer.

Tag 331, 1 pint in 100 gallons water, or 1 teaspoon per gallon.

Tepp 20 per cent emulsifiable liquid, ½ pint in 100 gallons, or ½ teaspoon per gallon.

Tepp 40 per cent emulsifiable liquid, ¼ pint in 100 gallons, ¼ teaspoon per gallon.

Zineb 70 per cent wettable powder, 2 pounds in 100 gallons water, or 2 tablespoons per gallon.

Zineb 7 per cent dust.

Ziram 70 per cent wettable powder, 2 pounds in 100 gallons, or 2 tablespoons per gallon.

Ziram 7 per cent dust.

14

GENERAL PARASITIC DISEASES

This chapter discusses some of the more prevalent diseases caused by bacteria and fungi, which occur on a wide variety of shade and ornamental trees.

Fire Blight

Although fire blight is known primarily as a disease of apple and pear trees, and consequently is of most concern to orchardists, it also occurs on many ornamental trees and shrubs in the Rose family. Commercial arborists, park and estate superintendents, and even the owner of a single susceptible tree must often cope with this very destructive disease. Besides attacking apple and pear trees, fire blight appears commonly on several species of cotoneaster, *Cotoneaster* spp.; cockspur thorn, *Crataegus crusgalli;* English hawthorn, *C. oxyacantha;* and mountain ashes, *Sorbus* spp. It occurs less commonly on fire thorn, *Pyracantha coccinea;* serviceberry, *Amelanchier canadensis;* flowering quince, *Chaenomeles japonica;* cultivated quince, *Cydonia vulgaris;* Christmas berry, *Photinia villosa;* flowering plum, *Prunus triloba* var. *plena;* spirea, *Spirea vanhouttei;* rose, *Rosa* spp.; and *Stransvaesia davidiana.* It also has been recorded on such widely different trees as walnut, *Juglans* spp., and persimmon, *Diospyros* sp.

Symptoms. The leaves near the growing tips and the flowers suddenly wilt, turn brown and black, and appear as though scorched by fire. Twigs also are blighted on most of the ornamental hosts. In some, the mountain ashes, for example, the infection spreads down to involve large branches. Extensive cankers on the trunk and main branches develop on the larger ornamental trees, as well as on apple and pear. The flowering quince is susceptible to blossom blight principally, the disease being rare on the woody parts. The presence of bloom infection and the absence of typical twig blight in such cases give the infected tree the appearance of having been injured by frost.

Cause. Fire blight is caused by a germ known as *Erwinia amylovora.* The bacteria are carried from oozing cankers on nearby pear, apple, or other infected hosts to the open blossoms by bees, flies, and other insects and by splashing rains. They multiply rapidly in the blossoms and are spread to other blossoms by bees in quest of nectar and pollen. In this manner hundreds of blossoms are contaminated in a relatively short period. The bacteria penetrate and infect the blossom tissues, then continue downward into the fruit spurs, twigs, and even branches, killing the various tissues infected.

Control. Blighted twigs should be pruned well below the infected areas and burned. This practice eliminates an important potential source of inoculum for subsequent epidemics. The pruning must be done carefully, all large wounds properly dressed, and the pruning tools carefully sterilized with alcohol to prevent spreading this highly contagious disease.

Bordeaux mixture, formerly used as a blossom spray, is no longer recommended. Far more effective are sprays containing the antibiotic streptomycin or mixtures of streptomycin and Terramycin.

The antibiotic should be sprayed when the trees are in the early-bloom and full-bloom stages. Poor control results when the sprays are delayed until the petal-fall stage. In the eastern United States three applications are needed for good control, in California as many as five may be necessary. The antibiotic is not capable of killing fire blight germs that have become deeply established inside the tree. It does prevent infections from becoming firmly established, provided the sprays are timed properly.

Less streptomycin is absorbed into a plant if rain falls within a few hours of application. A rain one or more days after application has little effect on the amount absorbed.

There are four formulations of antibiotics now widely available to large-scale users (orchardists, nurserymen, arborists, etc.) for fire blight control. Their active ingredient is streptomycin in either the sulfate or the nitrate form. Of these, Merck's Agri-Strep, Squibb's Phytomycin, and Lederle's Acco-Streptomycin contain only strepomycin. Pfizer's Agri-mycin 100 contains Terramycin in addition to streptomycin.

The preventive measures mentioned above usually suffice for most ornamental trees and shrubs. Where large ornamental trees and apple or pear develop cankers on the trunks and larger branches, additional treatments may be necessary. To eliminate an important source of contagion, the cankers should be removed by surgical methods or the bacteria in them destroyed by means of chemical paints.

The surgical removal of the canker involves cutting back into the healthy bark about 4 inches around the margin of the canker during the dormant season and removing all tissues inside the cut zone. The resulting wound is then disinfected with bichloride of mercury solution followed by the application of any good wound dressing. Such a treatment is effective for small cankers but is not recommended for large cankers extending around more than half of the affected branch. The best procedure in the latter case is to remove the branch completely, inasmuch as its chances of survival are small.

A more rapid method for destroying the bacteria in large cankers without surgical treatment involves the application of chemical paints to the cankered area. These paints penetrate the bark and destroy the harmful bacteria without severely injuring the tree. One of the best paints for such purposes, developed by Dr. H. E. Thomas of California, contains the following ingredients:

Cobalt Nitrate	3½ ounces
Commercial Glycerine	1.7 fluid ounces
Oil of Wintergreen	3.4 fluid ounces
Acetic Acid	1.7 fluid ounces
Ethyl Alcohol (or Denatured)	27.2 fluid ounces

The last four ingredients are mixed together by stirring, and the cobalt nitrate powder is added slowly until it is completely dissolved. This cobalt nitrate canker paint should be applied over large cankers with a brush, so that it covers not only the infected portions but several inches of the surrounding bark as well. It will not penetrate and kill the bacteria if the canker is covered with rough, corky bark. Consequently, best results are obtained if the rough layer of outer bark is scraped away before the paint is applied.

The mixture should be kept in a tightly stoppered bottle when not in use.

Crown Gall

Crown gall (Fig. 84) is a common disease of trees in the Rose family, such as apple, cherry, pear, plum, quince, and flowering almond. The disease also occurs on such widely different trees as chestnut, Arizona cypress, European juniper, incense cedar, sycamore maple, oleander, poplar, English walnut, willow, English yew, and several species of Euonymus including *E. patens*, *E. radicans* (*Fortunei*), and *E. radicans vegetus*.

Symptoms. Roughened swellings or tumor-like galls of varying sizes appear at the base of the tree or on the roots. On poplar and willow, these galls may also appear on the aerial parts. Growth of the tree may be retarded, the leaves may turn yellow, or the branches or roots may die as a result of the presence of these galls. Although old, established trees are rarely damaged appreciably, young trees may be killed.

FIG. 84. Crown gall on walnut stem.

Cause. Crown gall is caused by the bacterium *Erwinia tumefaciens*. The organism does not kill the parts attacked, but stimulates cell growth in them. It enters the susceptible plants through wounds and is primarily a parasite of young nursery stock.

Control. Nurserymen should discard all young fruit and ornamental trees showing galls on the stem or main roots. They should avoid wounding the stems and roots of young trees in the nursery. They should also be aware that irrigation water is capable of carrying the crown gall bacteria.

Where the disease is known to be present in the nursery, dipping the roots in a solution containing 200 parts per million of Agri-mycin 100 (4 tablespoons in 5 gallons of water) for half an hour is suggested as a possible preventive practice.

Powdery Mildews

The leaves and the twigs of many deciduous trees and shrubs are occasionally covered with a grayish-white dusty material known as powdery mildew. The mildew is made up of delicate, hyaline, cobweb-like strands of fungus tissue covered with microscopic colorless spores. Many minute, spherical, black bodies, fruiting structures of the sexual stage of the fungus, may be visible to the unaided eye on the grayish-white areas. The spores of the fungus overwinter in these bodies.

Although mildew fungi grow mainly over the surface, they also penetrate the leaf surface with fine sucking organs in many places and extract the juices of the host plant. Heavily mildewed leaves may turn yellow and fall prematurely. Mildews appear most commonly toward the end of summer and in late fall and are most prevalent on trees growing in shaded and damp locations.

Cause. There are a number of distinct species of powdery mildew fungi that can be distinguished only by microscopical examination. The following list includes most of the species, together with the trees or shrubs they parasitize:

Microsphaera alni on alder, birch, lilac, oak, plane tree, and walnut; *Phyllactinia corylea* on ash, beech, catalpa, chestnut, dogwood, elm, hawthorn, hornbeam, oak, and tulip; *Podosphaera oxyacanthae* on apple, hawthorn, pear, persimmon, plum, serviceberry, and spirea; *Sphaerotheca pannosa* on peach; *Uncinula salicis* on poplar and willow; *U. circinata* on maple; *U. flexuosa* on buckeye; *U. macrospora* on elm; *U. clintonii* on linden; and *Erysiphe aggregata* on alder.

Control. Lime sulfur 1-50 or wettable sulfurs were formerly recommended for powdery mildew control. These have been largely supplanted in recent years by Karathane and Mildex (p. 232). As a rule, however, only particularly valuable larger specimens or small nursery trees are sprayed to control mildew. Certain precautions listed on the package must be followed.

Shoestring Root Rot

Some of our most valuable shade and ornamental trees are susceptible to the so-called shoestring root rot disease. The name is derived from the long, round, black strands of tissue, closely resembling shoestrings, produced by the fungus underneath infected bark, over infected roots, or in the soil. Other

Fig. 85. Branch die-back and general decline in this oak are symptoms of shoestring root rot.

names often given to the same disease are Armillaria root rot, crown rot, and mushroom rot. Among shade and ornamental trees, oaks and maples appear to be most commonly infected, although the disease is occasionally destructive on apple, birch, black locust, chestnut, larch, mountain ash, pine, plane, poplar, spruce, and rhododendron. In all probability the disease may occur on almost any tree or shrub grown, if the necessary conditions for infection are present.

Symptoms. The above-ground symptoms cannot be differentiated from those produced by many other diseases or agents that cause root or trunk injuries. Probably the most striking external symptom is a decline in vigor of a part or the entire top of the tree (Fig. 85). Where the progress of the disease is slow, branches die back from time to time over a period of several years; where rapid, all the leaves wilt suddenly and dry up, presenting the appearance of severe manufactured illuminating gas injury. The latter symptom has been noted by the author during the last 15 years in many Norway, sugar, and red maples growing along city streets.

The most positive signs of this disease, however, are found at the base of the trunk at or just below the soil line, or in the main roots in the vicinity of the root collar. Here, fan-shaped, white wefts of fungus tissue closely appressed to the sapwood are visible when the bark is cut away or lifted. Scraping or lifting the white wefts of mycelia, which have a strong mushroom odor, will reveal water-soaked sapwood. Where the entire top has wilted, the fungus tissue will be found completely around the trunk. Where a large branch has died back, or one side of the tree shows poor vigor, the fungus will be found only on one or two main roots or on one side of the trunk base. Where the tree has been dead for some time, the dark brown to black 'shoestrings' may occur beneath the bark or in the soil near the infected parts. Clusters of light brown mushrooms, so-called honey mushrooms, may appear in the vicinity of the rotted wood in late autumn (Fig. 86). These rarely occur near infected street and shade trees, however, inasmuch as conditions are usually unfavorable for their development, or the infected trees are removed before the mushrooms have a chance to form.

Cause. Shoestring rot is caused by the fungus *Armillaria mellea.* The fruiting stage, the honey mushroom, somewhat resembles the mushroom sold in stores but is slightly larger, is yellowish brown, and the top of the cap is dotted with dark brown scales. The mushrooms usually grow in close clusters with their stem bases pressed together. Spores released from the underside of the cap are blown by the wind to bark injuries at the base of living trees or to dead tree stumps. Under favorable conditions, the spores germinate and produce mycelium, which penetrates and infects the tree tissues. Later, fungus strands (shoestrings) grow out from the tissues and through the soil. These strands may then contact healthy roots of nearby trees and penetrate and infect them. In fact, infection of living trees is thought to occur almost exclusively by means of these fungus strands. The author has seen infected trees in cities, however, that were isolated by whole blocks from other possible sources of infection, indicating that infection must have occurred by means other than fungus strands growing through the soil.

FIG. 86. The mushroom stage of the shoestring root rot fungus.

Some doubt exists as to the virulence of *Armillaria mellea*. Because the disease is almost constantly associated with trees that were previously in poor vigor, several investigators believe that the fungus is only weakly parasitic and cannot attack vigorously growing trees. Others, including the author, believe the causal fungus is capable of attacking living roots and can penetrate sound, healthy bark. It can also enter the tree through dead roots and through bark injuries. But no matter how it enters, the fact remains that once the fungus gains a hold in trees, it usually kills them in a relatively short while.

Control. Because much of the evidence at hand indicates that shoestring root rot is associated with weak trees, probably the best precautionary measures are fertilization, watering, and soil improvement.

An infected tree whose entire root system or trunk is diseased cannot be saved. The larger roots in the vicinity of the trunk as well as the trunk itself should be removed and burned. Soil in the immediate vicinity should also be removed.

Attempts to save partly infected valuable trees are occasionally successful, especially where only one or two main roots or one side of the trunk is involved. In such cases, the soil within a radius of about 2 feet around the trunk should be removed in the spring or summer to expose the root collar and the larger roots. Roots whose bark is completely rotted should be removed and the affected bark on partly diseased roots and on the trunk carefully cut away. All visible fungus tissue and water-soaked wood should be removed. The bark edges should be shellacked and the entire wounded surface covered with a good tree paint. Branches should be thinned out and small plants growing nearby removed to allow as much sunlight and air as possible in the vicinity of the treated parts. In addition, the trees should be fed with some quickly available plant food to stimulate additional growth. The exposed roots and root collar should be covered with fresh soil in late fall to prevent freezing.

Where the disease has killed one tree in a group, spread to nearby trees can occasionally be delayed or even completely checked by removing the tree and the main roots and by digging a circular isolation trench. The trench should be about 1 foot wide, 2 feet long, and about 5 or 6 feet from the former location of the trunk. All soil from the trench as well as that within the circle should be carted away. All roots within the trenched area should be removed, even if they belong to nearby healthy trees. Although this complete treatment is expensive, it is warranted where particularly valuable trees are growing near a tree that has succumbed to shoestring root rot. The remaining trees should, of course, be fertilized, watered, and given all care possible to ensure continued vigorous growth.

The problem of planting another tree in the site on which one has previously died from this disease occasionally confronts the shade-tree commissioner, arborist, or property owner. The author has known of cases where three successively planted trees have died from shoestring root rot under such conditions.

Where a tree must be planted in the same place, several precautions are necessary to ensure the safety of the new tree. The old soil, including the larger roots and as many as possible of the smaller roots of the dead tree, should be carted away. New soil should be used in its place but it should not be placed around the trunk. The immediate vicinity of the trunk should be filled with a layer a foot wide and 6 inches or so deep of ¾-inch crushed stone. This will help aerate the trunk base and provide a barrier between the soil and the trunk through which the fungus would have difficulty in spreading and becoming established.

Where such a procedure involves considerable expense or where it cannot be practiced because of obstructions, such as pavements and walks, an attempt may be made to destroy all traces of fungus tissue by disinfecting the soil and roots with some chemical. The best material for such purposes is formaldehyde solution, prepared by dissolving 1 gallon of 37 per cent formaldehyde in 50 gallons of water. Enough of this mixture should be poured over the root zone to penetrate to the subsoil. The treated area should be covered for a few days with canvas to enclose the fumes, after which the soil is uncovered and aerated for several weeks. All traces of formaldehyde fumes must be gone before a tree can be safely replanted. This treatment cannot be used where trees or other forms of vegetation are growing near by.

Carbon disulfide has been successfully used on the West Coast to control the shoestring fungus in citrus groves while infected trees are still standing. One and one-half ounce doses of the chemical are inserted 8 inches below the surface and 18 inches apart. The top of the soil is kept moist, or wet soil is applied to enclose the fumes. From 30 to 60 days are required for the chemical to kill the fungus in the soil and in diseased roots. The dead trees are removed and new trees are then set in their places.

Phenylmercury acetates have also been suggested as controls for *Armillaria mellea* by research workers at the Connecticut Agricultural Experiment Station. Drenching the soil with ½ pint of the acetate per 100 gallons of water is suggested.

Verticillium Wilt

Verticillium wilt, one of the most common fungus diseases of plants, is found on more than 300 kinds, including shade and ornamental trees, food and fiber crops, annual and perennial ornamentals. Among its more important suscepts are the many species of maples planted along city streets. In fact, this disease has caused the death of more maples during the last 20 years than any other disease. A recent survey in New Jersey revealed that the silver maple (*Acer saccharinum*) is the most susceptible of the streetside maples, with the Norway maple (*A. platanoides*), red maple (*A. rubrum*), and sugar maple (*A. saccharum*) decreasingly susceptible in that order.

The following additional trees known to be susceptible to Verticillium wilt are listed in alphabetical order of their botanical names: broad-leaved maple (*Acer macrophyllum*); painted maple (*A. mono*); boxelder (*A. negundo*); California boxelder (*A. Negundo* var. *Californicum*); black maple (*A. nigrum*); Japanese maple (*A. palmatum*); Japanese red maple (*A. palmatum* var. *rubrum*); Schwedler's maple (*A. platanoides* var. *Schwedleri*); sycamore maple (*A. pseudo-platanus*); Drummond maple (*A. rubrum* var. *Drummondi*); trident maple (*A. rubrum* var. *trilobum*); horsechestnut (*Aesculus hippocastanum*); tree-of-heaven (*Ailanthus altissima*); southern catalpa (*Catalpa bignonioides*); western catalpa (*C. speciosa*); carob (*Ceratonia siliqua*); redbud (*Cercis canadensis*); Judas tree (*C. Siliquastrum*); camphor tree (*Cinnamomum camphora*); yellowwood (*Cladrastis lutea*); flowering dogwood (*Cornus florida*); smoke tree (*Cotinus coggygria*); quince(*Cydonia oblonga*); Japanese persimmon (*Diospyros Kaki*); Texas persimmon (*D. texana*); persimmon (*D. virginiana*); beech (*Fagus* spp.); green ash (*Fraxinus pennsylvanica* var. *lanceolata*); ash (*Fraxinus* spp.); English walnut (*Juglans regia*); China tree (*Koelreuteria japonica*); goldenrain tree (*K. paniculata*); tulip tree (*Liriodendron tulipifera*); osage-orange (*Maclura pomifera*); apple (*Malus* spp.); banana (*Musa paradisiaca* var. *sapientum*); olive (*Olea europaea*); avocado (*Persea americana*); pistachio (*Pistacia vera*); European aspen (*Populus tremula*); almond (*Prunus communis*); apricot (*P. Armeniaca*); sweet cherry (*P. avium*); cherry plum (*P. cerasifera*); sour cherry (*P. cerasus*); Morello cherry (*P. cerasus* var. *austera*); garden plum (*P. domestica*); Mahaleb cherry (*P. mahaleb*); Japanese apricot (*P. mume*); prune (*Prunus* spp.); pear (*Pyrus* spp.); oak (*Quercus* spp.); black locust (*Robinia pseudo-acacia*); sassafras (*Sassafras variifolium*); Brazil pepper tree (*Schinus terebinthifolius*); Japanese pagoda tree (*Sophora japonica*); cacao (*Theobroma cacao*); American linden (*Tilia americana*); little-leaf linden (*T. cordata*); Crimean linden (*T. euchlora*); American elm (*Ulmus Americana*); English elm (*U. campestris*); Scotch elm (*U. glabra*); Siberian elm (*U. pumila*); and slippery elm (*U. fulva*).

Symptoms. Wilt symptoms vary somewhat with the kind of tree, its age, location, and other factors, but there are some symptoms common to all situations. A sudden wilting of the leaves on one limb, on several limbs, or even on the entire tree is a general symptom (Fig. 87). The severity of this symptom depends less on species of tree than on the amount of infection in the below-ground parts. Long before wilting becomes visible, the fungus has been at

FIG. 87. This tree is heavily infected by the fungus *Verticillium albo-atrum*.

work below ground. If most of the roots are infected, the tree will die quickly. On the other hand, if infections in the roots are confined to one side of the tree, only the parts above that side will show wilt symptoms. Periods of drought frequently accentuate the wilting symptoms and speed the death of the infected tree.

The wilted leaves may fall after they have dried completely, or they may hang on for the remainder of the season. Sometimes a few branches or even the entire top of the tree will die during the winter and hence fail to leaf out the following spring. Cases of this sort are usually attributed to winter injury.

Because many other agents can cause similar symptoms, positive diagnosis of Verticillium wilt can be made only by culturing sapwood tissue in the laboratory. The cultures are taken from discolored sapwood, which occasionally appears in the branches but more commonly and more reliably near the base of the main trunk or in the main roots.

The Verticillium fungus may enter through the branches, but this is the exception rather than the rule. It is a soil-inhabiting fungus primarily, and hence its most common avenue of entrance is via the roots. One would expect, therefore, that the discoloration of the sapwood would be more marked and more common closer to the point of entrance. This is actually the case, as the author has discovered in his studies on the cause of dying of streetside trees over the last three years.

When the Verticillium fungus is present, the outer sapwood rings (those nearest the bark) will show a discoloration. The color varies with the species of tree and the duration of the infection. In Norway maple it is a bright olive-green; in American linden, dark gray; in tree-of-heaven, yellowish-brown; in black locust, brown to black; in northern catalpa, purple to bluish-brown; in American elm, brown (like the Dutch elm disease fungus); and in Japanese pagoda tree, greenish-black. But the presence of discolored sapwood tissue cannot be used as an exact diagnostic symptom. It merely indicates that a fungus may be involved. For example, the author has isolated at least six different kinds of fungi from sapwood of Norway, silver, and red maples which showed the bright green discoloration! Moreover, he reproduced the same discoloration in young healthy maples by inoculating them with these fungi.

It is well to reiterate, therefore, that a positive diagnosis for Verticillium wilt, as for the Dutch elm disease, can be made only by laboratory isolation tests.

Cause. Wilt is caused by the fungus *Verticillium albo-atrum.* Many strains or segregates of this fungus are known to exist. The fungus can live in soil for some time without actually parasitizing roots. Although it is thought to enter roots and stems only through injuries, infection may occur via young, uninjured roots. Some research workers believe air currents may spread spores over long distances. When such spores lodge in tree wounds, above-ground infection may occur. Sapwood wounds are usually necessary for the start of an infection in the aerial portions of the tree. Contaminated pruning tools may serve to promote aerial infections.

FIG. 88. Greenish discoloration of sapwood may be caused by the Verticillium fungus. Several layers of healthy sapwood now cover the infected area in this branch.

Once inside the water-conducting tissues, the Verticillium fungus grows upward longitudinally; that is to say, the infection spreads upward from the point of inoculation. The actual discoloration which results may be due to the presence of the fungus itself or to a chemical change resulting from the action of the fungus.

Control. Trees showing general and severe infection by the Verticillium fungus cannot be saved. Such trees should be cut down at once and destroyed quickly and completely. As many of the roots as possible should also be removed. A species of tree satisfactory for the particular location but not susceptible to wilt should be planted. A few commercial arborists, faced with the request to replant the same kind of tree in infested soil, have reported good results from drenching the soil with formaldehyde solution, 1 part in 50 parts of water, several weeks to a month before replanting the new tree. Not only must formaldehyde be handled with care, however, but its reliability for doing the job under large-scale conditions is questionable.

Several investigators, including the author, have noted recovery of trees from mild cases of Verticillium wilt after liberal applications of a high nitrogenous soluble fertilizer such as Ra-pid-gro. The material must be used early in the growing season on trees which show only one or two wilted branches. The heavy fertilization apparently stimulates leaf growth, which in turn enables the rapid formation of a thick layer of sapwood that seals in the infected parts beneath (Fig. 88). In all probability the new Agrico All-soluble plant food would also be helpful.

Other investigators have found that heavy nitrogenous fertilizer treatment increased the spread of the fungus in the tree. A number of investigators have also tried injecting chemicals into infected trees or applying them to the soil around the roots as a curative measure. Such treatments are not reliably effective and hence must be kept in an experimental category for the present.

Other General Diseases

Many other general parasitic diseases affect a wide variety of trees. Among these are canker and die-back caused by the fungus *Botryosphaeria ribis* (p. 359), canker caused by several species of Cytospora (p. 418), and trunk and root rot caused by species of Ganoderma (p. 288).

15

DISEASES AND INSECT PESTS OF LOW-GROWING TREES

This chapter deals primarily with the diseases and insect pests * of the low-growing deciduous trees discussed in the early part of Chapter 8. Where both low-growing and tall-growing kinds occur in the same genus, as with ashes, magnolias, and maples, the disease and insect problems are also discussed here.

For each tree, the individual diseases are first presented in the logical order: symptoms, cause, and control. Next comes a description of the damage caused by each insect pest, followed by very brief descriptions of the various stages of the insect, and by the control measures.

The cultural requirements of some of the trees are discussed briefly before the more common diseases and pests. Where this information is omitted, it may be found in Chapter 8.

ASH (*Fraxinus*)

White, green, and black ash are unusually free from attacks by fungus parasites, but are frequently attacked by several insect pests.

A serious decline in the health of ash trees growing as shade and forest trees in the East and Middle West has been observed in recent years. The cause of the decline is not known.

Diseases

RUST. The leaves and twigs of green and red ash are occasionally attacked by the fungus *Puccinia peridermiospora*. The leaves are conspicuously dis-

* Although mites are not true insects, they are treated under this group for the remainder of the book.

torted and the twigs swollen. The spores of the fungus, yellow powder in minute cups, appear over the swollen areas. The spores produced on ash are incapable of reinfecting ash, but they do infect the so-called marsh and cord grasses, on which is formed another spore stage, which can return to ash and again produce rust on that host.

Control. The disease is rarely destructive enough to warrant special control measures. Eradication of the grass hosts in the vicinity of ash would disrupt the life cycle of the rust fungus and consequently prevent infections of ash. Sulfur sprays can be used where the disease is serious.

LEAF SPOT. A number of leaf spots occur on ash. The most serious is that produced by the fungus *Gloeosporium aridum.* Large areas of the leaf, especially along the edges, turn brown. During rainy springs, these spots may be numerous enough to blight the entire leaf and cause premature defoliation. Another leaf spot, especially prevalent in nurseries, is caused by the fungus *Piggotia fraxini.* White, green, and Oregon (*F. oregona*) ashes are most susceptible.

Control. Gather and burn fallen leaves. This practice is usually sufficient to keep leaf spot disease at a minimum. In recent years the disease has become increasingly prevalent in the New England states. Partial control has been achieved with bordeaux mixture or ziram sprays.

CANKER. A number of fungi including *Dothiorella fraxincola, Nectria cinnabarina, Cytospora annularis,* and *Diplodia infuscans* cause branch and trunk cankers.

Control. Prune out infected branches. Maintain trees in good condition by proper feeding, watering, and spraying.

INSECTS

SAWFLY. *Tomostethus multicinctus.** Trees may be completely defoliated in May or early June ** by yellow sawfly larvae. The adult is a bee-like insect that lays eggs in the outer leaf margins. Winter is passed in the pupal stage in the ground.

Control. Spray with malathion or arsenate of lead about the middle of May.

LILAC BORER. *Podosesia syringae fraxini.* Rough, knot-like swellings on the trunk and limbs and the breaking of small branches at the point of injury

* The scientific name of the insect is given after the common name.

** Unless specifically noted, the dates mentioned in the rest of this book for appearance of symptoms and time of control apply to the latitude of New York City.

are indications of the presence of the lilac borer, a brown-headed, white-bodied larva, ¾ inch long. The adult female appearing in early fall is a moth with clear wings having a spread of 1½ inches, the front pair of which is deep brown. The larvae pass the winter underneath the bark.

Control. Prune out severely infested parts. Spray the trunk and branches with malathion or DDT around May 1, and repeat twice at 2-week intervals.

CARPENTER WORM. *Prionoxystus robiniae.* Large scars along the trunk, especially in crotches, and irregularly circular galleries about ½ inch in diameter, principally in the heartwood, are produced by a 3-inch pinkish-white caterpillar, the carpenter worm. The adult moth, with a wingspread of nearly 3 inches, deposits eggs in crevices or rough spots on the bark during June and early July. A period of 3 years is necessary to complete the life cycle.

Control. Larvae feeding in the wood can be killed by injecting Bortox into the tunnels and sealing the openings with chewing gum, putty, or grafting wax. Spraying the trunks with malathion or DDT when the females are depositing greenish-white sticky eggs in summer will also provide control.

FALL WEBWORM. *Hyphantria cunea.* In fall webworm infestations the leaves are chewed in August and September, and the branches covered with webs or nests, which enclose skeletonized leaves and pale yellow or green caterpillars, 1 inch long. The adult moth has white- to brown-spotted wings with a spread of 1½ inches.

Control. Cut out or remove nests, or spray the tree thoroughly with arsenate of lead, malathion, or DDT when the webs are first visible.

OYSTER SHELL SCALE. *Lepidosaphes ulmi.* Masses of brown bodies shaped like an oyster shell and about 1/10 inch long, covering twigs and branches, are characteristic of this insect. The pests overwinter in the egg stags under the scales. The young crawling stage appears in late May.

Control. Spray in late spring, before the buds open, with 1 part concentrated lime sulfur in 10 parts of water, or with winter-strength miscible oil or oil emulsion according to the directions of the manufacturer. Malathion sprays when the young are crawling about in May and June also give control.

SCURFY SCALE. *Chionaspis furfura.* A grayish scurfy covering on the bark indicates the presence of this pest. The scale covering the adult female is pear-shaped, gray, and about 1/10 inch long. The pest overwinters as purple eggs under the female scale.

Control. Spray in late spring, before the leaf-buds open, with winter-strength miscible oil or oil emulsion, or in mid-May and again in early June with malathion.

FIG. 89. Ash flower gall caused by mites.

LILAC LEAF MINER. *Gracilaria syringella.* Light yellow, ¼-inch larvae first mine the leaves of ash, deutzia, privet, and lilac and then roll and skeletonize the leaves. Small moths emerge from overwintering cocoons in the soil in May to deposit eggs on the underside of the leaves. A second brood emerges in July.

Control. Spray with malathion before larvae curl the leaves.

FLOWER GALL. *Eriophyes fraxiniflora.* The staminate (male) flowers of white ash are transformed into galls by the feeding and irritation of this mite (Fig. 89).

Control. Malathion spray applied after the buds swell and before the new growth emerges in spring will provide control.

BIRCH (*Betula*)

The birches are graceful as well as beautiful (Fig. 90). Extremely particular about soil conditions, they are unable to adapt themselves as street trees and are used principally on lawns.

FIG. 90. The bark and growth habit of the gray birch (*Betula populifolia*) are very interesting.

The canoe birch (*B. papyrifera*) has attractive bark and a single trunk. Gray birch (*B. populifolia*) is gray-barked, and many specimens have several trunks.

Ornamental birches are susceptible to several fungus parasites and insect pests. Of these, the bronze birch borer is mainly responsible for the death of most of the trees used in ornamental plantings.

Diseases

LEAF SPOT. Several leaf spot fungi attack birch. The fungus *Gloeosporium betularum* produces brown spots with a dark brown to black margin. The fungus *Cylindrosporium betulae* forms smaller spots with no definite margin. Both fungi may become sufficiently prevalent to cause some premature defoliation.

Control. Gathering and burning fallen leaves usually suffices for practical control. Bordeaux mixture or any other copper spray may be used in late spring as a preventive.

LEAF BLISTER. Three fungi cause leaf blister on many species of birch. *Taphrina bacteriosperma* and *T. carnea* produce red blisters and curling of the leaves on many of the birch species, and *T. flava* forms yellow blisters on gray and canoe birches.

Control. Gather and burn all fallen leaves. Spray with 1 part concentrated lime sulfur in 10 parts of water just before buds open in spring.

LEAF RUST. Birches are subject to a leaf rust that rarely does any appreciable damage.

Symptoms. In the summer, small, round, reddish-yellow pustules form on the lower leaf surfaces. Later dark brown to black pustules appear on the same leaves.

Cause. The fungus *Melampsoridium betulinum* causes leaf rust. The spores from the darker pustules appearing in late summer or fall are capable of infecting larch.

Control. The disease rarely becomes destructive enough to warrant special control measures.

CANKER. Black, paper, sweet, and yellow birches are particularly susceptible to canker.

Symptoms. Young cankers are usually not readily visible. Upon close examination, however, they are seen to be darker in color than the adjacent healthy tissue and appear water-soaked. Later the edge of the diseased area cracks, thus exposing the canker (Fig. 91). Callus tissue forms over the cracked area, but becomes infected and dies. The process is repeated annually, forming concentric rings of dead callus. The bark within the cankered zone falls away, leaving the wood exposed. When the canker completely girdles the stem or trunk, the distal portion dies.

FIG. 91. Canker on black birch.

Cause. Canker is believed to be caused by the fungus *Nectria galligena* or a closely related species. Small, globose, dark red fruit bodies of the fungus are barely visible to the unaided eye on the dead bark.

Control. Because canker on birches is always associated with trees in poor vigor, fertilization and watering of the more valuable specimens is suggested. Avoidance of wounds, removal of dead branches, and treatment of all bark injuries as soon as they occur also help to ward off infections.

WOOD DECAY. A number of wood-decay fungi attack birches. One, *Polyporus betulinus,* attacks dying or dead birches and produces shelf- or hoof-

FIG. 92. These birches are dying back because of infestations of the bronze birch borer.

shaped, gray, smooth, fungus bodies along the trunk. Others, such as *Torula ligniperda, Fomes fomentarius, F. igniarius,* and *F. applanatus,* are associated with decay of living trees.

Control. Wood decays cannot be checked once they have become extensive. Avoidance of wounds and maintenance of the trees in good vigor by fertilization are the best preventive practices.

DIE-BACK. The progressive dying back of the upper branches of ornamental birches is usually caused by severe infestations of the bronze birch borer (Fig. 92). In a few instances, however, it is associated with the fungus *Melanconium betulinum,* which attacks trees weakened by drought.

Control. Prune affected branches to sound wood, and fertilize and water heavily to help revitalize the tree.

INSECTS

CASE BEARER. *Coleophora salmani.* Leaves are mined and shriveled and small cases are formed under the leaves by the case bearer, a light yellow to green caterpillar, ⅕ inch long with a black head. The adult is a brown moth with a wingspread of ⅖ inch. The pest overwinters as a larva in a case attached to the bark.

Control. Spray in late spring, before growth begins, with 1 part concentrated lime sulfur in 8 parts of water, or spray with standard malathion solution in late July or early August.

LEAF MINER. *Fenusa pusilla.* Gray, canoe, and cut-leaf birches are especially susceptible to attacks by the leaf miner, a small white worm which causes leaves to turn brown in late spring or early summer (Fig. 93). The adult is a small black sawfly which overwinters in the soil as a pupa.

The first brood begins to feed anywhere from very early to late May, depending on the season and the location of the trees. The first brood causes most damage because it attacks the tender, spring foliage. Other broods hatch during the summer but these cause less damage because they do not attack mature foliage but confine their feeding to leaves on sucker growths and to newly developing leaves in the crowns of the trees.

Control. Both lindane and malathion sprays give excellent control. The first spray should be applied around the first of May. If the spring is a cold one, the first application can be delayed a week or so. For best control, two additional applications should be made at 10-day intervals after the first. To control the second brood of leaf miners, spray again about July 1 and July 10.

LEAF SKELETONIZER. *Bucculatrix canadensiella.* The lower leaf surface is chewed and the leaf is skeletonized and may turn brown as a result of feeding by the leaf skeletonizer, a yellowish-green larva, ¼ inch long. The adult moth has white-lined, brown wings with a spread of ⅜ inch.

Control. Spray with malathion or arsenate of lead about mid-August.

FIG. 93. The leaf miner causes brown blotches on birch leaves. The tiny adult flies and the larvae are also shown.

BRONZE BIRCH BORER. *Agrilus anxius.* This pest annually destroys many ornamental birches, particularly the weeping cut-leaf European birch (*B. alba laciniata*); the canoe birch (*B. papyrifera*) is more resistant. Additional details are presented under Poplar on page 366.

Control. Spray the trunk and branches thoroughly and the leaves lightly with DDT about June 1 and again 3 weeks later. Keep the trees in good vigor by feeding and watering when needed.

CHERRY (FLOWERING) (*Prunus*)

Flowering cherries are used rather extensively as ornamentals. In latitudes north of southern New Jersey, they are occasionally damaged by low winter temperatures, as is evidenced by longitudinal cracks (Fig. 94) on the south or west side of the trunk or in the branch crotches.

Flowering cherries are susceptible to some of the fungi and insect pests that attack the common cherry. These parasites, however, have not been studied in detail on flowering cherries, and any treatment must be based on the control measures suggested for ordinary cherry. Information on the diseases and insects of ordinary cherry is readily available from the state agricultural experiment stations.

Diseases

LEAF SPOT. Several leaf spots occur on flowering cherry. One, believed to be caused by the fungus *Higginsia hiemalis*, is destructive during wet seasons as in 1958, causing almost complete defoliation.

Control. Wettable sulfur, ferbam, or ziram sprays will control fungus leaf spots of this host. Recently a new fungicide sold under the name Cyprex has given excellent control of *Higginsia hiemalis* on edible cherries. In all probability this material will also control the same fungus on flowering cherry.

YELLOWING. Very often yellowing and premature defoliation of flowering cherry occur without previous spotting of the leaves. These symptoms are associated with excessively wet or dry soils or with low-temperature injuries to the crown and roots. As a rule, a second set of leaves is formed following the premature defoliation in the spring or early summer. The new leaves are normal to all outward appearances.

Sometimes leaf yellowing is due to other factors including virus disease, borer infestation in the trunk and branches, and diseases in the roots. These are more serious and may even be fatal.

Insects

ASIATIC GARDEN BEETLE. *Autoserica castanea.* The leaves are chewed during the night by a brown beetle, ¼ inch long. During the day, the insect hides just below the soil surface.

Control. Arsenate of lead, DDT, or malathion sprays will keep this pest in check.

FIG. 94. Winter injury to flowering cherry. This species is particularly susceptible in eastern areas north of Philadelphia.

PEACH BORER. *Sanninoidea exitiosa.* Masses of gum exuding from the base of the tree result from the work of the peach borer within the trunk. The adult is a clear-winged moth, which appears and deposits eggs in July and August.

The lesser peach borer, *Synanthedon pictipes,* may also attack flowering cherry. The adult, a metallic, blue-black, yellow-marked moth, emerges in June in the vicinity of New York City and in May in the south, to deposit eggs on the bark higher up in the tree than the peach borer.

Control. Four applications of malathion to the trunk and branches will control both types of borers. The first application should be made about June 15 in lattitudes of New York City, a few weeks earlier in the South. The other applications should follow at about 2-week intervals.

WEST INDIAN PEACH SCALE. *Pseudaulacaspis pentagona.* The presence of West Indian peach scale is indicated by white incrustations on the bark, masses of circular white scales, 1/10 inch in diameter, which suck juices from below the bark.

Control. While the tree is dormant, spray with 1 part concentrated lime sulfur in 10 parts of water, or with dormant strength oils. Malathion sprays applied when the young are crawling about in spring will also control these pests.

ORIENTAL FRUIT WORM. *Graphiolitha molesta.* Wilting of the tips of twigs may be due to boring by the oriental fruit worm, a small pinkish-white larva, about ½ inch long. The adult female is about ½ inch long and is gray with chocolate-brown markings on the wings. Larvae overwinter in the soil.

Control. Where only a few trees are involved, removal and burning of wilted tips as they appear, is usually sufficient. Large numbers of trees can be protected by periodic applications of DDT sprays to which a compatible mite killer has been added, starting as soon as the leaves begin to emerge and repeating twice at 10- to 12-day intervals.

SAN JOSE SCALE. *Aspidiotus perniciosus.* Gray, closely appressed masses of circular scales, 1/10 inch in diameter, with a raised nipple in the center, are characteristic of the San Jose scale. The winter is passed in the immature stage on the bark.

Control. Dormant oil sprays or lime sulfur 1-10 on the trunk, branches, and twigs before the buds open in the spring is effective. A more complete control can be obtained by following with malathion sprays in May and June.

PEAR-SLUG SAWFLY. *Caliroa cerasi.* Dark green to orange, tadpole-shaped, ½ inch long larvae covered with slime skeletonize the leaves of this host, as well as pear, plum, hawthorn, mountain ash, and quince (Fig. 95).

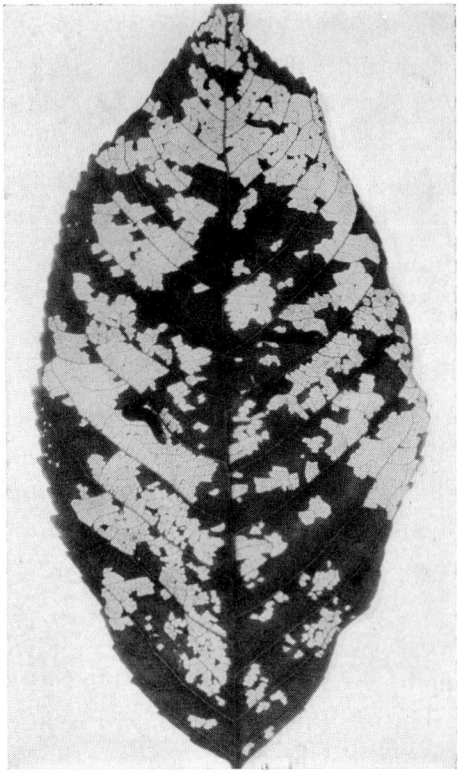

FIG. 95. Pear-slug sawfly larval skeletonizing cherry leaf.

Control. Spray with lead arsenate or malathion when the slugs begin to feed.

JAPANESE BEETLE. See under Linden (Chapter 16).

CRABAPPLE (FLOWERING) (*Malus*)

Flowering crabapple trees used for ornamental purposes are subject to attack by many of the fungi and insects that occur on the common apple.

DISEASES

Of the fungi attacking the leaves, the rusts and scab occur most frequently (Fig. 96). Spray schedules have not been devised specifically for these ornamentals, because of their relatively few numbers as compared with the fruit-bearing trees, but information on the control of apple pests is available from state agricultural experiment stations. The general spray schedules listed below and adapted from apple schedules will probably control all the more important diseases of flowering crabapples.

FIG. 96. Rust spots on flowering crabapple leaves.

Control. If practicable, eliminate any red cedars (*Juniperus*) within a mile radius. Or spray with ½ pound of 76 per cent ferbam and 3 pounds of elemental sulfur per 100 gallons of water. For a few trees the dilution rate per gallon of water is 2 teaspoons of ferbam and 2 tablespoons of sulfur. Four or 5 applications at 7- to 10-day intervals are needed, starting when the orange fungus masses appear on junipers. For most effective results, the sprays should be applied before rainy periods.

INSECTS

Control. Most of the insects which infest the leaves and twigs, such as aphids, alder lace bug, leafhoppers, and several kinds of caterpillars, are easily controlled with malathion if applied when the young are feeding. Oyster shell and Putnam scales can be controlled with a dormant oil spray, or with malathion when the young scales are crawling about in late spring. Malathion can be safely combined in the ferbam-sulfur spray recommended above for disease control.

CRAPE MYRTLE (*Lagerstroemia*)

This beautiful shrubby tree is reliably hardy only in the South, although fine specimens are growing in protected places in the latitude of New York City.

Diseases

POWDERY MILDEW. Young shoots in early spring and leaves and flowers later are affected by this fungus throughout the South. Tissues are distorted and covered with a white coating of the fungus.

Cause. Erysiphe lagerstroemiae. This fungus lives over the winter as so-called mycelium in dormant buds.

Control. A dormant lime sulfur spray just before buds burst in the spring, followed by more dilute wettable sulfur sprays (1 tablespoon per gallon of water) at 2-week intervals will control this disease.

BLACK SPOT and TIP BLIGHT. The former caused by a *Cercospora* and the latter caused by *Phyllosticta lagerstroemia* are two less common fungus diseases.

Control. Copper fungicides applied several times at periodic intervals after growth starts in spring will provide control.

Insects

APHID. *Myzocallis kahawaluokalani.* This species attacks only crape myrtle and exudes a great amount of 'honeydew' on which the sooty mold fungus (*Capnodium* sp.) thrives.

Control. Spray with lindane or malathion early in the growing season.

FLORIDA WAX SCALE. *Ceroplastes floridensis.* This reddish- or purplish-brown scale covered with a thick, white waxy coating tinted with pink attacks a wide variety of trees and shrubs in the South.

Control. Spray with malathion when the crawling stage is about.

DOGWOOD (*Cornus*)

The dogwoods, among our best ornamental low-growing trees, are subject to several important fungus diseases and insect pests.

Diseases

CROWN CANKER. The most serious disease of flowering dogwood (*C. florida*), crown canker, is prevalent in several northeastern states.

Symptoms. An unthrifty appearance is the first general symptom. The leaves are smaller and lighter green than normal, and turn prematurely red in late summer. At times, especially during dry spells, they may curl and shrivel. Later, twigs and even large branches die. At first the diseased parts occur principally on one side of the tree, but within a year or two they may appear over the entire tree.

The most significant symptoms and the cause for the weak top growth is the slowly developing canker on the lower trunk or roots, at or near the soil level. Although the canker is not readily discernible in the early stages, it can be located by careful examination. Cutting into it will reveal that the inner bark, cambium, and sapwood are discolored. Later, the cankered area becomes sunken, and the bark dries and falls away, leaving the wood exposed. When the canker extends completely around the trunk base or the root collar, the tree dies.

Cause. Crown canker is caused by the fungus *Phytophthora cactorum*. The same parasite apparently causes the so-called bleeding canker disease of maples (p. 284) and canker of American beech (p. 307). It also parasitizes a large number of other plants. *P. cactorum* can live over in the soil in partly decayed organic matter, and its spores may be washed to nearby uninfested areas. It appears to gain entrance primarily through wounds, and then invades the tissues in all directions. Thus far the disease has not been found on dogwoods growing in the woods, but only on transplanted trees in ornamental plantings.

Control. Crown canker cannot be controlled after the fungus has invaded most of the trunk base or root collar. Control is possible if the infection is confined to a relatively small area at the trunk base. After the cankered area is outlined, a strip of healthy bark approximately 1 inch wide should be removed all around the edge of the canker. The bark removal should be deep enough to uncover the cambium. Within the area thus left bare, all discolored bark and sapwood should be carefully cut out with a gouge. The edge of the wound should be painted with orange shellac and the area between with any good tree paint. Success with this procedure will depend on early detection of the infection and thorough and complete surgical treatment.

Because the disease has been found only on transplanted trees, and because the fungus appears to enter the host plants more readily through wounds, some observers have suggested that infection occurs primarily through unavoidable injuries inflicted during transplanting. Another avenue of entrance suggested, especially on long-established trees, is wounds made by lawn machinery and cultivating tools. Consequently, all wounds should be painted immediately with shellac and then with a good tree paint, after they have been properly shaped. The possibility of injury to valuable lawn specimens can be decreased by providing some protection around the trunk base, such as a No-Trim Tree Guard.

FIG. 97. Blight on flowering dogwood leaves caused by the fungus *Botrytis cinerea*.

FLOWER AND TWIG BLIGHT. In rainy seasons such as 1958 in the eastern United States, the white flower bracts fade and rot and infect leaves on which they fall (Fig. 97). In some cases twigs are also blighted.

Cause. The fungus *Botrytis cinerea* causes blight on dogwoods.

Control. Spray the entire tree lightly with zineb (Parzate or Dithane Z-78), 1½ pounds to 100 gallons of water, early in the flowering period. For making up small quantities, use 1 tablespoon per gallon of water. Repeat the spraying if rain follows within 3 or 4 days.

LEAF SPOT. Small brown spots may also be prevalent on the leaves in wet seasons.

Cause. Septoria cornicola is a common cause for leaf spotting. The fruiting bodies of this fungus can be detected in the browned lesions with a powerful lens.

Control. A number of fungicides are effective in combating leaf spot. Among these are maneb (Manzate) or captan, or phenyl mercury compounds (Puratized Agricultural Spray or Tag 331). Applications should be made once a month starting in April when the flower buds are in the cup stage, and continuing until the flower buds for the following year are formed in late summer.

SPOT ANTHRACNOSE. This fungus disease is becoming increasingly prevalent on wild and cultivated dogwoods.

Symptoms. Small, slightly raised spots with reddish-gray to gray centers appear on the leaves (Fig. 98). More than 100 such spots can appear on a single leaf. Spots with light tan centers and purple to brown borders may also appear on the white flower bracts (Fig. 99). Cankers may also develop on the current season's stems.

Cause. The fungus *Elsinoe corni* causes spot anthracnose.

Control. The spray schedule suggested for leaf spot control above, will also control spot anthracnose.

INSECTS

BORERS. Dogwoods are frequently attacked by the flat-headed borer and the dogwood borer (*Thamnosphecia scitula*). The cambial area in the branches may be chewed by whitish caterpillars with brown heads and up to ½ inch long, causing girdling and death of the branches. The adult females are small moths, with blue-black margins on their clear wings. They deposit eggs on roughened bark from late spring to mid-summer.

Control. Paint or spray the trunk and branches with DDT or dieldrin, 4 to 6 teaspoons in 1 gallon of water, three times at 20-day intervals starting in late May. In recent tests in Connecticut, a single application of aldrin or lindane to the trunk and branches in early May gave good control of overwintering larvae.

FIG. 98. Small slightly raised reddish-gray spots on dogwood leaves are caused by the fungus *Elsinoe corni.*

FIG. 99. Spots on the white flower bracts are also caused by *Elsinoe corni.*

WHITE FLY. *Tetraleurodes mori.* Leaves of dogwood, mulberry, holly, maple, and plane infested by white fly are usually sticky from honeydew secreted by tiny, black, oval, scale-like young, 1/35 inch long, with a prominent white border. The adults are tiny, white flies, which dart away when the leaves are disturbed.

Control. Spray with malathion in midsummer and again in late summer.

SCALES. A number of scales including dogwood, cottony maple (Fig. 100), obscure, oyster shell, and San Jose occasionally infest dogwood.

FIG. 100. Cottony maple scale on dogwood leaves.

Control. A dormant oil or lime sulfur spray, followed by one or two applications of malathion during the growing season will control scales.

CLUB GALL. *Mycodiplosis alternata.* Club-shaped galls or swellings, ½ to 1 inch long, on twigs of flowering dogwood are caused by a reddish-brown midge which attacks the twigs in late May. Small orange larvae develop in the galls until early fall, when they drop to the ground.

Control. Pruning and burning twigs with the club galls during the summer will provide control.

TWIG GIRDLER. See under Elm (Chapter 16).

EMPRESS TREE (*Paulownia*)

The empress tree is prized most for its showy clusters of violet flowers in early spring. It is relatively free of pests, but it is quite sensitive to extremes in the weather.

Diseases

LEAF SPOT. Two fungi, *Ascochyta paulowniae* and *Phyllosticta paulowniae*, are known to produce brown spots on the leaves in rainy seasons.

Control. Gather and burn fallen leaves. Preventive sprays containing copper or dithiocarbamate fungicides can be used on extremely valuable specimens in rainy seasons.

WOOD DECAY. The fungi *Polyporus spraguei* and *P. versicolor* are associated with a wood decay of this host.

Control. As with other wood-decaying fungi, control is difficult. Avoiding wounds near the trunk base and keeping the tree in good vigor by feeding and by watering when necessary, are suggested.

Insects

No important insects are known to attack the empress tree.

FRANKLIN TREE (*Franklinia*)

Diseases

LEAF SPOT. *Phyllosticta gordoniae.* Leaves may be spotted as a result of infection by this fungus.

Control. A copper fungicide will probably control this leaf spot on valuable specimens.

BLACK MILDEW. *Meliola cryptocarpa.* Leaves of the Franklin tree in the deep South are occasionally attacked by this fungus.

Control. Measures are rarely adopted to control this fungus.

FRINGE TREE (*Chionanthus*)

Diseases

LEAF SPOT. Four species of fungi are known to cause spotting of leaves on this host: *Cercospora chionanthi, Phyllosticta chionanthi, Septoria chionanthi,* and *S. eleospora.*

Control. Spray with a copper fungicide or with one of the dithiocarbamates.

POWDERY MILDEW. The fungus *Phyllactinia corylea* occasionally infects the leaves of fringe tree.

Control. Spray with sulfur or Mildex (Karathane) where the mildew is severe.

GOLDENCHAIN (*Laburnum*)

Diseases

TWIG BLIGHT. A species of *Fusarium* has been reported as the cause of twig blight on this host. Another species *F. lateritium* also causes twig blight on goldenchain and the Japanese pagoda tree (p. 279).

Control. Prune affected branch tips to sound wood and apply a copper fungicide several times at 2-week intervals during rainy springs.

LEAF SPOT. *Phyllosticta cytisii.* This fungus produces light gray spots, which later turn brown. They have no definite margin. Another fungus, *Cercospora laburni,* also causes leaf spotting.

Control. Spray as for twig blight where these leaf spots are prevalent.

Insects

APHIDS. The cowpea aphid (*Aphis medicaginis*), black with white legs, clusters at the tips of the branches. The bean aphid (*Aphis fabae*) also infests goldenchain.

Control. Spray with lindane or malathion when the aphids appear in late spring.

GOLDENRAIN TREE (*Koelreuteria*)

Diseases

LEAF SPOT. A species of *Cercospora* has been reported on the leaves of this tree in Florida.

Control. When the disease is serious, copper fungicides should be used as a preventive measure.

CANKER. The fungus *Nectria cinnabarina* occasionally produces cankers on the branches.

Control. Same as for maple cankers (p. 287).

WILT. The fungus *Verticillium albo-atrum* (p. 251) occasionally infects and kills this tree.

HAWTHORNS (*Crataegus*)

Hawthorns, sometimes called thorns, are members of the Rose family, to which the apple and pear belong. As such, they are frequently attacked by the same fungi, bacteria, and insects that attack those trees.

Diseases

LEAF BLIGHT. In late summer English hawthorns may be completely defoliated by the leaf blight disease. Cockspur thorn and Washington thorn appear to be much more resistant to the trouble.

Symptoms. Early in the spring small, angular, reddish-brown spots appear on the upper leaf surface. As the season advances, the spots enlarge, then coalesce, and the leaves finally drop.

Cause. The fungus *Fabraea theumenii* causes leaf blight. It is primarily a parasite of hawthorns, although some investigators believe it is also responsible for a similar disease of pear and quince.

Numerous black, flattened, orbicular bodies are visible, with the aid of a hand lens, in the discolored spots on both leaf surfaces. These bodies contain spores that initiate numerous infections during rainy springs.

Control. Gather and burn all the leaves in the fall to reduce infections the following spring, inasmuch as the fungus survives from one season to the next primarily in diseased leaves.

There is some evidence that the fungus also overwinters as mycelium in small cankers on the bark. Because such sources of inoculum cannot be completely eliminated, sprays containing zineb, maneb, or ferbam are needed. These fungicides should be used at the rate of 1 pound in 100 gallons of water and applied three times at 10-day intervals. The first application should be made immediately after the leaves have unfolded. Recently applications of the antibiotic cycloheximide (Actidione) have also proved effective in controlling leaf blight.

Trees that have been subjected to several consecutive years of severe leaf blight infections may be considerably weakened. A good tree fertilizer, applied in fall or spring, will help such trees to regain their vigor. Fertilizers high in nitrogenous materials should be avoided, inasmuch as they increase the tree's susceptibility to fire blight.

FIRE BLIGHT. Hawthorns, especially the English hawthorn, are subject to the bacterial disease fire blight, discussed on page 238.

CEDAR-HAWTHORN RUST. Orange to yellow pustules on the lower leaf surface are produced by the rust fungus *Gymnosporangium globosum* (Fig. 101).

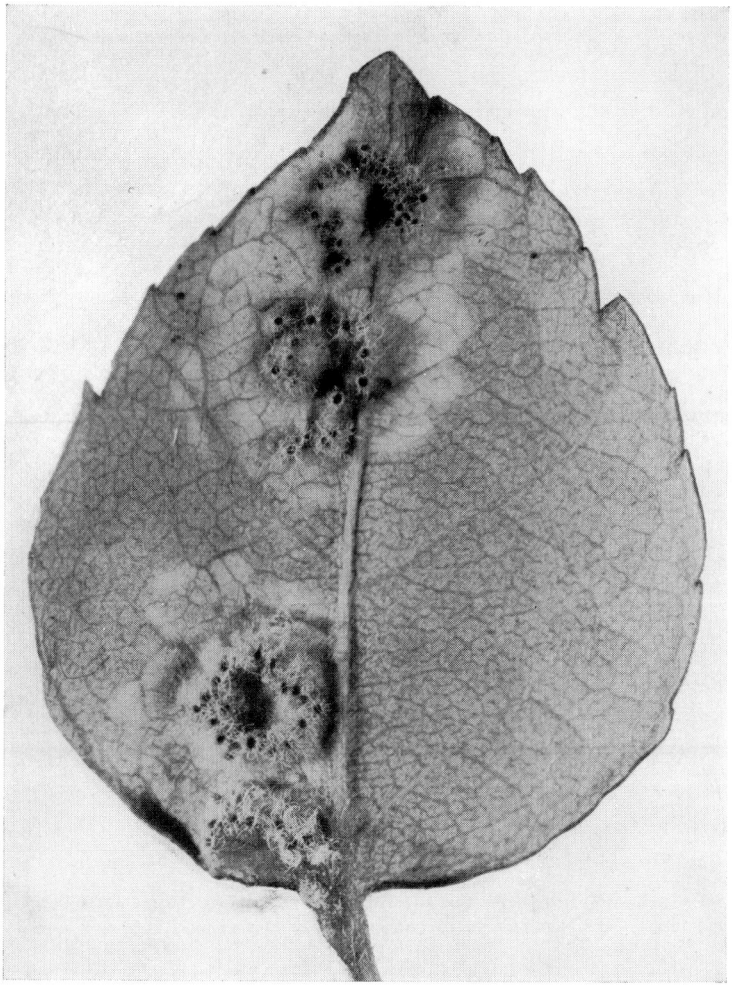

FIG. 101. The fruiting bodies of the rust fungus are shown on this hawthorn leaf.

The alternate stage of this fungus develops on eastern red cedar and on dwarf, prostrate, and Rocky Mountain junipers.

Control. If practicable, eliminate susceptible junipers within one mile of hawthorns. If not practicable, spray the hawthorns with ferbam or a mixture of ferbam and wettable sulfur four to five times at 7- to 10-day intervals when orange masses appear on junipers.

Recent research shows that this disease can also be controlled by spraying junipers with cycloheximide. See under Juniper (Chapter 17).

LEAF SPOT. A large number of fungi are known to cause leaf spots on hawthorns. Among the more common ones are: *Cercospora confluens, C. apiifoliae, Cercosporella mirabilis, Cylindrosporium brevispina, C. crataegi, Gloeosporium crataegi, Hendersonia crataegicola, Septoria crataegi,* and *Monilinia Johnsonii.* The last also causes spots on the fruits, as does the scab fungus *Venturia inaequalis.*

Control. The sprays suggested for leaf blight control will also control most of the fungi listed above.

POWDERY MILDEW. Two species of powdery mildew fungi, *Phyllactinia corylea* and *Podosphaeria oxyacanthae,* attack hawthorns.

Control. The dithiocarbamates which control leaf blight and rust are ineffective against powdery mildew. For these use either sulfur or Mildex (Karathane).

Insects

HAWTHORN LEAF SKELETONIZER. *Hemerophila pariana.* Leaves attacked by this pest, a yellowish-green larva, ½ inch long, are skeletonized, only the veins and a thin layer of tissue being left. The adult female is a reddish-brown moth with a wingspread of half an inch. The moths pass the winter in sheltered places near the thorn trees.

Control. Spray with arsenate of lead, DDT, or malathion when the larvae start to feed in the spring.

WESTERN TENT CATERPILLAR. *Malacosoma pluviale.* Tawny or brown caterpillars with a dorsal row of blue spots feed primarily on this host as well as wild cherry and alder, making tents at the same time as the fall webworm. The moths are smaller and somewhat lighter than the eastern tent caterpillar adults.

Other caterpillars that may feed on hawthorns include the eastern tent, the forest tent, the red-humped, the variable oak leaf, and the walnut caterpillar. The caterpillar stages of the gypsy moth and the western tussock moth also occur on this host.

Control. Spray with arsenate of lead, DDT, or malathion when the caterpillars begin to chew the leaves.

PEAR-SLUG SAWFLY. See under Flowering Cherry.

HORNBEAM (*Carpinus*)

Diseases

TWIG BLIGHT. The fungus *Fusarium lateritium* (p. 280) causes a twig blight on this host in the South.

Control. Prune infected twigs on valuable ornamental specimens. A protective fungicide containing copper might also help.

CANKER. Cankers on the bark may be caused by *Pezicula carpinea, Solenia ochraceae,* or *Nectria galligena.* The felt fungus, *Septobasidium curtisii,* occurs on a number of trees in the South. The felt is purplish-black and covers the insects it parasitizes.

Control. Pruning out twigs and branches infested with these fungi is warranted on valuable specimens. The use of fungicides containing copper may also be justified in some situations.

LEAF SPOT. *Gloeosporium robergei, Gnomoniella fimbriata,* and *Septoria carpinea* are three fungi known to cause leaf spots on hornbeam.

Control. These leaf spots are rarely serious enough to warrant control measures. Copper, ferbam, or phenyl mercury sprays are effective in preventing heavy outbreaks.

Insects

SCALE. *Phenacoccus acericola.* Although this scale is more common on sugar maple, it occasionally infests hornbeam, where it forms large white cottony masses on the lower leaf surface.

Control. Malathion sprays directed to the lower leaf surfaces in late spring are effective.

JAPANESE PAGODA TREE (*Sophora*)

Diseases

CANKER. Oval, 1- to 2-inch cankers with definite, slightly raised, dark red-brown margins with light tan centers occasionally appear on this host. When the cankers completely girdle the stem, the distal portion dies.

Cause. The fungus *Fusarium lateritium* causes this disease. Pink spore structures of the fungus develop in wet weather. Another fungus, *Cytospora sophorae,* is also found on dead branches.

Both fungi are usually associated with frost damage in late fall or early spring.

Control. Pruning out diseased or dead branch tips to sound wood and several sprays of a copper fungicide during the growing season are usually sufficient to keep these fungi in check.

TWIG BLIGHT. Two fungi, *Nectria cinnabarina* and *Diplodia sophorae,* occasionally cause a die-back of twigs.

Control. Same as for canker.

INSECTS

No insects of any importance have been reported on this host.

KENTUCKY COFFEE TREE (*Gymnocladus*)

This species is remarkably free of parasites and insect pests.

DISEASES

LEAF SPOT. Three fungi, *Cercospora gymnocladi, Phyllosticta gymnocladi,* and a species of *Marssonia,* have been reported on this host.

Control. Special control measures are rarely required, but periodic application of copper or dithiocarbamate fungicides will protect valuable specimens.

MAGNOLIA (*Magnolia*)

Magnolias are subject to a number of fungus diseases and insect pests. Only the more important ones are treated here.

DISEASES

LEAF BLIGHT. The fungus *Pellicularia koleroga* causes a leaf blight on *Magnolia grandiflora.* The same fungus also affects apple, citrus, dogwood, Japanese persimmon, pear, pecan, quince, and many shrubs and vines in the South.

Control. Valuable specimens can be protected with two or three applications of copper fungicides.

LEAF SPOT. More than a dozen fungi cause leaf spots on magnolias but those belonging to the genus *Phyllosticta* (*P. cookei*, *P. glauca*, and *P. magnoliae*) are the most common. In the deep South a parasitic alga, *Cephaleuros virescens*, also affects the leaves.

Control. Fungus leaf spots on valuable specimens can be controlled by periodic applications, early in the growing season, of copper or dithiocarbamate fungicides.

DIE-BACK. A dying back of the tops frequently occurs on magnolias. Infected areas on the large limbs and trunk show cankers that are bordered by longitudinal cracks in the bark. The bark is uniformly dark brown under the external gray. The wood is discolored a bluish-gray.

Cause. The cause of die-back has not been definitely established. A fungus belonging to the genus *Phomopsis* is constantly associated, however, with the cankers.

Control. No control measures, aside from pruning infested branches to sound wood, can be recommended until the cause of the trouble is clearly established.

NECTRIA CANKER. *Nectria magnoliae.* This fungus produces symptoms similar to those of *N. galligena* (p. 259) but infects only magnolia and tulip trees.

Control. Same as for *N. galligena.*

WOOD DECAY. A heart rot, associated with the fungi *Fomes geotropus* and *F. fasciatus*, has been reported on magnolia. Affected trees show sparse foliage and die-back of the branches. In early stages the rot is grayish-black, with conspicuous black zone lines near the advancing edge of the decayed area. The mature rot is brown. The causal fungi gain entrance through wounds.

Control. No control measures are effective, once the rot has become extensive. Avoid trunk wounds and maintain good vigor by fertilization and watering.

INSECTS

MAGNOLIA SCALE. *Neolecanium cornuparvum.* Underdeveloped leaves and generally weak trees may result from heavy infestations of the magnolia scale, a brown, varnish-like hemispherical scale, ½ inch in diameter, with a white waxy covering (Fig. 102). The young scales appear in August and overwinter in that stage.

TULIP-TREE SCALE (p. 369) also infests magnolias and lindens at times.

Control. Dormant oil sprays are very effective in combating this pest. The crawling stages of both the magnolia and tulip tree scales, unlike most scales, appear in late summer. Hence the use of malathion sprays for the crawling stages should be delayed until late August or early September.

FIG. 102. Magnolia scale.

MAPLE (*Acer*)

DISEASES

WILT. Already discussed in great detail on page 247, wilt is perhaps the most important vascular disease of maples. Positive diagnosis of this disease is possible only by collecting adequate specimens of discolored sapwood beneath the bark of the trunk near the soil line and culturing the material in the laboratory. The Verticillium fungus, however, is not the only fungus which is capable of discloring the sapwood of maples. Several others have been found to do so. Some of these, however, have not killed trees that were artificially inoculated with them, while others have. The parasitic kinds have not yet been identified.

BASAL CANKER. The author was the first to discover this disease on Norway maples in New Jersey about 20 years ago.

Symptoms. An early symptom is a thin crown resulting from a decrease in the number and size of the leaves. Trees die within a year or two following this period of weak vegetative growth. Another, more striking symptom is the presence of cankers at the base of the trunk near the soil line. The inner bark, the cambium, and in many instances the sapwood are reddish-brown in the cankered area (Fig. 103). Death occurs when the entire root system decays or when the cankers completely girdle the trunk.

FIG. 103. Bark cut away to show the reddish-brown discoloration of the sapwood produced by the fungus *Phytophthora cinnamomi.*

Cause. Basal canker is caused by the fungus *Phytophthora cinnamomi.* This fungus was known as *P. cambivora* when the author first isolated it from diseased Norway maples. The fungus appears to be most destructive on trees growing in poorly drained or shallow soils.

Control. No effective control measures are known. Diseased trees should be removed and burned. New plantings of Norway maples should be made in well-drained soils, high in organic matter. Frost cracks and mechanical injuries near the base of the trunk should be properly treated and dressed and trees fertilized and watered to maintain good vigor.

BLEEDING CANKER. This highly destructive disease was first reported from the New England states centering around Rhode Island. The author found the first case of this disease in New Jersey in October 1940. Norway, red, sycamore, and sugar maples; oaks, elms, and American beech are susceptible to this disease. The severity of the disease is in some way associated with hurricane damage.

Symptoms. The disease is named for its most characteristic primary symptom, the oozing of sap from fissures overlying cankers in the bark. Infected inner bark, cambium, and sapwood develop a reddish-brown necrotic lesion, which commonly exhibits an olive-green margin. These symptoms differ markedly from those produced in trees affected with basal canker. A secondary symptom, the wilting of the leaves and dying back of the branches, is said by one investigator to be due to a toxic material secreted by the causal fungus.

Cause. Bleeding canker is caused by the fungus *Phytophthora cactorum*, which is closely related to the fungus that causes basal canker. *P. cactorum* is also responsible for the crown canker disease of dogwoods (p. 268). The author has produced infection in dogwood with the fungus isolated from Rhode Island maples. The same organism kills the growing tips of rhododendrons during rainy seasons and is known to attack a large number of other trees including apple, apricot, cherry, peach, and plum.

Control. As with basal canker, no effective control measures are known.

LEAF SPOT. Several species of fungi produce leaf spotting of maples. Most of the spots involve relatively small areas of the leaf and therefore do not reduce the efficiency to any measurable degree. A few types of spots are so large, or may coalesce to such an extent, as to involve a large part of the leaf or even result in premature defoliation. In such instances, considerable harm to the tree may result, especially if the spotting occurs annually.

Though caused by distinctly different fungi, many of the spots resemble one another so closely that only a specialist can determine the species of fungus involved. For practical purposes, however, the causal agent is of little importance, because nearly all types of spots are controllable by much the same practices.

Leaf spots are abundant only after prolonged periods of cool, wet weather.

They were unusually abundant in the Northeast early in the summer of 1958 because of the long wet spell in late May and early June of that year.

The following are the most common leaf spots of maple:

Anthracnose on maple leaves produces lesions similar to those found on planes (p. 354). The spots may coalesce, and the entire leaf die and fall. Silver and sugar maples are the most susceptible species. Anthracnose is caused by the fungus *Gloeosporium apocryptum*.

Purple eye spot (Fig. 81) is found on most species of maple. Large, irregularly circular, brown spots surrounded by a broad purple border are produced by the fungus *Phyllosticta minima*. Black pin-point bodies, the fruiting structures of this fungus, are usually visible in the browned center. This leaf spot rarely impairs the functions of the leaf, although a few leaves may drop prematurely.

Bull's-eye spot is a name coined by the author to describe a spot that occurs on sycamore, red, sugar, and silver maples growing in particularly shaded spots in the eastern United States. The spots show a distinct target pattern with layers of concentric rings (Fig. 104). The causal organism was originally described as *Cristulariella depraedens*, but the author has some evidence that indicates that it may belong to the genus *Sclerotinia*.

FIG. 104. Bull's-eye spot of Sycamore maple.

Tar spot is an appropriate description of a common trouble on red and silver maples. Black, thickened, raised, tar-like spots, ⅛ to ½ inch in diameter, on the upper leaf surface, are actually the overwintering bodies or sclerotia of the fungus *Rhytisma acerinum*.

A *bacterial leaf spot* has been reported on the broad-leaved maple (*A. macrophyllum*) from the West Coast. Spots vary from pin-point dots to areas ¼ inch in diameter. They first appear as though water-soaked and are surrounded by a yellow zone; later they turn brown and black. The petioles may be attacked, resulting in defoliation. The causal organism is *Pseudomonas aceris*.

OTHER LEAF SPOTS. A number of other leaf spots that occur occasionally are said to be caused by fungi belonging to the genera *Septoria*, *Cylindrosporium*, *Phyllosticta*, and *Monochaetia*. Although the ocellate leaf spot (Fig. 105) on red maple resembles spots produced by fungi, it is actually caused by a tiny transparent maggot *Cecidomyia ocellata*.

FIG. 105. Ocellate leaf spot of maple produced by the feeding of the larvae of *Cecidomyia ocellata*.

Control. Most of the fungi and bacteria causing leaf spots are known to pass the winter on fallen diseased leaves. Consequently the first step in any control program is the gathering and burning of all diseased leaves in the fall. Where the spotting is particularly severe, preventive sprays should be employed. Fungicides containing copper are most effective. Several applications at 2-week intervals should be made in spring, starting when the leaves are unfolding from the buds. Copper sprays may cause damage to Norway maple leaves when they are applied during cool, rainy weather. Recently it has been found that a single delayed dormant application of phenyl mercury acetate gives as good control of some leaf diseases as do several applications of copper fungicides.

LEAF BLISTER. Several species of maples are commonly attacked by a group of fungi belonging to the genus *Taphrina*. These fungi produce blisters, curling, and blighting of the leaves, symptoms often attributed erroneously to the Anthracnose fungus. The following are the more common species of *Taphrina* together with their hosts:

Taphrina sacchari, on sugar and black maple.

T. lethifer on mountain maple, *A. spicatum*.

T. aceris on Rocky Mountain hard maple, *A. grandidentatum*.

T. dearnessii on red maple.

T. carveri on silver maple.

Control. Leaf blisters on valuable specimens can be prevented by early spring applications of copper fungicides as suggested for leaf spot control. Copper sprays, particularly bordeaux, may cause some spray burn.

CANKER. Maples often show sunken and discolored bark areas on the trunk and larger limbs, especially in the vicinity of pruning wounds, bark injuries, and dead branch stubs. The wood beneath these areas is colored green to black. The causal fungus, a species of *Nectria*, is responsible in many instances, and its presence can be detected by close inspection of the sunken tissue, where tiny, reddish-brown, globose fruiting bodies are visible (Fig. 81).

Boxelder and red and sugar maples occasionally show another type of canker, which differs strikingly from that produced by the nectria canker. The cankers are irregularly circular and contain broad, slightly raised concentric rings of callus tissue. The cankered tissue is firmly attached to the wood with heavy, white to buff, fan-shaped wefts of fungus tissue under the bark near the margins. Tiny black fungus bodies are present in the centers of the old cankers. The fungus responsible for this type of canker is known as *Eutypella parasitica*. It was recently reported on a Norway maple growing along a Syracuse, New York, street.

Control. Cankers on trunks can rarely be eradicated by surgical methods once they have become very extensive. Those on branches can be destroyed by removing and burning the affected members. Removal of dead branches, avoidance of unnecessary injuries, and maintenance of trees in vigorous condition by fertilization and watering are probably the best means known of preventing canker formation.

DIE-BACK. A die-back of Norway maple branches occurs commonly in trees growing along city streets in the northeastern United States. Often only the upper end of the branch loses its foliage, while the portion attached to the main stem remains perfectly normal. Inspection of the stem between these two areas will reveal a sunken canker completely girdling the branch. Usually a crack ¼ inch wide separates the healthy from the diseased tissue. The canker is associated with winter injury followed by invasion of weak fungus parasites. The wood of infected branches does not show green discoloration, as does that of some wilt-infected trees.

Control. The author has checked this trouble in a large number of trees during the last 15 years merely by pruning the branches well below the visibly infected area. No reinfections have occurred in such trees.

GANODERMA ROT. Rapid decline and death of many streetside trees is associated with a fungus that forms large, reddish, and varnish-like coated fruit bodies at the base of the infected tree or on its surface roots. Although the fungus is most frequently found on dead trees, the author, in an intensive survey of street trees in New York and New Jersey several years ago, found the fungus actively invading living trees, particularly red (A. *rubrum*) and Norway (A. *platanoides*) maples (Fig. 106). Moreover, he was able to infect young maples with pure cultures of the fungus (Fig. 107) and to kill young trees whose roots were surrounded with fresh, chopped fruiting bodies of the Ganoderma.

Cause. The fungus *Ganoderma lucidum*, or a closely related species, causes decay of living trees. The determination that G. *lucidum* was involved in the author's studies was made by Dr. Donald Rogers during his tenure as Mycologist for The New York Botanical Garden.

Control. No control is known for diseases of this type. Planting trees in deep, fertile soil, avoiding bark and root injuries, proper fertilization and watering will reduce chances of infection.

INSECTS

NORWAY MAPLE APHID. *Periphyllus lyropictus.* Stunted, wrinkled, or blotched leaves and the dropping of a sticky secretion to the pavement below, suggest the presence of the Norway maple aphid, a large, hairy, yellowish-green insect.

FIG. 106. The author is examining a dying Norway maple infected with the fungus *Ganoderma lucidum*. He is holding a live branch cut from the healthy side of this tree. Note the fungus bodies at the base of the tree.

FIG. 107. INSET: A culture of the Ganoderma fungus growing on a synthetic medium in the laboratory.

Control. When the pests become numerous, spray the undersides of the leaves with lindane or malathion.

LEAFHOPPERS. *Cicadellidae.* The production of tiny white specks on the leaves and the appearance of narrow, yellowish-green insects, ⅛ inch long, which seem to hop but actually fly, are typical signs of this pest.

Control. Sprays containing DDT or a mixture of DDT and a miticide will control leafhoppers.

COTTONY MAPLE SCALE. *Pulvinaria innumerabilis.* Soft maples are weakened by the sucking of the undersides of branches by the cottony maple scale, an insect ¼ inch long, having cottony egg masses protruding from the brown scale. The adult female appears in late May or June. Winter is passed in the adult stage.

Control. Malathion spray applied in spring when young scales are crawling about is particularly effective.

TERRAPIN SCALE. *Lecanium nigrofasciatum.* Twigs may be killed as a result of the sucking of sap by numerous reddish-brown, terrapin-shaped female scales, ⅛ inch long. A sooty mold develops on the secretions of this pest. The winter is passed in the partly grown stage.

Control. Spray with 1 part concentrated lime sulfur solution in 10 parts of water while the tree is dormant, or with a standard malathion spray about mid-May. The latter should be repeated a month later. Dormant oils are unsafe for some maples, including sugar and silver.

WOOLLY MAPLE SCALE. *Phenacoccus acericola.* This scale, which closely resembles the cottony maple scale, infests maples, especially the sugar. It forms long white cottony masses on the undersides of the leaves.

Control. A malathion spray in spring just as the buds burst and a second early in August give good control.

BLADDER GALL MITE. *Vasates quadripedes.* Globular, bladder-like galls, $1/10$ inch in diameter, on the upper leaf surfaces are produced by white, pink, or red mites, $1/125$ inch long. The galls are first green, then red, and finally black. The mites live in the galls and leaf tissues. They overwinter in scars and wounds on the bark.

The mite *Paratetraynchus aceris,* occasionally infests maple leaves, causing mottling and premature yellowing.

Control. Excellent control of both species of mites is attained by spraying the trees thoroughly, while they are dormant, with 1 part concentrated lime sulfur solution in 10 parts of water. Spraying with malathion when the leaves are just emerging and again when they are half-grown will also control most species of mites.

LEOPARD MOTH. *Zeuzera pyrina.* The trunks and branches are tunneled by the 2-inch long, pinkish-white, black-spotted larvae of the leopard moth. Many of the branches are completely girdled. The adult female has white, metallic-blue spotted wings with an expanse of 1½ inches. Eggs are deposited in bark crevices throughout the summer. The life cycle is completed in 2 years.

Control. Prune and burn severely infested branches. Borers inside branches can be destroyed by inserting a flexible wire into their tunnels to crush them, or by injecting a nicotine paste such as Bortox and then sealing the opening with chewing gum or putty. The females, which are unable to fly, may be collected and destroyed before they have laid their eggs.

SUGAR MAPLE BORER. *Glycobius speciosus.* Wide channels in the inner bark and sapwood of the trunk and larger branches are made by the sugar maple borer, a pinkish-white larva 2 inches long. The bark above the infested area dies, leaving large scars. The adult beetle is an inch long and is velvety black with yellow markings. The eggs are deposited in bark crevices in July and August. The pest has a 2-year life cycle. The first year is spent beneath the bark; and the second, in the heartwood.

Control. Prune and burn severely infested branches by June 1, and inject a nicotine paste such as Bortox into the tunnels in the remaining parts of the trunk and branches.

MAPLE AND OAK TWIG PRUNER. *Elaphidion villosum.* Branches an inch or less in diameter are girdled by the twig pruner, a white, cylindrical, conspicuously segmented grub. The girdled branches harboring the grubs fall to the ground when the wind blows. The adult, a slender, grayish-brown, long-horned beetle, ½ inch long, deposits eggs inside the twigs in July, and the larvae overwinter in the fallen twigs.

Control. Collect and burn all fallen twigs in summer and autumn.

FOREST TENT CATERPILLAR. *Malacosoma disstria.* The leaves are chewed by the forest tent caterpillar, a 1½-inch long, bluish-black pest with white spots down the back. The caterpillars mass in large numbers on the trunk during the day. The adult female is a brown moth with two dark lines across the wings, which have a spread of 1½ inches. Eggs are laid on the twigs in rings or bands of 200 or more in June and July.

Control. Spray with arsenate of lead, DDT, or malathion in April or May when the caterpillars are young.

BOXELDER BUG. *Leptocoris trivittatus.* The leaves of boxelder are injured by the sucking of young bright-red bugs. The adult stage is a stout, ½-inch long, grayish-black bug with three red lines on the back. All stages of this pest are clustered on the bark and branches in the early fall. Eggs are deposited in the bark crevices in the spring. The winter is passed in the adult stage. Like the elm leaf beetle (p. 323), this pest frequently enters homes in the fall to hibernate.

Control. Spraying in late May or June with pyrethrum or malathion is effective in combating this pest.

FLAT-HEADED BORER. *Chrysobothris femorata.* Trees in poor vigor are attacked by the flat-headed borer, a light yellow larva, an inch long, which builds flattened galleries below the bark, often girdling the tree. The adult is a dark coppery-brown beetle, ½ inch in length, which deposits eggs in bark crevices in June and July. The pest overwinters underneath the bark in the larval stage.

Control. Increase the tree's vigor by feeding and watering, or protect recently transplanted trees with crepe wrapping paper. Spraying the trunk and branches with malathion in late May and repeating twice more at 2-week intervals should reduce the adult population. Borers active in the tree may be killed with a flexible wire or a knife.

ALDER BLIGHT APHID. See under Alder (Chapter 16).

BAGWORM. See under Arborvitae (Chapter 17).

CARPENTER WORM. See under Ash.

JAPANESE BEETLE. See under Linden (Chapter 16).

SPRING AND FALL CANKERWORM. See under Elm (Chapter 16).

TUSSOCK MOTH. See under Elm.

WHITE FLY. See under Dogwood.

MOUNTAIN ASH (*Sorbus*)

Because mountain ashes belong to the Rose family, they are subject to most of the same diseases and insects common in this family.

Diseases

LEAF RUST. The alternate stages of the rust fungi *Gymnosporangium aurantiacum, G. globosum, G. juniperinum, G. Nelsoni, G. nootkatense,* and

G. libocedri occur on *Juniperus* or *Librocedrus* species. Circular, light yellow, thickened spots first appear during the summer. Later, orange cups develop on the lower surface of these spots.

Control. Where mountain ashes are highly prized, and where practicable, remove the alternate hosts. If such removal is impracticable, valuable trees can be protected by sprays recommended for rust control of flowering crabapples (p. 267).

CANKER. Two fungi, *Cytospora chrysosperma* and *C. massariana*, occur occasionally on mountain ash, especially on weakened trees.

Control. Prune severely affected branches, and fertilize. The latter practice may increase the tree's susceptibility to fire blight, however, if high nitrogenous fertilizers are used.

CROWN GALL. See page 240.

INSECTS

PEAR LEAF BLISTER MITE. *Eriophyes pyri.* Tiny brownish blisters on the lower leaf surface and premature defoliation result from infestations of the pear leaf blister mite, a tiny, elongated, 8-legged pest, 1/125 inch long. The pests overwinter beneath the outer bud scales. Eggs are deposited in the spring in leaf galls, which develop as a result of feeding and irritation by the adult.

Control. Spray the trees in late fall or early spring with 1 part concentrated lime sulfur in 10 parts of water. Malathion sprays applied just after the leaves begin to unfurl in spring will also control mites.

ROUND-HEADED BORER. *Saperda candida.* Trees are weakened and may be killed by the round-headed borer, a light yellow, black-headed, legless larva an inch long. Galleries in the trunk near the soil level, frass at the base of the tree, and round holes of the diameter of a lead pencil in the bark are typical signs of this pest. The adult is a beetle, ¾ inch long, brown, with two white longitudinal stripes on the back. It emerges in April and deposits eggs on the bark.

Control. Spray trunk and branches thoroughly with either DDT or malathion starting in mid-May and repeating three times at 2-week intervals.

SCALES. As with other members of the Rose family, mountain ashes are markedly susceptible to scale insects. Among the common ones are black, cottony maple, oyster shell, San Jose, and scurfy scales.

Control. Dormant oil or lime sulfur sprays, followed by malathion sprays during the growing season will control these insects.

SAWFLY. *Pristiphora geniculata.* Green larvae with black dots feed on mountain ash leaves from early June to mid-July, leaving only the larger veins and midribs. The adults, yellow with black spots, deposit eggs on the leaves in late May.

Control. Spray with arsenate of lead, DDT, or malathion when the leaves are fully expanded.

WOOLLY APPLE APHID. See under Elm (Chapter 16).

REDBUD (*Cercis*)

This tree, extremely popular in some of the states in the middle South, is subject to several serious fungus diseases but few insect pests.

Diseases

LEAF SPOT. A conspicuous leaf spot disease occurs on redbud throughout its range in the eastern half of the United States.

Symptoms. Scattered circular to angular brown spots, about ¼ inch in diameter, appear on the leaves early in the summer. The spots increase in size until they attain a diameter of nearly ½ inch by fall.

Cause. The fungus *Mycosphorella cercidicola* produces circular to angular spots with dark brown borders on the leaves. Spores from overwintered fallen leaves initiate spring infections. The fungi *Cercosporella chionea* and *Phyllosticta cercidicola* also cause leaf spots on this host.

Control. Valuable specimens can be protected by periodic applications of copper or dithiocarbamate fungicides in late spring and early summer.

BOTRYOSPHAERIA CANKER. Redbud is an important host of this widespread fungus disease discussed in detail on page 359. On redbud the cankers begin as small sunken areas and increase slowly in size. The bark in the center of the canker blackens and cracks along the edges. The wood beneath the cankered area becomes discolored. The causal fungus is easily recovered from discolored wood by standard tissue culture techniques in the laboratory.

Insects

TWO-MARKED TREEHOPPER. *Enchenopa binotata.* The leaves and stems of redbud are occasionally punctured by a dusky brown insect 3/10 inch long.

A horn-like projection from the upper part of its body makes the insect resemble, in outline, a bird at rest.

Locust, sycamore, hickory, and willow are also infested by this treehopper.

Control. Spray the leaves and stems with malathion when the insects are young.

CATERPILLARS. The larval stage of the California tent caterpillar (*Malacosoma californicum*) and the grape leaf folder (*Desmia funeralis*) occasionally infest redbuds.

Control. Spray with lead arsenate, DDT, or malathion when the young caterpillars begin to chew the leaves.

SCALES. Three species of scales—greedy, oleander, and terrapin—also infest the twigs and branches of this host.

Control. Dormant lime sulfur or oil sprays usually are sufficient to control scales. Severely infested trees should also be sprayed in May and June with malathion.

RUSSIAN OLIVE (*Elaeagnus*)

Russian olive, an excellent tree for very difficult situations, is relatively free of fungus parasites and insect pests.

Diseases

LEAF SPOT. Several fungi including *Cercospora carii, C. elaeagni, Phyllosticta argyrea, Septoria argyrea,* and *S. elaeagni* cause spots on leaves of this host.

Control. These fungi are rarely severe enough to warrant the use of protective fungicides containing copper or dithiocarbamates.

CANKERS. *Nectria cinnabarina* and a species of *Fusarium* and one of *Phytophthora* are known to cause cankers on the branches and trunk of Russian olive.

Control. No control measures have been developed.

RUST. *Puccinia caricis-shepherdiae* with the alternate host, *Carex,* and *P. coronata* var. *elaeagni* with the alternate host, *Calamagrostis,* occur on Russian olive in the midwestern United States.

CONTROL. These diseases are not important enough to warrant control measures.

Insects

OLEANDER-THISTLE APHID. *Capitophorus braggii.* This pale yellow and green aphid overwinters on Russian olive and lives on thistles during the summer.

Control. Control measures are unnecessary.

SCALES. The olive parlatoria (*Parlatoria oleae*) and the purple scale (*Lepidosaphes beckii*) occasionally infest this host.

Control. Malathion sprays during the summer will control heavy infestations of these scale insects.

SASSAFRAS (*Sassafras*)

Sassafras, an excellent tree for landscape plantings, is relatively free of fungus parasites but is extremely susceptible to Japanese beetle infestations. Like American holly, it is dioecious, that is the pistillate and staminate flowers grow on different trees. The slate-blue fruits are ornamental and provide food for the birds.

Diseases

CANKER. Two fungi, *Nectria galligena* and *Physalospora obtusa,* are frequently associated with branch cankers.

Control. Pruning affected branches to sound wood is suggested.

LEAF SPOT. Five fungi cause leaf spotting on this host: *Actinothyrium gloeosporioides, Phyllosticta illinoensis, P. sassafras, Gloeosporium fructigenum,* and a species of *Septoria.*

Control. These fungi are rarely serious enough to warrant control measures.

YELLOWS. *Virus.* Bunching and fasciation of the branch tips, leaf rolling, and leaf dwarfing are believed to be caused by a virus.

Control. No control measures are known.

Insects

PROMETHEA MOTH. *Callosamia promethea.* The leaves are chewed by the promethea moth larva, a bluish-green caterpillar that grows to 2 inches in length. The adult female has reddish-brown wings, near the tips of which eye-like spots are visible; the wingspread is nearly 3 inches. The cocoons of this pest are suspended in the trees.

The caterpillars of the hickory horned devil (*Citheronia regalis*), the io

(*Automeris io*), and the polyphemus moth (*Antheraea polyphemus*) also chew sassafras leaves.

Control. Spray with arsenate of lead, DDT, or malathion when the caterpillars are small.

WEEVIL. *Odonopus calceatus.* This small, snout weevil feeds on the leaves as soon at the buds break and before the leaves have expanded, resulting in numerous holes in the leaves. The female adults then deposit eggs on the midribs of the leaves, and the larvae which hatch from these mine into the leaves to cause blotches. Tulip tree and magnolia are also susceptible.

Control. Malathion sprays applied when the adults begin to feed, or before the eggs hatch, should provide control.

SERVICEBERRY (*Amelanchier*)

Serviceberry, also known as Juneberry and shadbush, is an excellent low-growing tree of the Rose family. As such it is subject to many of the diseases and insects common to members of that family.

Diseases

WITCHES'-BROOM. Several species of serviceberry are subject to the witches'-broom disease, which rarely causes much damage.

Symptoms. The production of many lateral branches from a common center results in a bunching or broom-like effect. The lower surfaces of leaves attached to such twigs are covered with sooty fungus growth.

Cause. The fungus *Apiosporina collinsii* causes witches'-broom. It penetrates the twigs in spring and stimulates the production of the many lateral branches. The black growth of the fungus on the leaves contains small, round fruiting bodies in late summer. These produce spores for the following season's infections.

Control. Prune and burn affected leaves and twigs.

LEAF BLIGHT. The fungus *Fabraea maculata* causes a leaf spotting and blighting similar to that produced on hawthorns.

Control. Dithiocarbamate sprays applied at 2-week intervals during the early growing season will control this fungus on valuable specimens.

RUST. More than 10 species of rust occur on serviceberry with alternate hosts on juniper, chamaecyparis, and incense cedar.

Control. Same as for hawthorn rusts.

MILDEWS. Three species of powdery mildew fungi also occur on this host: *Erysiphe polygoni, Phyllactinia corylea,* and *Podosphaera oxyacanthae.*

Control. Sulfur sprays or Mildex (Karathane) will control mildews on particularly valuable specimens.

INSECTS

BORERS. A number of borers including the lesser peach (*Synanthedon pictipes*), apple bark (*Thamnosphecia pyri*), round-headed apple tree (*Saperda candida*), and the shot-hole (*Scolytus rugulosus*) occasionally attack this host.

Control. Keep trees in good vigor by feeding and watering. Spraying the trunk and branches with DDT or malathion at 2- to 3-week intervals during late spring and early summer will control these borers.

PEAR LEAF BLISTER MITE. See under Mountain Ash.

PEAR-SLUG SAWFLY. See under Flowering Cherry.

WILLOW SCURFY SCALE. See under Willow (Chapter 16).

SILK TREE (*Albizzia*)

Silk tree (*Albizzia julibrissin*), also known as mimosa, is used as a street and lawn tree in the South and as an ornamental in protected places along the Atlantic Coast as far north as Boston. Its most destructive disease is wilt, and its most destructive insect the webworm.

DISEASES

WILT. This highly destructive disease is extremely prevalent from Maryland to Florida, where the silk tree is planted most extensively. The author was first to find it in New Jersey at Cape May some years ago and has since found it at The New York Botanical Garden in New York. This is the northernmost location of the disease at this time.

Symptoms. Wilting of the leaves, which soon die, is the most striking symptom. A brown ring of discolored sapwood, usually in the current annual ring of the stem and branches, is another positive symptom. Bleeding along the stems occurs occasionally. Infected trees usually die within a year or so of infection.

Cause. The fungus *Fusarium oxysporium* var. *perniciosum* is known to cause wilt. The fungus lives in soil and spreads in this medium. A closely related species was reported from the West Indies a few years ago.

Control. The only control known is to use wilt-resistant trees. These have been selected by research workers of the Division of Forest Pathology of the United States Department of Agriculture after many years of painstaking research. Resistant individuals are available from several nurseries in the South.

CANKER. A canker of silk tree caused by *Nectria cinnabarina* has been reported from several Middle Atlantic states.

Control. Pruning infected branches is the only control measure known.

Insects

WEBWORM. *Homadaula albizziae.* The larvae of this pest appear on the foliage in early spring. They are also highly destructive to honey locust. The larvae are ⅗ inch long when full grown and are gray-brown, sometimes pinkish, with 5 narrow white lengthwise stripes. At first they feed together in a web, but later spread throughout the tree, tying the leaves in conspicuous masses and skeletonizing them. When mature, they drop to the ground on silken threads and spin cocoons in various kinds of cracks and crevices. The moths, gray with a wingspread of about ½ inch, appear in June to lay eggs on the silk tree flowers and leaves.

Control. When the young caterpillars are just beginning to feed, spray with lead arsenate or with malathion. DDT sprays are not recommended because they occasionally cause severe defoliation.

SCALE. *Pinnaspis minor.* This white, semitransparent scale occasionally infests silk trees in addition to avocado, citrus, hackberry, palms, and other trees and shrubs in the deep South.

Control. Spray with malathion when the young are crawling about.

SILVERBELL (*Halesia*)

Silverbell is subject to few diseases and insect pests.

Diseases

LEAF SPOT. In rainy seasons, brown circular spots are caused by the fungus *Cercospora halesiae.*

Control. This leaf spot rarely becomes serious enough to warrant periodic applications of a copper fungicide.

WOOD DECAY. A decay of the wood, caused by *Polyporus halesiae,* has been reported from Georgia.

Control. Wood-decay fungi are difficult to control. Avoid bark wounds and keep the tree in good vigor.

INSECTS

No insects of importance are reported on silverbell.

SORREL TREE (*Oxydendrum*)

DISEASES

TWIG BLIGHT. *Sphaerulina polyspora.* This fungus occasionally causes blighting of leaves at the tips of the branches. Trees injured by fire or in poor vigor appear to be most subject to this disease.

Control. Pruning infected twigs is usually sufficient to keep this disease under control. Feed and water when necessary to keep trees in good vigor.

LEAF SPOT. Two fungi, *Cercospora oxydendri* and *Mycosphorella caroliniana,* occasionally spot the leaves of sorrel tree. The latter produces reddish or purple blotches with dry, brown centers in midsummer.

Control. Spray with a copper fungicide, starting when the leaves are fully expanded and repeating in 2 weeks.

INSECTS

No insects of any consequence attack this tree.

SMOKE TREE (*Cotinus*)

The smoke tree, a close relative to ordinary sumac, is occasionally used in ornamental plantings. When it is grown in soils where nitrates tend to become deficient, its leaves show red spots or blotches. The application of a commercial fertilizer high in nitrogen usually helps to overcome this trouble. Only one serious disease, wilt (p. 247) affects this tree.

DISEASES

RUST. Two rusts have been reported on this tree: *Puccinia andropogonis* var. *onobrychidis* and *Pileolaria cotinicoggyriae.* The latter, found in Georgia several years ago, produces conspicuous spots with hypertrophied centers surrounded by dead tissue.

Control. These diseases are not serious enough to warrant control measures.

LEAF SPOT. Leaves are spotted by the fungus *Cercospora daleae*. The fungus has been found only in the Middle West.

Control. Valuable specimens can be protected with copper fungicides.

Insects

OBLIQUE-BANDED LEAF ROLLER. *Archips rosaceana*. The leaves may be mined and rolled in June by pale yellow larvae. The adult moth is reddish-brown and has a wingspread of 1 inch, with the front wings crossed by 3 distinct bands of dark brown.

Other trees subject to this pest are apple, apricot, ash, birch, cherry, dogwood, hawthorn, horsechestnut, linden, maple, oak, peach, pear, plum, and poplar.

Control. Spray early in June with lead arsenate or malathion.

TREE-OF-HEAVEN (*Ailanthus*)

Tree-of-heaven is never used as a streetside tree but thrives in the larger cities where no other tree will grow. It frequently sprouts in parks and yards of large cities. Only the seed-bearing kinds should be planted, as the male or pollen-bearing form has an offensive odor when it blooms.

Diseases

WILT. *Verticillium albo-atrum* (p. 247) is the most destructive fungus parasite of this host. Many trees have been killed by the fungus in the Philadelphia-New York environs in the last 15 years.

SHOESTRING ROOT ROT. *Armillaria mellea* (p. 243) is another fungus that has been destroying this tree in the northeastern United States in recent years.

LEAF SPOT. Three leaf-spotting fungi, *Cercospora glandulosa*, *Phyllosticta ailanthi*, and *Gloeosporium ailanthi*, occasionally attack this tree.

Control. Control measures are never applied for leaf spots.

TWIG BLIGHT. *Fusarium lateritium*. This fungus also causes branch and twig cankers on Japanese Pagoda Tree (page 279).

Insects

CYNTHIA MOTH. *Samia cynthia*. The larva of the Cynthia moth is one of the few insects known to feed on the tree-of-heaven. Trees may be completely defoliated. The mature larva is about 3½ inches long, light green, and covered with a glaucous bloom. The adult moth is beautifully colored and has a wingspread of 6 to 8 inches (Fig. 108). A crescent-shaped white marking is present in the center of each of the grayish-brown wings.

FIG. 108. Cynthia moth. The larval stage of this insect feeds voraciously on tree-of-heaven foliage.

Control. Spray with arsenate of lead or malathion in June when the caterpillars begin to chew the leaves.

AILANTHUS WEBWORM. *Atteva aurea.* Olive-brown caterpillars with 5 white lines feed in webs on the leaves in August and September. Adult moths have bright orange forewings, with 4 crossbands of yellow spots on a dark blue ground.

Control. Spray with arsenate of lead or malathion when the young larvae appear in August.

16

DISEASES AND INSECT PESTS OF TALL-GROWING TREES

The diseases and insect pests of tall-growing deciduous trees are treated here. As discussed in Chapter 8 most of these are best planted along wide streets, in parks and larger properties.

ALDER (*Alnus*)

Most species of alder grow best in moist soils. Those grown in ornamental plantings are less subject to diseases than to insect pests. The few diseases that occasionally appear produce little permanent damage.

Diseases

LEAF CURL. The leaves of red alders grow to several times their normal size, are curled and distorted, and turn a decided purple when attacked by the fungus *Taphrina macrophylla*. Infection is apparent on the leaves as soon as they appear in the spring.

Three other species, *T. amentorium*, *T. occidentalis*, and *T. robinsoniana*, are known to cause enlargement and distortion of the scales of female catkins. These project as curled reddish tongues, which are soon covered with a white glistening layer of fungus tissue.

Control. Spray the trees in late fall or while they are still dormant in the spring with 1 part of concentrated lime sulfur in 10 parts of water. This will destroy most of the spores, which overwinter in the bud scales and on the twigs, and which are largely responsible for early infections. Sprays are suggested only for valuable specimens.

POWDERY MILDEW. Several species of powdery mildew fungi attack the female catkins of alder. The most common, *Erysiphe aggregata,* develops as a white powdery coating over the catkin. Two other species, *Microsphaeria alni* and *Phyllactinia corylea,* occur less frequently.

Control. These fungi rarely cause enough damage to warrant control measures. Where justified, Mildex or Karathane sprays will give control.

CANKER. A number of fungi cause cankers and die-back of the branches. Among the most common are *Nectria coccinea, Solenia anomala,* and *Physalospora obtusa.* Trunk cankers may be caused by *Diatrypella oregonensis* and *Hymenochaete agglutinans.*

Control. No simple control measures are known. Cankered branches should be pruned to sound wood and the trees fed and watered to increase their vigor.

INSECTS

ALDER BLIGHT APHID. *Prociphilus tessellatus.* The downward folding of leaves, in which are found large woolly masses covering bluish-black aphids, is typical of this pest (Fig. 109). Eggs are deposited in bark crevices and pass the winter in these locations. Maples are also infested by this species.

Control. Malathion or lindane sprays when the young begin to develop in spring give excellent control.

ALDER FLEA BEETLE. *Altica ambiens.* The leaves are chewed during July and August by dark brown larvae with black heads. The adult, a greenish-blue beetle ⅕ inch long, deposits orange eggs on the leaves in spring.

Control. Arsenate of lead, DDT, or malathion sprays when the larvae begin to feed will control this heavy feeder.

ALDER LACE BUG. *Corythucha pergandei.* This species infests not only alder, but occasionally birch, elm, and crabapple.

Control. Spray with lindane or malathion in late spring.

ALDER PSYLLID. *Psylla floccosa.* This sucking insect is common on alders in the Northeast. The nymphal stage produces large amounts of wax. When psyllids are massed on the stems they resemble piles of cotton.

Control. Spray with malathion when the nymphs or adults are seen.

BEECH (*Fagus*)

Beeches are among our most beautiful trees, their dormant shapes and winter bark contribute much to any landscape.

The American beech, *F. grandifolia,* is a handsome, low-branched, slow-growing tree. Its dense foliage and shallow rooting habit, however, make diffi-

FIG. 109. Alder blight aphid.

cult the maintenance of a lawn under it. It grows best in cool, moist soils and cannot withstand constant trampling over the soil nor does it withstand city conditions very well. Under ordinary conditions it is extremely difficult to transplant.

The European beech, *F. sylvatica,* is similar to the American species except that its bark is darker gray and its foliage more glossy. Its many horticultural varieties include purple beech, variety *atropunicea,* with deep purple leaves; weeping beech, variety *pendula,* one of the world's most beautiful weeping trees; weeping purple beech, variety *purpureo-pendula;* and cut-leaf, variety *laciniata,* with very beautiful, narrow, almost fern-like leaves. Other new or interesting kinds are Darwyck beech, *F. sylvatica fastigiata;* golden-leaved beech, variety *aurea;* a red fern-leaved beech, variety *Rohanni;* and finally the tricolor beech with pink, white, and green leaves that turn a coppery bronze.

Diseases

The dying of a large number of American beeches is believed by some investigators to be associated with drought and subsequent death of the roots. Associated with the progressive weakening of some of the trees, however, is a trouble known as leaf mottle or scorch.

LEAF MOTTLE. This disease was first reported by Dr. W. H. Rankin in 1928 as being prevalent on American beech in the vicinity of Philadelphia. Since that time it has appeared somewhat sporadically, although it has been unusually abundant in the eastern United States during the last 15 years.

SYMPTOMS. In the spring, small translucent spots surrounded by yellowish-green to white areas appear on the young unfurling leaves. These spots turn brown and dry, and by the first of June the mottling is very prominent, especially between the veins near the midrib and along the outer edge of the leaf. Within a few weeks the brown areas increase in number until the entire leaf presents a scorched appearance. A considerable part or, in some instances, all of the leaves then drop prematurely. Where complete defoliation occurs, new leaves begin to develop in July. The second set of leaves in such cases appears quite normal and drops from the trees at the normal time in the fall.

The extensive loss of foliage early in the season exposes the branches to the direct rays of the sun. Because beeches are rather sensitive to such rays, considerable scalding of the bark occurs. Such injury predisposes the tree to attacks by the two-lined chestnut borer and possibly by other insects. Heavy infestations of borers will kill large branches or even the entire tree.

Cause. The cause of leaf mottle is not known. Investigations by the author many years ago failed to reveal definite clues. Site, age of tree, water supply, or atmospheric conditions do not appear to be correlated with the incidence or severity of the trouble. For example, trees growing in open lawns appear to be as subject as trees in the woods; young trees seem to contract the trouble as readily as old ones; trees growing in light, well-drained soils apparently are as susceptible as those in heavy moist soils; and the disease appears to be prevalent in wet seasons as in dry ones. The symptoms are not typical of any known nutrient deficiency, and injections of magnesium nitrate and ferrous sulfate into the trunks have failed to produce any improvement.

Dr. Spencer Davis of Rutgers University has reported that leaf mottle is actually a form of leaf scorch. Trees that exhibit mottling have a different protoplasm or genetical background than those that are not mottled, according to Dr. Davis. Hence one tree responds differently from the other to the same soil conditions.

Control. Until more is known about the causal nature, the only recommendation that can be made is to provide adequate fertilization. The bark of particularly valuable specimens that are completely defoliated might well be protected from the midsummer sun with burlap or some other material until the second set of leaves has developed sufficiently to provide the necessary shade.

BLEEDING CANKER. This disease, first reported on maples in Rhode Island about 1939, is known to occur on elm, oak, almond, apricot, cherry, plum, and peach. The name suggests the most common symptom visible on beech, maple, and elm, an oozing of a watery light brown or thick reddish-brown liquid from the bark.

Cause. The fungus *Phytophthora cactorum* causes bleeding canker. On dogwood it causes the disease crown canker, and it causes trunk cankers on fruit trees.

Control. No effective preventive measures are yet known. Infected specimens should be cut down and burned to prevent spread to nearby trees. Mildly affected trees have been known to recover.

BEECH BARK DISEASE. The natural stands of beech in eastern Canada and northern Maine have been severely damaged by this disease in the last 30 years. Although it is primarily a disease of forest beech, it is potentially dangerous to ornamental beech. In 1956 the disease was known to exist in Ulster and Sullivan counties of New York State.

Cause. The combined attack of the woolly beech scale, *Cryptococcus fagi* and the fungus *Nectria coccinea* var. *fagi* cause this disease.

Infestations of the scale on the bark always precede those of the fungus. During August and September, countless numbers of minute, yellow, crawling larvae appear over the bark. By late autumn they settle down and secrete a white fluffy material over their bodies. This substance is very conspicuous, and the trunks and branches appear as though coated with snow. Through the feeding punctures of the insect, the Nectria fungus then penetrates the bark and soon kills it. The insects soon die because of the disappearance of their source of food.

Death of the infected bark is followed by drying; infected areas on the trunk thus are depressed and cracked. Eventually, deeply sunken cankers are formed, which assume a more or less circular or oval shape. The destruction of the bark and other tissues of the tree leads to a progressive dying of the top. Late in the season the leaves usually curl and turn brown, the twigs die, and new buds fail to form. The dead leaves remain attached throughout the winter. In the spring, affected branches fail to produce foliage, and other branches, lacking reserve food materials, produce small yellow leaves that usually die during the summer. Eventually, the entire tree dies.

Control. Because of infestations of the woolly beech scale must precede fungus penetration, the eradication of the insect pest will prevent the start of the disease. Malathion sprays applied to the trunk and branches of valuable ornamental trees growing in the infested areas in early August and in September should control the young scales. A dormant lime sulfur spray applied to the trunk and branches will control overwintering adult scales. Oil sprays are also effective but are not reliably safe on beech trees.

Insects

BEECH BLIGHT APHID. *Prociphilus imbricator.* The bark is punctured and juices are extracted by the beech blight aphid, a blue insect covered with a white cottony substance. The pests also feed on the leaves.

Control. Spray with lindane or malathion when the pests first appear in the spring.

WOOLLY BEECH APHID. *Phyllaphis fagi.* Leaves are curled and blighted by the woolly beech aphid, a cottony-covered insect, the cast skins of which adhere to the lower leaf surface. The purple beech is more commonly infested than the American beech.

Control. Before the leaves are curled, spray with the same formula as that recommended for the beech blight aphid.

BROWN WOOD BORER. *Parandra brunnea.* Winding galleries in the wood, made by the white-bodied, black-headed borers, 1¼ inches long, and tiny holes in the bark, made by emerging shiny, brown beetles ¾ inch long, are typical signs of brown wood borer infestation. The eggs are deposited in bark crevices or in decayed wood.

Control. Infestations can be prevented to a large extent by avoiding mechanical injuries to bark and wood and by treating open wounds.

BEECH SCALE. *Cryptococcus fagi.* White masses on the bark of the trunk and lower branches, composed of tiny circular scales, 1/40 inch in diameter, are the common signs of beech scale infestation. The eggs are deposited on the bark in late June and early July, and the young crawling stage appears in August and September. The pest overwinters as a partly grown adult scale. Beech scale is primarily a pest of forest trees and is mentioned here only because of its association with the beech bark disease. In 1956 this pest was present in 12 New York State counties.

Control. See under beech bark disease.

TWO-LINED CHESTNUT BORER. *Agrilus bilineatus.* Trunks and branches of trees defoliated by the leaf mottle disease or of those weakened by other causes are subject to attacks by the two-lined chestnut borer. Cankers of varying sizes appear on the sides of the branches exposed to the sun's rays. A dark red liquid exudes from a small puncture in the center of such cankers. The formation of several cankers along a branch reduces the sap flow and weakens the distal portions of the infested branch. A description of the pest is presented under Oaks on page 352.

SCALES. In addition to the scale mentioned above, beech trees are subject to the following scales: black, cottony-cushion, European fruit lecanium, oyster shell, Putnam, and San Jose.

Control. Spray with malathion when the young scales are crawling about and with dormant lime sulfur in early spring.

APPLE TREE TENT CATERPILLAR. See under Willow.

DATANA CATERPILLAR. See under Oak.

GYPSY MOTH. See under Elm.

CATALPA (*Catalpa*)

Two species of catalpa, common and western, are used in ornamental plantings. Common catalpa (*C. bignonioides*) should be planted only where its flowers and fruits are not objectionable. Western catalpa (*C. speciosa*) is hardy and grows rapidly. Several leaf spots and powdery mildews in addition to the Verticillium wilt disease (p. 247) occur on these hosts. Insects are usually more of a problem than the diseases.

Diseases

LEAF SPOT. A common disease of catalpa, leaf spot is most prevalent during rainy seasons. It was prevalent in the eastern United States in 1958 because of the frequent rains in late spring and early summer.

Symptoms. Tiny water-soaked spots, scattered over the leaf, appear in May. The spots turn brown and increase in size until they attain a diameter of about ¼ inch. Holes in the leaves are common as a result of dropping out of the infected tissue. Where the spotting is unusually heavy, the leaves may drop prematurely.

Cause. Three fungi, *Phyllosticta catalpae*, *Alternaria catalpae*, and *Cercospora catalpae*, are frequently associated with these spots. Injury by the catalpa midge and infection by bacteria are believed to increase the susceptibility to leaf spots.

Control. Gather and burn the leaves in the fall. Spray valuable susceptible trees three times with bordeaux mixture or any other copper fungicide, first, as the leaves unfurl, then when the leaves are half-grown, and again when full grown.

POWDERY MILDEW. Two species of mildew, *Microsphaeria alni* var. *vaccinii* and *Phyllactinia corylea*, attack catalpas.

Control. Spray valuable trees with either wettable sulfur or Mildex (Karathane).

WOOD DECAY. The western catalpa is extremely susceptible to heartwood decay caused by the fungus *Polyporus versicolor*. The heartwood becomes straw-yellow, light, and spongy. Another fungus, *Polyporus catalpae*, causes decay of the trunk near the soil line. The wood becomes tough, brown, brittle, and full of cracks.

Control. Avoid wounds, inasmuch as the spores of both fungi enter through bark injuries. Keep the trees in good vigor by fertilizing and watering.

INSECTS

MEALY BUG. *Pseudococcus comstocki.* Distorted growth of twigs, limbs, and trunk may be produced by the sucking of a small, elliptical, waxy-covered insect, the mealy bug. The leaves may be covered with black, sooty mold, which develops on so-called honeydew secreted by this pest. Winter is passed as eggs in masses of white, waxy secretions on the bark.

Control. Before the buds open, spray with 1 part concentrated lime sulfur in 10 parts of water, or with winter strength miscible oil. A malathion spray applied to the trunk and branches in late spring and again in early summer will provide additional control.

MIDGE. *Itonida catalpae.* Leaves are distorted and circular areas inside the leaves are chewed, leaving a papery epidermis, as a result of infestation by tiny yellow maggots. The adult, a tiny fly, with a wingspread of $\frac{1}{16}$ inch, appears in late May or early June, to lay eggs on the leaves. Winter is passed in the pupal stage in the soil.

Control. Cultivate the soil beneath the trees to destroy the pupae, and spray in late May with malathion or a mixture of malathion and DDT.

SPHINX. *Ceratomia catalpae.* Leaves may be completely stripped from a tree by a large yellow and black caterpillar, the sphinx, which attains a length of 3 inches. The adult female is a grayish-brown moth with a 3-inch wingspread. The winter is passed as the pupal stage in the ground.

Control. Spray the foliage with lead arsenate, DDT, or malathion early in May and again in mid-August.

CHESTNUT (*Castanea*)

DISEASES

BLIGHT. The rapid disappearance of one of our best forest, ornamental, and nut trees, the American chestnut, as a result of infection by one of the most virulent tree parasites, is too well known to warrant much discussion in this book. Despite tireless effort and tremendous monetary expenditures, dead and diseased chestnut trees are all that remain of the losing battle man has waged to check this invader. Despite the lack of many optimistic signs, a few individuals maintain that it is yet too early to predict the ultimate fate of this valuable tree. They point out that cankers on diseased trees appear to develop less rapidly than in former years and that more and more partly healed cankers are visible. Other hopeful signs, they assert, are the fact that sprouts continue to develop at the base of dead stumps and from the roots, some attaining a height of more than 30 feet and a trunk diameter of 8 or more inches, and the fact that thousands of other sprouts, though not attaining so large a size, are able to produce nuts before they succumb.

No one will dispute the statement that the chestnut blight disease has done more than any other single factor in American history to make the public tree-conscious. Within the span of 50 years many persons have witnessed the passing of this irreplaceable tree. Believed to be of minor importance when first reported by the late Hermann Merkel, who found a few infected trees in the New York City Zoological Park in 1904, the disease proceeded to wipe out the chestnut stands in New England forests and along the eastern slope of the Allegheny and Blue Ridge mountains, the principal range of this host. Today some chestnuts still stand in the extreme southern and western parts of this tree's natural range: in Tennessee, Georgia, and South Carolina. It is safe to say, however, that they too will soon suffer the same fate as their northern kin, for blight has been reported in these states, as well as in many midwestern and several far-western states.

Though blight is essentially a disease of the American chestnut, it also occurs commonly on the chinquapin (*Castanea pumila*). The causal fungus has been found growing on red maple, shagbark hickory, and staghorn sumac (*Rhus typhina*), and on dead and dying white, black, chestnut, and post oaks.

Symptoms. The only evidences of the once magnificent chestnut trees are the tremendous barkless trunks that still stand among living trees of other species, or the rotting stumps around whose edges vigorous living sprouts and small trees continue to appear. Many of the young trees seem to escape the disease for the first half dozen years of their lives. Whether this is owing to their smooth bark, which is relatively free from injury during this period, or whether they have some degree of resistance in infancy, is not definitely known. Once the destructive fungus gains a strong foothold, however, the leaves on one branch, or several branches, or even on the entire young tree suddenly wilt, turn brown, and hang dry on the branches (Fig. 110). The death of the leaves and of the stems on which they are attached results from the death and girdling of some portion below. When the infection occurs rather high in the tree, only the distal parts are affected. When it develops at the base near the soil line, the entire top of the tree dies.

Close inspection of the lower parts of dying twigs, branches, or trunks reveals the presence of cankers. These are discolored, slightly sunken areas in the bark. Layers of flat, fan-shaped, buff-yellow wefts of fungus tissue are usually revealed between the bark and the sapwood, when the bark in the cankered area is lifted or peeled carefully.

FIG. 110. Chestnut blight on young sprouts. The leaves hang dry and brown on girdled stems. Note the stump of the original tree.

Cause. Blight is caused by the fungus *Endothia parasitica.* When a considerable bark area has been invaded, tiny, pin-point fruiting bodies, the tips of which barely protrude from the bark, are developed. In damp weather, sticky, yellowish-orange masses of spores ooze from the openings in these bodies. The spores are splashed by rain or are carried by birds and by crawling and flying insects to wounds in the bark below or to nearby trees, where they cause new infections. A second type of fruiting structure is also developed. This is a small, black, flask-shaped body embedded in the bark with a relatively long beak protruding above the surface. The spores from this type are shot into the air and are blown for many miles by the wind. In fact, spore traps placed on high buildings in New York City, when the epidemic was at its peak years ago, caught many spores of this type. The wind-blown spores lodge in open wounds on chestnut trees and produce new infections when conditions are favorable. The mycelium, the vegetative stage of the fungus, penetrates the bark, cambium, and sapwood with considerable rapidity and soon completely girdles the infected twig, branch, or trunk. Chestnut blight has been unchecked by man because of his inability to prevent the dissemination of spores by wind, birds, and insects.

Control. Although many suggestions and recommendations have been published, none has proved effective in controlling chestnut blight. Persistent efforts have been made to find some chemical, that, when injected into the tree, would check the development of cankers. Numerous claims for such chemicals have been advanced, but the author is unaware of the existence of any really effective material at the present time.

After 40 years of research in this country, no American chestnut has been found with sufficient resistance to be of practical value. Large blight-free American chestnuts were known to exist in 36 states in 1957. Only time will reveal whether any of these are definitely resistant.

When research scientists finally realized that the chestnut blight fungus was uncontrollable, substitutes were sought. The answer seemed to be the introduction of Asiatic chestnuts which were known to be resistant to blight.

Although a few of these chestnuts were introduced as long ago as 60 years, the greatest numbers were introduced in the late 1920's. But the introduction of these has resulted in the appearance of other diseases such as the blossom-end rot of the nuts caused by the fungus *Glomerella cingulata* and twig canker discussed below.

Asiatic chestnuts (*C. japonica* and *C. mollissima*) are now widely available from many nurseries. Most of them are best suited for ornamental plantings and nut production, and not as forest trees. Many have a shrub-like growth habit, with multiple trunks arising near the ground level. They do not grow so tall and straight as the American chestnut and cannot compete in wooded areas when interplanted with other trees.

TWIG CANKER OF ASIATIC CHESTNUTS. The increased use of Asiatic species of chestnuts, because of their resistance to blight, has revealed that these species are susceptible to a less destructive disease, twig blight.

Symptoms. Cankers on trunks, limbs, or twigs result from fungus invasion of the bark, cambium, and sapwood. Cankers on the branches may cause complete girdling, followed by death within a single season. When complete girdling does not take place in one season, the canker may callus over temporarily, and the infection then may continue the following season, to cause death of the parts above.

Cause. Several fungi have been associated with this disease, the most common and virulent one being *Cryptodiaporthe castanea.* It has been found to kill young trees in nurseries and older trees in permanent plantings, but more commonly it kills individual branches, thus decreasing growth and deforming the trees.

Control. Twig blight is most prevalent on trees in poor vigor. Maintaining good vigor will do much to ward off attacks. Planting sites should be carefully selected, and fertilization and watering practiced to ensure vigorous growth. All unnecessary injuries to the trees should be avoided, since the causal organisms penetrate and infect most readily through bark wounds. Trees should be carefully inspected in early summer, when symptoms are most apparent, and affected twigs should be pruned to sound wood and burned. Large cankers on the trunk or larger branches should be removed by surgical methods.

INSECTS

BROWN WOOD BORER. See under Beech.

DATANA CATERPILLAR. See under Oak.

JAPANESE BEETLE. See under Linden.

LECANIUM SCALE. See under Oak.

MAPLE AND OAK TWIG PRUNER. See under Maple (Chapter 15).

SPRING CANKERWORM. See under Elm.

TUSSOCK MOTH. See under Elm.

TWO-LINED CHESTNUT BORER. See under Oak.

ELM (*Ulmus*)

Diseases

DUTCH ELM DISEASE. Much has been written in the last 25 years both in the United States and in Europe on the Dutch elm disease, the most destructive fungus disease attacking elms. The misleading name given the disease merely refers to the Netherlands, where it was first identified in 1919. The disease is believed to have entered the United States in the late 1920's on burled elm logs from Europe. After killing literally thousands of elms in the eastern United States, it has spread to the Middle West and to Canada. A few diseased trees were discovered in Denver, Colorado, in 1948 but the disease is still largely concentrated in the northeastern United States.

Symptoms. Wilting leaves on one or more branches followed by yellowing, curling, and dropping of all but a few of the leaves at the branch tips about midsummer are the first outward symptoms. The disease is often recognizable in winter by the tuft of dead brown leaves adhering to the tips of curled twigs. When diseased twigs or branches are cut, spots or flecks are visible in the sapwood near the bark (Fig. 111). A longitudinal section of the diseased twig shows long brown streaks following the grain of the wood.

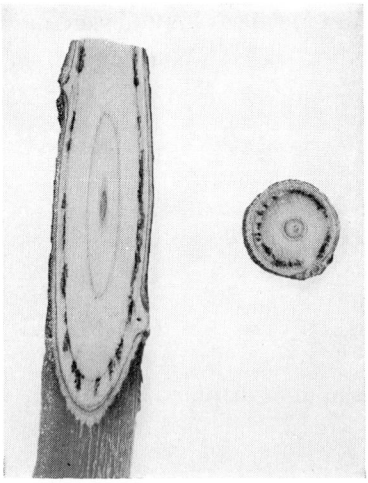

FIG. 111. LEFT: Discoloration in twig resulting from infection by the Dutch elm disease fungus, *Ceratocystis ulmi*.

FIG. 112. RIGHT: The Dutch elm disease fungus as it appears in pure culture when isolated from infected elm twigs.

Discoloration in the wood also occurs in several other less destructive elm diseases. Consequently, a positive diagnosis can be made only after the fungus responsible for the discoloration has been isolated. Plant pathologists at the state agricultural experiment stations are equipped to make a diagnosis, provided adequate twig specimens showing internal discolorations are submitted to them.

Cause. The Dutch elm disease is caused by the fungus *Graphium ulmi* (Fig. 112), the perfect or sexual stage of which is *Ceratocystis ulmi.*

Several species of insects are largely responsible for the spread of the fungus from diseased to healthy trees. Among the most common vectors in this country are the smaller European elm bark beetle, *Scolytus multistriatus*, and the native elm bark beetle, *Hylurgopinus rufipes.* A number of other boring insects are also known to transmit the causal fungus, or are suspected of transmitting it. The fungus may occasionally be spread by rain water and through grafting of roots of a diseased with a healthy elm. Other suspected means of spread are wind-blown spores, birds, and pruning tools, although no clear-cut evidence is available.

The fungus penetrates the tree only through wounds. Once inside the wood, it can spread rapidly either by spores developed in the wood vessels and carried in the sap stream, or by growth of the fungus hyphae. Death of affected branches is believed to be due to toxins produced by the fungus or to lack of water as a result of the plugging of the vessels by materials formed in the process of fungus invasion.

Control. After almost 25 years of research, no positive control has been developed. Control is indirect, that is, against the insects, bark beetles, which are the principal disseminators of the fungus.

One way to control the insect carriers is to search out and destroy their breeding places. These include all dead, dying, or devitalized elm material such as sick trees, hurricane-damaged ones, broken limbs, elmwood piles, and elm fence posts. To do a good job such material must be removed from a relatively large area because the smaller European elm bark beetle, the principal disseminator in this country, can fly more than 3 miles and frequently does. The long flights are usually made in search of suitable breeding places, but these same beetles may also feed on living trees. Such beetles have been found to carry viable spores of the Dutch elm disease fungus for more than 2 miles. The elm bark beetles breed and feed in all species of elms that grow in this country.

The best way to prevent the movement of these beetles is to use a residual-type contact insecticide on elm trees. The emulsion-type sprays containing DDT are most effective. But coverage is fully as important as spray concentration. Good coverage can be achieved only with adequate equipment handled by skilled operators under proper weather conditions.

For hydraulic sprayer applications, 3 gallons of 25 per cent emulsifiable DDT in 100 gallons is the standard dosage. For mist-blower application, 1 gallon of 25 per cent emulsifiable DDT plus 1 gallon of water is the usual ratio.

Two applications are necessary, one before the leaves appear (by late March or early April) and the second about mid-July. The prefoliar (or dormant spray) is by far the more important because the elms are most susceptible to the disease in spring and early summer. Both sprayings should cover the entire outside surface of the tree, especially the crotches of small twigs and branches. The second spraying should contain half as much DDT as the prefoliar spray.

The excessive use of DDT has led to other problems such as an increase in the population of Putnam scale, mites, and aphids. These pests can be overcome, however, by combining other insecticides and miticides with the DDT.

Frequent reports appear in the press about a chemical which will prevent infection or cure diseased trees. According to research scientists most closely associated with the problem, nothing has been developed, as of October 1958, which could be considered either as a cure or as a definite control of carriers. For all practical purposes as thorough a sanitation program as possible is perhaps the most important control practice. Spraying with DDT during the late dormant period provides a degree of protection to extremely valuable trees, but it is not recommended for community-wide projects because of the very high costs and the additional pest and injury complications.

Greatest hope for eventual control lies in the discovery of American elms that have natural immunity to the disease. A number of such individuals are being watched carefully by United States Department of Agriculture scientists.

Several other species of elms are known to be resistant to the disease. The Christine Buisman, a selection of the European smooth-leaved elm, *U. carpinifolia*, is said to be resistant to both Dutch elm disease and phloem necrosis (p. 319). It was released to nurseries for further propagation and sale in 1949, 1950, and 1951 and is now available from the Willis Nursery Company, Ottawa, Kansas, which specializes in shade trees. It has good foliage and growth characters but grows more slowly than the American elm. It is rather susceptible to elm leaf beetles and is not reliably hardy in very cold areas like northern Iowa and Minnesota. The Hansen Manchurian elm, like most Asiatic elms, is decidedly resistant to the disease. A close relative of the American elm, Japanese keaki (*Zelkova serrata*) is somewhat more resistant to the disease than is the American elm. It is vase-shaped and has bark resembling beech and elm-like foliage that turns red in the fall.

PHLOEM NECROSIS. Even more deadly to elms is the disease known by the odd name phloem necrosis. First identified as a virus disease in 1938 following investigations of the dying of many elms in Ohio, the disease is now present throughout the Middle West. Its eastern limits are along a line connecting Marietta, Ohio; Charleston, West Virginia; and Chattanooga, Tennessee. With the appearance of the disease in the Tennessee city, arborists in the Northeast, where the disease has not yet appeared, are becoming concerned. Chattanooga is beyond an area of sparse elm growth which, it had been hoped, would serve as a barrier.

Symptoms. The earliest symptoms appear in the extreme top of the tree, at the outer tips of the branches. Here, the foliage becomes sparse, and the leaves droop, because of downward curvature of the leaf stems. Individual leaves curl upward at the margin, producing a trough-like effect that makes them appear narrow and grayish-green. They are often stiff and brittle. The leaves then turn yellow, and fall prematurely. These symptoms appear throughout the tree and are not confined to one or several branches as are wilt disease infections.

In advanced stages of the disease, the small fibrous roots and eventually the larger ones die. One of the most typical signs is the discoloration of the phloem tissue which precedes the death of the larger roots. The discoloration frequently extends into the trunk and branches. The cambial region becomes light to deep yellow, and the adjacent phloem tissue turns yellow, then brown with small black flecks scattered throughout. Following this, the phloem tissues are browned and killed. Moderately discolored phloem has an odor resembling wintergreen.

In chronic cases of phloem necrosis, there is a gradual decline over a 12- to 18-month period before the tree dies. In acute cases, an apparently healthy and vigorous tree may wilt and die within 3 to 4 weeks.

Cause. Phloem necrosis is caused by an infectious principle known as a virus. It can be transmitted experimentally by grafting patches of diseased bark, scions, or roots on healthy trees. In nature the infectious principle is transmitted by the elm leafhopper *Scaphoideus luteolus*. Actually there are two strains of this insect in the central states where the disease is most concentrated, but only one is capable of transmitting the virus.

Control. DDT sprays will control the insect vector. Unfortunately even with nearly perfect control of the insect, only a few viruliferous ones which are missed by the spray may spread the disease.

The Christine Buisman elm mentioned under Dutch elm disease control has thus far proved to be highly resistant to phloem necrosis. As with the other disease, claims of a preventive or cure for the virus disease have no sound experimental evidence to back them.

MOSAIC VIRUS. This virus disease, which causes yellow mottling of leaves, is rather rare and relatively harmless. It can be transmitted from diseased to healthy trees through both pollen and pistils. No control is known.

DOTHIORELLA WILT. Two common diseases of elms are indistinguishable in the field from the Dutch elm disease. They are known only by their scientific names, Dothiorella wilt, sometimes called Cephalosporium wilt, and Verticillium wilt. The latter is more severe on maples (see p. 282). The former is confined largely to elms. Though not so destructive as the Dutch elm disease, Dothiorella wilt occurs over a much larger area.

Symptoms. The early symptoms are a drooping and wilting of the leaves, preceded or accompanied by a distinct yellowing. The leaves soon become brown, roll upward along the midrib, and drop prematurely. The disease progresses rapidly above the point of infection, but advances very slowly in a downward direction, resulting in extensive formation of suckers below the infected regions.

The internal symptoms usually precede the external. The wood, particularly the sapwood, of affected twigs and branches, shows brown streaks resembling those produced by the Dutch elm disease. Because both external and internal symptoms so closely resemble those of the Dutch elm disease, positive differentiation can be made only by isolating and determining the causal fungus.

*Cause.** Wilt and die-back are caused by the fungus *Dothiorella ulmi*. Small black fruiting bodies of the fungus develop on affected twigs and branches during the summer. Sticky masses of spores are released from these

* Several other fungi are also associated with wilting and dying back of elms. Among these are fungi belonging to the genera *Coniothyrium*, *Cytosporina*, *Cytospora*, *Phomopsis*, and *Sphaeropsis*.

bodies the following spring and are spread chiefly by rain and wind, though insects may also be involved in dissemination. Infection usually occurs through wounds made by insects or other agents. The fungus then moves down the leaf into the petiole and eventually into the wood vessels. Trees can be infected also through the roots, and, when such infections occur, the trees die very rapidly. Diseased leaves falling to the ground may provide inoculum for such root infections.

Control. Severely infected trees should be removed and burned. Mildly infected ones should be pruned heavily to remove as much of the diseased wood as possible. Because the fungus may develop internally well beyond the area of the external symptoms, pruning does not always produce the desired results. Heavy fertilization may help mildly diseased trees to recover.

The number of leaf infections can be reduced by applying combination sprays containing a fungicide and an insecticide. Fertilization is suggested as a general precautionary measure, despite the fact that there seems to be no correlation between the vigor of the tree and its susceptibility to the disease.

CANKER. A number of types of canker occur on the bark of elms. They are associated with fungi belonging to the genera *Diaporthe, Coniothyrium, Diplodia, Phoma,* and *Botryosphaeria.*

Many small cankers can be eradicated by surgical means. The cuts should extend well beyond the visibly infected area to ensure complete removal of fungus-infested tissue. The edge of the wound should be painted with shellac and the entire exposed area finally dressed with a good tree paint.

LEAF SPOT. Several leaf spot diseases occur commonly on elms. The fungus *Gloeosporium inconspicuum* produces brown, irregular spots on the leaves, and *G. ulmicolum* kills the leaf tissue along the veins and margins of the leaves. Probably the most prevalent leaf spot, however, is that caused by the fungus *Gnomonia ulmea,* the first symptom of which appears early in spring as small white or yellow flecks on the upper leaf surface. The flecks then increase in size, and their centers turn black (Fig. 113). Where infections occur early and are heavy, the leaf may drop prematurely. Usually, however, the disease becomes prevalent in late fall about the time the leaves drop normally, and consequently, little damage to the tree occurs.

Control. Gather and burn the fallen leaves in autumn. Bordeaux mixture, or phenyl mercury sprays such as Puratized Agricultural Spray, applied three times at 10- to 12-day intervals starting when the leaves are half-grown, will give good control of leaf spots.

OTHER LEAF DISEASES. Several powdery mildew fungi and one leaf-blister fungus occasionally occur on elms, but rarely cause any damage. These can be checked by following the recommendations made for similar diseases on other trees.

FIG. 113. Leaf spot of elm caused by the fungus *Gnomonia ulmea*.

WETWOOD. A term applied to certain forms of wilt, branch die-back, and internal and external fluxing of elms. It occurs in American, Moline, Littleford, English, Siberian, and slippery elms.

Symptoms. Elms affected by wetwood have dark, water-soaked malodorous wood. The condition is usually confined to the inner sapwood and heartwood in trunks and large branches. Little or no streaking occurs in the outer sapwood, and no discoloration is seen in the cambial region or phloem.

Cause. Wetwood is caused by the bacterium *Erwinia nimipressuralis*. In the course of their development the bacteria produce gas, which often creates pressures ranging from 5 to 20 pounds in the tree.

Control. Injection of such chemicals as mercuric chloride, copper sulfate, silver nitrate, and 8-hydroxyquinoline benzoate into infected trees has failed to control the disease. Although no effective control for the disease has been found, some relief to fluxing trees is possible by installing drains in the trunk. Such drains do not reduce the infection but provide a way for the sap and toxic gas to escape. (See under slime flux, p. 165).

WOOD DECAY. A number of fungi are associated with decay of elm wood. They cannot be checked once they have invaded large areas of the trunk. Many can be prevented from gaining access to the interior of the tree, however, by avoiding bark injuries, by properly treating injuries that occur, and by keeping the tree in good vigor by fertilization and watering.

INSECTS

LEAF BEETLE. *Galerucella xanthomelaena.* Two distinct types of injury are produced by this pest. Soon after the leaves unfurl in spring, rectangular areas are chewed in them by the adult beetles, brownish-yellow insects, ¼ inch

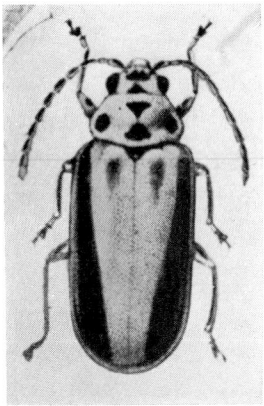

FIG. 114. LEFT: Adult elm leaf beetle, which occasionally becomes a household pest.
FIG. 115. RIGHT: Eggs of elm leaf beetle are deposited on the lower leaf surface of elms. Note some have already hatched.

long (Fig. 114). Later in the season, the leaves are skeletonized, and curl and dry up as a result of feeding on the lower surface by the larvae, black grubs with yellow markings. The eggs (Fig. 115) are deposited on the lower leaf surface by the beetles.

Control. Spray with 3 pounds of arsenate of lead and ½ pint of fish oil in 50 gallons of water, after the leaves are partly out, and again about 3 weeks later. DDT sprays, one pound of 50 per cent wettable DDT powder in 50 gallons of water, also provide control. Mites may increase on leaves sprayed too often with this material. The masses of grubs or pupae around the base of the tree can be destroyed by wetting the soil with a mixture containing 1 tablespoon of 15 per cent wettable lindane powder in a gallon of water.

TUSSOCK MOTH. *Hemerocampa leucostigma.* The leaves are chewed by the tussock moth larva, a hairy caterpillar, 1½ inches long, with a red head, longitudinal black and yellow stripes along the body, and a tussock of hair on the head in the form of a Y. The adult female is a wingless, gray, hairy moth that deposits white egg masses on the trunk and branches.

CONTROL. Spray the foliage in May with arsenate of lead and a sticker, or with DDT.

SPINY ELM CATERPILLAR. *Nymphalis antiopa.* The leaves are chewed by this pest, a 2-inch long, spiny black caterpillar with a row of red spots on its back. The adult, a large butterfly with yellow-bordered, purplish-brown wings, overwinters in bark cavities and other protected places. It deposits masses of eggs around small twigs in May.

Control. Spray with arsenate of lead, DDT, or malathion when the caterpillars are small. Prune and burn infested twigs and small branches.

SPRING CANKERWORM. *Paleacrita vernata.* When infested by spring cankerworms, the leaves are chewed by looping worms of various colors and about 1 inch in length. The adult female moth, which is ½ inch long and is wingless, climbs up the trunk to deposit eggs in the spring.

Control. Spray the leaves with arsenate of lead, DDT, or malathion while the worms are small. In spring the trunk may be banded with a sticky material (Tree Tanglefoot) to trap the adult females as they crawl up the tree (Fig. 116).

FALL CANKERWORM. *Alsophila pometaria.* The leaves are chewed by the fall cankerworm, a black worm about 1 inch in length. The adult female is a wingless moth that deposits eggs in late fall on the twigs and branches.

Control. The same sprays recommended for the spring cankerworm will control this pest. Bands of sticky material should be applied by late September to trap the adult females as they climb the tree.

ELM CASE BEARER. *Coleophora limosipennella.* When elms are infested by the case bearer, small holes are chewed in the leaves and angular spots mined between the leaf veins by a tiny larva. The adult is a small moth with a

FIG. 116. Tree banded with sticky material to trap crawling cankerworm moths.

½-inch wingspread. The pest overwinters in the larval stage in small, cigar-shaped cases made from leaf tissue.

Control. Lead arsenate, DDT, or malathion sprays in May, or a dormant spray before growth starts, will control this pest.

COCKSCOMB GALL. *Colopha ulmicola.* Elongated galls which resemble the comb of a rooster are formed on the leaves as a result of feeding and irritation by wingless, yellowish-green aphids. Eggs are deposited in bark crevices in the fall.

Control. A malathion spray applied as the buds open in spring will destroy the so-called stem-mother stage.

WOOLLY APPLE APHID. *Eriosoma lanigerum.* Stunting and curling of the terminal leaves result from infestations of bluish-white aphids. The pest overwinters in the egg stage in bark crevices.

Control. In spring just after the buds burst, spray with malathion. Repeat in 10 days to 2 weeks.

EUROPEAN ELM SCALE. *Gossyparia spuria.* Yellowing and premature defoliation may result from heavy infestations of the European elm scale, an elliptical, greenish-brown scale, 1/12 inch in diameter, with a woolly fringe around the edge. The young scales appear on the foliage in June and move to the bark in fall, after they mature.

Control. Spray all surfaces with a dormant superior spray oil before the buds open in spring, or with malathion in late May and again in mid-June.

GYPSY MOTH. *Porthetria dispar.* The leaves of a large number of forest, shade, and ornamental trees are chewed by the gypsy moth larva, a hairy, dark gray caterpillar with pairs of blue and red dots down its back, and ranging up to 3 inches in length. Among the most susceptible trees are apple; speckled alder; gray, paper, and red birch; hawthorn; linden; oaks; poplars; and willows. Trees that are not favored as food include ash, balsam fir, butternut, black walnut, catalpa, red cedar, flowering dogwood, sycamore, and tulip tree. The average annual damage caused by feeding of this insect amounts to nearly $2 million. More than 800,000 acres of woodlands have been 25 to 100 per cent defoliated in a single year. During 1957 several million acres of woodland in Connecticut, New York, and New Jersey were sprayed with DDT by means of airplanes. The practice received much publicity, and some citizens started a law suit to prevent the authorities from applying DDT in the future. The Federal judge before whom the case was tried ruled in favor of the authorities.

Control. Large-scale control of this pest can be achieved only by applying DDT via airplanes. Valuable individual ornamental trees can be protected by spraying with arsenate of lead when the caterpillars are small. Collecting and burning the egg masses during winter and early spring also helps to protect valuable ornamental trees.

FIG. 117. Typical galleries in wood produced by the elm bark beetle.

EUROPEAN BARK BEETLE. *Scolytus multistriatus.* Trees in weakened condition are most subject to infestation by the European bark beetle. The adult female, a reddish-black beetle, 1/10 inch long, deposits eggs along a gallery in the sapwood. The small white larvae hatching from the eggs tunnel out at right angles to the main gallery (Fig. 117). Tiny holes are visible in the bark when the adult beetles finally emerge. The beetle is of interest because it is one of the principal vectors of the Dutch elm disease fungus (p. 316). Adult beetles will feed to a slight extent on buds and bark of twigs during the summer.

Control. Remove and burn severely infested branches or trees, and fertilize and water weakened trees. High concentrations of DDT are needed to keep this pest from transmitting the Dutch elm disease.

ELM BORER. *Saperda tridentata.* Weakened trees are also subject to attack by the elm borer, a white grub, 1 inch long, which burrows into the bark and sapwood and pushes sawdust out through the bark crevices. The adult is a grayish-brown beetle, ½ inch long, with brick red bands and black spots. Eggs are deposited on the bark in June. The larvae overwinter in tunnels beneath the bark.

Control. Spray the bark of the trunk and branches in late June with either malathion or DDT.

TWIG GIRDLER. *Oberea tripunctata.* The appearance of dead twigs on the ground in late May or early June may result from girdling by a white, cylindrical larva, ½ inch long, the twig girdler. The adult is a beetle ½ inch long, with an orange-reddish body and a reddish-brown head. The larvae overwinter in the twig burrows.

Control. To destroy the larvae, gather and burn all fallen twigs as soon as they are noticed; to destroy the adults, spray the trees with arsenate of lead, DDT, or malathion in June.

LEAF MINER. *Fenusa ulmi.* Leaves of American, English, Scotch, and Camperdown elms are mined and blotched in May and June by white, legless larvae, sometimes as many as 20 in a single leaf. Adult females are shiny black sawflies which deposit eggs in slits in the upper leaf surfaces.

Control. Spray the leaves in early and mid-May with either lindane or malathion.

LACE BUG. *Corythucha ulmi.* The leaves of American elms lose their color as early as mid-June as a result of this sucking insect. The adults overwinter in leaves on the ground.

Control. Spray with lindane, malathion, or DDT when the young bugs appear in late spring.

MITES. *Eotetranychus carpini borealis* and *Metatetranychus ulmi.* Elm leaves turn yellow prematurely as a result of infestations of these mites.

Control. The inclusion of a miticide such as Aramite or Ovex in the regular insecticide spray will curb these pests.

CARPENTER WORM. See under Ash (Chapter 15).

FALL WEBWORM. See under Ash.

LEOPARD MOTH. See under Maples (Chapter 15).

SCURFY SCALE. See under Ash.

HACKBERRY (*Celtis*)

Diseases

WITCHES'-BROOM. In the eastern and central states, the American hackberry is extremely susceptible to the witches'-broom disease.

Symptoms. The early symptoms are visible on the buds during the winter. Affected buds are larger, more open, and hairier than normal ones. Branches that develop from such buds are bunched together, producing a broom-like effect, which is most evident in late fall or winter. The witches' brooms are more unsightly than harmful. Their presence, however, causes branches to break off more readily during wind storms, and the exposed wood is then subject to decay.

Cause. The exact cause of the witches'-broom disease is not definitely known. An Eriophyes mite and the powdery mildew fungus, *Sphaerotheca phytoptophyla,* are almost constantly associated with the trouble and are believed to be responsible for the deformation of the buds that results in the bunching of the twigs.

Control. No effective control measures are known. Pruning back all infected twigs to sound wood and possibly spraying with 1 part of lime sulfur in 10 parts of water in early spring might help. The Chinese hackberry is less susceptible than the common hackberry and should be substituted in areas where the disease is prevalent. The southern hackberry (*Celtis mississippiensis*) is also less subject to witches'-broom.

LEAF SPOT. Fungi belong to the genera *Cercosporella, Cylindrosporium, Phleospora, Phyllosticta,* and *Septogloeum* will cause leaf spots on hackberry in rainy seasons.

Control. The leaf spots are rarely serious enough to require control practices, but a copper or dithiocarbamate fungicide would probably control them.

GANODERMA ROT. See under Maple (Chapter 15).

Insects

NIPPLE GALL. *Pachypsylla celtidis-mamma.* Small round galls opening on the lower leaf surfaces, and resembling nipples, are caused by a small jumping louse or psyllid.

Control. A malathion spray when the leaves are one-quarter grown should control this pest.

SCALES. Several species of scales including camphor, cottony-cushion, cottony maple, oyster shell, Putnam, and San Jose occasionally infest this host.

Control. Malathion sprays when the young are crawling about in spring will provide control. Where infestations are heavy, a dormant spraying before leaves emerge in spring might also be applied.

PAINTED HICKORY BORER. See under Hickory.

SPINY ELM CATERPILLAR. See under Elm.

HICKORY (*Carya*)

Hickories are rarely used as street trees because of their large taproots, which render transplanting difficult, and their highly prized nuts, which are gathered by persons with little regard for damage to the tree. The shagbark hickory, with its shaggy bark, is the most desirable species for lawn planting.

Diseases

LEAF SPOT. Several leaf-spotting fungi occur on hickory. Of these, the fungus *Gnomonia caryae* is the most destructive. It produces large, irregularly circular spots, which are reddish-brown on the upper leaf surface and brown on the lower. The margins of the spots are not sharply defined, as are those of many other leaf spots. The minute brown pustules on the lower surface are the summer spore-producing bodies. Another spore stage develops on dead leaves and releases spores the following spring to initiate new infections.

The fungus *Monochaetia desmazierii* occasionally produces a leaf spot on hickory, but is more prevalent on oaks.

The fungus *Marssonina juglandis* also attacks hickory (Fig. 118) but is more destructive to black walnut. Two species of *Septoria*, *S. caryae* and *S. hicoriae*, also cause spots on hickory.

Control. Gather and burn leaves in the fall to destroy the fungi they harbor. To protect valuable specimens, spray the leaves with 1½ pounds ferbam or ziram per 100 gallons of water when the leaves open, when half-grown, and again after they are full-grown.

CANKER. Several canker diseases occasionally occur on hickory. These are associated with the fungi *Strumella coryneoidea, Nectria galligena,* and *Rosellinia caryae.*

FIG. 118. Leaf spot of hickory caused by the fungus *Marssonina juglandis.*

Control. Prune dead or weak branches and paint the wounds with a good dressing. Avoid bark injuries and keep the trees in good vigor by fertilizings and by watering during dry spells. Keep borer and other insect infestations under control.

WITCHES'-BROOM. *Microstroma juglandis.* This fungus, which causes a leaf spot of butternut and black walnuts, is capable of causing a witches'-broom disease on shagbark hickory. The brooms, best seen when the trees are dormant, are composed of a compact cluster of branches. Early in the growing season the leaves on these branches are undersized and curled, later they turn black and fall.

Control. No effective control measures have been developed, but a dormant lime-sulfur spray, followed by ferbam or ziram sprays during the growing season, is suggested.

CROWN GALL. The crown gall bacterium, *Erwinia tumefaciens,* has been found occasionally on hickory (see p. 240).

INSECTS

HICKORY GALL APHIS. *Phylloxera caryaecaulis.* Hollow green galls in June, which turn black in July, on leaves, stems, and small twigs of hickory are caused by the sucking of this small louse. In June the insides of the galls are lined with minute shiny lice of varying sizes. Galls range in size from a small pea to more than ½ inch in diameter.

Control. A dormant spray in late spring just as growth begins should destroy many overwintering lice. Malathion sprays early in the growing season and repeated several times at 2-week intervals also give control. Sprays are not effective once the galls begin to form.

BARK BEETLE. *Scolytus quadrispinosus.* Young twigs wilt as a result of boring by the bark beetle, a dark brown insect, ⅕ inch in length. The bark and sapwood are mined, and the tree may be girdled by the fleshy, legless larva, ¼ inch long. The larvae overwinter under the bark.

Control. Spray the foliage with arsenate of lead, DDT, or malathion when the beetles appear in July. Remove and burn severely infested trees, and peel the bark from the stump. Increase the vigor of weak trees by fertilization and watering.

CIGAR CASE BEARER. *Coleophora caryaefoliella.* The leaves are mined, turn brown, and fall when infested by the cigar case bearer, a larva, ⅕ inch long, with a black head. The adult female is a moth with brown wings that have fringed hairs along the edge and a spread of ⅖ inch. The larvae overwinter on twigs and branches in cigar-shaped cases, ⅛ inch long.

Control. Spray with lead arsenate, DDT, or malathion as soon as the leaves are fully developed.

PAINTED HICKORY BORER. *Megacyllene caryae.* The sapwood of recently killed trees is soon riddled by painted hickory borers, creamy white larvae that attain a length of ¾ inch. The adult beetle is dark brown, has zigzag lines on the back, and is ¾ inch long. Eggs are deposited in late May or early June.

Control. Remove and burn dead trees immediately. Inject a nicotine paste such as Bortox into the tunnels of live trees and then seal the openings with chewing gum, grafting wax, or putty. Spraying the bark with DDT or malathion in early June might also help.

TWIG GIRDLER. *Oncideres cingulata.* Twigs girdled by this pest, a larva ½ inch long, are then broken off by the wind and fall to the ground. The adult is a reddish-brown beetle, ¾ inch long. The larvae overwinter inside the twigs on the ground.

Control. Gather and burn severed branches and twigs in the autumn or early spring.

JUNE BUGS. *Phyllophaga* spp. The leaves may be chewed during the night by light to dark brown beetles, which vary from ½ to ⅞ inch in length. The beetles rest in nearby fields during the day. The larva is ¾ to 1 inch long, white, and soft-bodied with a brown head. Three or more years are required for completion of the life cycle of most of the species.

Control. Spray the leaves with arsenate of lead, DDT, or malathion during late May or early June, depending on the locality. The larval stage, at which time the larva feeds on grasses in lawns and golf courses, can be controlled by treating the lawns with dieldrin or chlordane.

MITE. *Eotetranychus hicoriae.* This pest occasionally infests hickory leaves, causing them to turn yellow and drop prematurely.

Control. Spray with malathion or a miticide such as Aramite, Dimite, or Ovex.

HORSECHESTNUT (*Aesculus*)

Diseases

LEAF BLOTCH. The most common leaf disease of the horsechestnut and related trees such as the Ohio and yellow buckeye is known as leaf blotch. The latter species are less susceptible than the common horsechestnut.

Symptoms. Small, irregular, slightly discolored, water-soaked spots appear on the leaves in the spring (Fig. 119). Later the centers of the spots become a reddish-brown, surrounded by a yellow zone that merges into the healthy green of the unaffected portion. Numerous minute black specks appear in the center of the spot on the upper surface of the leaf. The spots often enlarge to involve a good position of the leaf (Fig. 120), and the tree appears as if scorched by fire. Many of the affected leaves drop prematurely.

Cause. Leaf blotch is caused by the fungus *Guignardia aesculi.* In the spring, spores of the fungus are forcibly ejected into the air from spore sacs developing on overwintered diseased leaves on the ground. Air currents carry them to the newly developing leaves, where they penetrate and infect during a rainy period. The black specks that later appear in dead tissue, resulting from early spring infections, exude millions of so-called summer spores, which are washed and scattered by rain and wind to other leaves. Additional infections then occur. By August a tree may show infections on almost every leaf.

FIG. 119. LEFT: Early stage of horsechestnut leaf blotch disease.
FIG. 120. RIGHT: Advanced stage of the disease.

Control. In the fall all leaves should be gathered and burned, to eliminate the overwintering stages of the fungus in these tissues. Protective sprays should be applied on the young foliage. Ziram (Zerlate, Karbam White) sprays at the rate of 2 pounds of the 76 per cent material in 100 gallons of water are effective. The first application of the fungicide should be made soon after the buds open in the spring, followed by at least two applications at intervals of 2 weeks. If the month of May is unusually rainy, as it was in the northeastern United States in 1958, a fourth application may be profitably made 2 weeks after the third spray.

LEAF SCORCH. Another common trouble often confused with leaf blotch, is discussed in great detail on page 194. There are no black pin-point fungus bodies in the browned areas affected by scorch, however.

Control. The application of an antidesiccant such as Wilt-Pruf as soon as the leaves are fully expanded may help to reduce the amount of scorch.

POWDERY MILDEW. In late summer, the lower leaf surface may be covered with white patches of the fungus *Uncinula flexuosa.*

Control. Where mildew is unusually severe, sprays such as Mildex (Karathane) will provide control.

SAPWOOD ROT. A white spongy rot of the sapwood is caused by the toadstool *Collybia velutipes.* The same fungus produces a similar rot in linden and American elm. The toadstools, which often grow out of the decaying tissue in clusters, have a yellow or brown cap, and the bases of the stems are covered with dark brown velvety hairs.

Control. Because the spores of the causal fungus gain access through openings, all bark injuries should be treated and dressed as soon as they occur. Where considerable decay already is present, extensive cavity treatments are rarely justified because horsechestnuts are relatively short-lived trees.

INSECTS

COMSTOCK MEALY BUG. *Pseudococcus comstocki.* This small, elliptical, waxy-covered insect attacks umbrella catalpa primarily, but it is also found on apple, boxwood, holly, horsechestnut, magnolia, maples, osage orange, poplar, and Monterey pine. After hatching in late May, the young crawl up the trunk to the leaves, where they suck out the juices and devitalize the tree. Twigs, leaves, and trunks may be distorted as a result of heavy infestations. The eggs winter in bark crevices or in large masses hanging to twigs.

Control. Malathion sprays are effective in controlling this pest.

SCALES. Several scales including the cottony maple, maple phenacoccus, oyster shell, Putnam, scurfy, and walnut also infest horsechestnut.

Control. Dormant oil sprays, or malathion sprays early in the growing season will control these insects.

BAGWORM. See under Arborvitae (Chapter 17).

FLAT-HEADED BORER. See under Maple (Chapter 15).

JAPANESE BEETLE. See under Linden.

TUSSOCK MOTH. See under Elm.

WALNUT SCALE. See under Walnut.

LARCH (*Larix*)

Larches grow in almost any type of soil including clay and limestone. They do best, however, in moist, loamy soils and in full sunshine. They do not tolerate dry soils or sandy hillsides where the summers are hot.

The varieties most commonly used in ornamental plantings are the American larch or tamarack, the European larch, and the Japanese larch.

Diseases

NEEDLE RUST. Three rust fungi attack larch needles. They develop principally on the needles nearest the branch tips. Affected needles turn yellow and have pale yellow fungus pustules on the lower surfaces.

The fungus *Melampsora bigelowii* occurs on American, European, Western, and Alpine larches. Its alternate hosts are several species of willows. Larches can be infected only by spores developing on willows, but the spores on willows can reinfect the willow.

The fungus *Melampsora medusae* attacks American larch and its alternate host, poplar. As with *M. bigelowii,* spores on larch cannot reinfect larch, but must come from poplars.

Melampsoridium betulinum affects American larch and several species of birches. The spore stage on birch can infect birch as well as larch.

Control. Where larches are the more valuable specimens in an ornamental planting, the removal of the alternate host will prevent infection. Infected needles should be submitted to a rust specialist for determination of the exact species involved, before attempts to eradicate the alternate hosts are made.

LEAF CAST. The needles of American and Western larches and the spur shoots that bear them may be killed by the fungus *Hypodermella laricis.* Early symptoms are yellowing, followed by browning of the needles. Very small, elliptical, black fruiting bodies of the fungus appear on the dead leaves during the winter.

Several other fungi, including *Lophodermium laricis, L. laricinum,* and *Meria laricis,* also cause leaf cast or leaf blights of larch.

Control. Gather and burn the needles in late fall or winter to eliminate the most important source of inoculum. This usually ensures satisfactory control.

EUROPEAN LARCH CANKER. This disease has been particularly destructive on larches in Europe for a long time. It is discussed here to acquaint tree men with its symptoms so that suspicious cases can be reported to the proper authorities immediately.

A few localized outbreaks arising from the importation of diseased plants from Europe occurred in the New England states from 1927 to 1935. Fortunately these were stamped out as soon as discovered. Because of the similarity of symptoms to several other less destructive canker diseases, larch canker has been erroneously reported as existing on various trees.

European and American larches are known to be very susceptible to canker, whereas the Japanese larch is relatively resistant.

Symptoms. Stems of young trees and branches of small diameter die suddenly as the result of girdling by cankers. On trunks and branches of large diameter, cankers that increase in size each year are formed. A heavy flow of resin is evident from the cankers. The slowly developing cankers on the trunk or large branches result in considerable distortion and swelling of the affected members. White, hairy, cup-shaped fruiting bodies, about ¼ inch in diameter, develop in the cankered tissues.

Cause. The fungus *Dasyscypha willkommii* causes larch canker. It enters through wounds and destroys the inner bark and cambium.

Control. European investigators have found that selection of favorable planting sites and maintenance of trees in good vigor discourage severe outbreaks of the disease. It is hoped that any future outbreaks in the United States will be discovered in time and eradicated so as to prevent extensive spread of the disease.

WOOD DECAY. Several species of fungi are constantly associated with the various types of wood decay of larch. Those most common in the eastern United States are *Fomes annosus, F. roseus, F. pini,* and *Polyporus schweinitzii.* They are found mainly on older, neglected trees.

Control. Little can be done to check wood decays by the time they are discovered. Maintaining valuable specimens in good vigor by periodic fertilization and by watering during dry spells will do much to prevent initial infections. Wounds on the trunks of such trees should be treated promptly.

INSECTS

LARCH CASE BEARER. *Coleophora laricella.* Leaves first turn yellow, then brown, and finally die from infestations by the larch case bearer, a dark brown-bodied, black-headed larva, ⅕ inch long.The adult female, a silvery gray moth with a wingspread of ⅓ inch, deposits cinnamon-colored eggs on the leaves in June. The larvae overwinter in tiny cases on the branches.

Control. To destroy the overwintering larvae in the cases, spray with 1 part concentrated lime sulfur solution in 10 parts of water, when the trees are dormant. Spray in late June with malathion.

SAWFLY. *Pristiphora erichsonii.* In sawfly infestations, the needles are chewed by ¾ to 1 inch long, olive-green larvae covered with small brown spines. The adult is a wasp-like fly with a wingspread of ⅘ inch. Eggs are deposited in incisions on twigs in late May and June. The larvae overwinter in brown cocoons on the ground.

Control. Gather and burn fallen needles beneath the tree. If necessary, spray with lead arsenate, DDT, or malathion in June or early July.

WOOLLY LARCH APHID. *Chermes strobilobius.* White woolly patches adhering to the needles are typical signs of this pest. The adult aphids are hidden beneath the woolly masses. The winged adults migrate to pines, and another generation returns to the larch the following season. Eggs are deposited at the bases of the needles in the spring. Young aphids overwinter in bark crevices.

Control. Spray with 1 part concentrated lime sulfur in 10 parts of water before growth starts, or with malathion when the young are hatching in May.

LINDEN (*Tilia*)

Diseases

LEAF BLOTCH. European lindens are occasionally affected by the leaf blotch disease.

Symptoms. Elongated, light brown spots occur along the veins in the various parts of the leaf but chiefly near the tip. A conspicuous, narrow black band appears between the dead and the healthy tissue. In severe infections, the tree may be completely defoliated.

Cause. Leaf blotch is caused by the fungus *Gnomonia tiliae,* which is thought to overwinter on diseased fallen leaves.

Control. Gather and burn fallen leaves. Spray with bordeaux mixture 4-4-50, or other copper fungicides, first when the leaves begin to grow, then when they are half-grown, and again after they are fully expanded. Ziram and phenyl mercury sprays, which are effective against leaf blotch diseases of other shade trees, might be effective on linden.

LEAF SPOT. Small circular brown spots surrounded by a darker border occur commonly on American and European lindens. When the spots become numerous, or when a number of them coalesce, the leaves drop prematurely. Young trees appear to be most susceptible.

Cause. The fungus *Cercospora microsora,* whose perfect stage is *Mycosphorella microsora,* is responsible for the leaf spotting.

Control. The same as for leaf blotch.

POWDERY MILDEW. Lindens are susceptible to powdery mildew fungi including *Uncinula clintonii, Microsphaeria alni,* and *Phyllactinia corylea.* A discussion of powdery mildew is presented on page 242.

NECTRIA CANKER. The American linden in forest plantings is extremely susceptible to nectria cankers, but the disease is rarely serious on ornamental or streetside plantings. It is discussed under birch on page 259. Several other cankers are produced by the fungi *Aleurodiscus griseo-canus,* and *Strumella* sp.

INSECTS

JAPANESE BEETLE. *Popillia japonica.* The leaves are chewed and skeletonized by the Japanese beetle, a metallic-green insect, ⅜ inch long (Figs. 121 and 122). The adult females deposit their eggs in the soil, where the pest overwinters in the larval stage (Fig. 123). It is in this stage that the insect feeds on grass roots.

Control. Spray with either malathion or DDT, or a mixture of both, in early July and again 10 days later. The spray should be directed to the upper leaf surfaces because the beetles feed high upon the sunny side of the tree.

APHID. *Myzocallus tiliae.* Sap is sucked from the leaves and a sticky substance is exuded by a yellow and black aphid with clouded wings.

Control. Spray with lindane or malathion when the young aphids appear on the leaves in spring.

LEAF BEETLE. *Calligrapha scalaris.* Ragged holes remain in leaves chewed by creamy white larvae with yellow heads. The adult beetle is ⅜ inch long, oval, yellow, with green spots on the wing covers and with a broad, irregular, coppery-green stripe down the back. Lemon-yellow eggs are deposited on the lower leaf surface in late June or early July. The beetles hibernate in the ground.

Control. Spray the foliage with arsenate of lead, DDT, or malathion as the larvae appear.

CECROPIA MOTH. *Hyalophora cecropia.* Leaves infested by the cecropia moth are devoured by 4-inch long, bluish-green caterpillars with rows of red, yellow, and blue tubercules along the body. The adult moth has a wingspread of

FIG. 121. Japanese beetles on linden (normal size) and the characteristic appearance of chewed leaves.

FIG. 122. UPPER INSET: Japanese beetle enlarged.

FIG. 123. LOWER INSET: Larval stage of Japanese beetle.

nearly 7 inches. Its wings are brown with a red-bordered white cross-band. Winter is passed as pupae in cocoons on the branches.

Control. Spray with arsenate of lead, DDT, or malathion when the caterpillars are small.

LINDEN BORER. *Saperda vestita.* Broad tunnels beneath the bark near the trunk base or in roots are made by the linden borer, a slender, white larva, 1 inch long. The adult, a yellowish-brown beetle, ¾ inch long, with 3 dark spots on each wing cover, feeds on green bark. Eggs are deposited in small bark crevices made by the beetle.

Control. Dig out the borers with a flexible wire, or inject a nicotine paste such as Bortox into the tunnels and seal the openings. Spraying the trunk base with DDT might control many of the young larvae after they hatch out of their eggs.

SCALES. Lindens are subject to many scales including cottony maple, oyster shell, European fruit lecanium, Putnam, San Jose, terrapin, tulip tree, walnut, and willow.

Control. Dormant oil sprays control many of these scales. Where infestations are heavy, malathion sprays should be applied when the young scales are crawling about—around mid-May for most kinds. Young of tulip-tree scale crawl about in late summer.

SPRING AND FALL CANKERWORMS. See under Elm.

GYPSY MOTH. See under Elm.

TUSSOCK MOTH. See under Elm.

TWIG GIRDLER. See under Hickory.

LOCUST AND HONEY LOCUST (*Robinia* and *Gleditsia*)

Diseases

WOOD DECAY. The greatest majority of older black locusts growing along roadsides and in groves in the eastern United States harbor one of several wood-decay fungi. Nearly all these decays have followed infestations of the locust borer (p. 342).

The fungus *Fomes rimosus* causes a spongy, yellow rot of the heartwood. It infects the trunk through tunnels made by the locust borer or through dead older branches. Following extensive decay of the woody tissues, the fungus grows toward the bark surface where it produces hard, woody, bracket- or hoof-shaped fruiting structures nearly a foot wide. The upper surface of the structure is brown or black and is cracked; the lower surface is reddish-brown.

The fungus *Polyporus robiniophilus* produces a soft, white rot of the heartwood. Like the Fomes fungus, it infects the tree through locust borer tunnels and dead branches. The fruiting structures, which form on dead bark, are at first white and firm, and later brown and corky.

Control. Because wood decays of black locust become established mainly through locust borer tunnels, their prevention rests primarily on freedom from the insect pests. Effective control of the latter, however, cannot easily be attained, and for this reason heartwood decays will continue to be prevalent. Cavity treatments should never be attempted on black locusts.

LEAF SPOT. Honey locust is subject to a leaf spot which occurs commonly in the South but which has not yet been reported from the Northeast. The fungus *Linospora gleditsiae* produces small black structures, scattered over the lower leaf surface.

Control. Gathering and burning of all fallen leaves should give practical control.

WITCHES'-BROOM. Honey locust and black locust are subject to one of the few virus diseases that affect trees. The disease, though common on the sprouts, rarely occurs on the large trees. It is characterized by production in late summer of dense clusters or bunches of twigs from an enlarged axis. The bunched portions ordinarily die during the following winter.

Control. Infected trees have appeared to recover naturally. No definite control measures are known.

INSECTS

BORER. *Megacyllene robiniae.* White, cylindrical larvae feed on the inner bark and sapwood of black and yellow locusts. Young trees may be so weakened by the galleries produced that they break off at the point of injury. Adult beetles are black with bright yellow, transverse, zigzag lines (Fig. 124).

Control. Spraying the trunks with malathion just before growth starts in spring will kill the young larvae which have been hibernating in small cells just beneath the corky layer of bark.

LOCUST LEAF BEETLE. *Chalepus dorsalis.* Holes in leaves may result from feeding in late April by the locust leaf beetle, a black-headed, orange-red beetle, ¼ inch long, with a black stripe down the back. The leaves turn brown and fall in summer as a result of mining by the black-headed, yellow larvae. The eggs are deposited on the under sides of the leaves, and the adult beetle overwinters in bark crevices.

Control. Spray with arsenate of lead and a sticker, or with DDT in April to poison the adults, or with lindane or malathion early in July to kill the young larvae as they begin to mine the leaves.

LOCUST GALL MAKER. *Ecdytolopha insiticiana.* Elongated, gall-like swellings, 1 to 3 inches long, on the twigs are produced by feeding and irritation of the pale yellow larvae of the locust gall maker. The adult female is a grayish-brown moth with a wing expanse of ¾ inch. The pest overwinters in the pupal stage among fallen leaves.

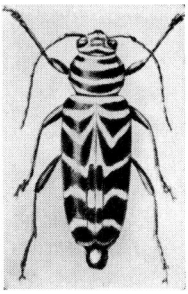

FIG. 124. Adult stage of the locust borer, *Megacyllene robiniae.*

Control. Prune and burn infested twigs in August, and gather and burn fallen leaves in autumn.

MITE. *Eotetranychus multidigituli.* Leaves of honey locust drop prematurely as a result of infestations of this mite. Infested leaves are stippled yellow prior to dropping. Defoliated trees usually leaf out again in late summer but are considerably weakened.

Control. Malathion or specially made miticides such as Aramite and Ovex, applied early in June, give good control.

WEBWORM. *Homadaula albizziae.* This pest is more common on silk tree, but it occasionally feeds on the leaves of honey locust, causing them to turn brown.

Control. Lead arsenate or DDT sprays applied in mid-June and mid-August will control this pest.

POD GALL. *Cecidomyia gleditsiae.* This midge causes globular galls ⅛ inch in diameter at the growing tips. Adults emerge in early July.

Control. A DDT spray applied in early July and repeated 10 days later will provide good control.

LEAFHOPPER. *Empoasca pergandei.* Honey locust leaves may be attacked by this sucking insect, which causes the leaves to turn yellow and drop.

Control. Malathion or DDT sprays will give control, although the latter might serve to increase the mite population unless combined with a miticide.

HONEY LOCUST BORER. *Agrilus difficilis.* A flat-headed borer burrows beneath the bark of honey locust and eventually may girdle the tree. Large quantities of gum exude from the bark near the infested nodes. The adult beetles, which emerge in June, are elongate, ½ inch long, black with a metallic luster.

Control. No really effective controls are known, although it is likely that DDT or malathion sprays applied to the trunks during the egg-laying period in mid-June and early July might prove effective.

SCALES. In addition to the two scales listed below, a number of other species including *Diaspidiotus osborni, Phenacaspis spinicola, Chionaspis gleditsiae, Chionaspis ortholobia,* and *Lecanium caryae* also infest locusts.

The larvae of the Imperial moth, *Eacles imperialis,* also feed on locust trees.

BAGWORM. See Under Arborvitae (Chapter 17).

CARPENTER WORM. See under Ash (Chapter 15).

COTTONY MAPLE SCALE. See under Maple (Chapter 15).

SPRING AND FALL CANKERWORMS. See under Elm.

SAN JOSE SCALE. See under Flowering Cherry (Chapter 15).

WALNUT DATANA. See under Walnut.

MAIDENHAIR TREE (*Ginkgo*)

The maidenhair tree is unusually resistant to fungus and insect attack.

LEAF SPOT. The only leaf disease reported on this host is anthracnose, caused by the fungus *Phyllosticta ginkgo.* The damage by this fungus is negligible. Several wood decay fungi, including *Polyporus hirsutus, P. lacteus, P. tulipiferus, P. versicolor,* and *Fomes meliae,* have also been reported, but these are of rare occurrence.

Few insects attack this tree. Among those occasionally found are the omnivorous looper, *Sabulodes caberata;* the grape mealybug, *Pseudococcus maritimus;* and the white-marked tussock moth, *Hemerocampa leucostigma.*

OAK (*Quercus*)

Diseases

WILT. This most highly publicized disease of oaks in recent years is causing some concern to arborists, nurserymen, tree owners, and lumber interests in the Middle West. At this writing, the disease has been found in 20 states, from Kansas and Nebraska eastward to Pennsylvania, and from Minnesota southward to Arkansas. It has not yet been found east of the Susquehanna River.

Symptoms. Wilt is characterized by wilting and death of the leaves, followed usually by death of the entire tree within the same year. Brown to black discoloration is often present in the sapwood of wilted branches, although this is not always a dependable diagnostic symptom.

Oaks belonging to the red oak group appear to be most susceptible, but the disease is not confined to oaks. It has been found in Chinese chestnuts and related genera.

Cause. The fungus *Ceratocystis fagacearum* is known to cause wilt. When first described in the literature in 1943 it was known as *Chalara quercina.* The fungus causes infections primarily through wounds made by man or other agents. It has been transmitted rather easily on tools used by arborists and foresters. The most susceptible period appears to be during springwood development in the tree.

Very little is known on how the oak wilt fungus spreads. Some spread occurs by root grafting of diseased and healthy trees but the means of long-range spread is still unknown. Several species of insects, including the fruit fly (*Drosophila melanogaster*) and Nitidulid beetles, are known to be capable of spreading it. The fungus has also been recovered from several other insects including a bark beetle, ambrosia beetle, and the two-lined chestnut borer.

So-called ascospores of the oak wilt fungus were found to survive more than 150 days in or on dead or dormant Nitidulid beetles. It is concluded, therefore, that insects coming in contact with such spores on mycelial mats in the fall can still carry viable spores at the end of their winter hibernation. Such spores could probably cause infection during the spring period when the oaks are resuming cambial activity—the period when they are most prone to infection.

The fungus has been recovered from infected roots, stems, twigs, and leaves of oaks but never from acorns of diseased trees.

Oak trees were experimentally infected by spraying their flowers with spores of the fungus.

The chinquapin, *Castanea pumila*, and two exotic species of oaks, *Quercus brutia* and *Q. lusitanica*, were experimentally infected with the oak wilt fungus.

The oak wilt fungus was capable of infecting apple trees under experimental conditions. Jonathan and Delicious apple trees were infected with the fungus, and some died as a result. The fungus was recovered from the roots and stem tips one year after inoculation. Apple is the first genus outside the oak-beech family that appears to be genuinely susceptible to the oak wilt fungus. The fungus was found to survive and spread to a limited degree in ash, birch, dogwood, hop hornbeam, hickory, sassafras, sourwood, and wild cherry.

Control. Because the oak wilt fungus appears to be most infectious early in the growing season when the new spring wood vessels are developing, University of Wisconsin researchers suggested that pruning operations in oaks be delayed until July or later.

It is wise not to store logs from oaks that have died from the wilt disease, because the fungus can develop in such logs and produce viable spores.

In February 1958, T. W. Jones and T. W. Bretz reported in Research Bulletin 657 of the Missouri Agricultural Experiment Station that the several treatments suggested earlier as control measures for oak wilt disease failed to provide control. These included isolation of diseased trees by poisoning surrounding healthy oaks, sanitation by cutting and burning diseased trees, and a combination of both treatments. Their report stated: 'The number of trees infected with oak wilt each year on untreated plots remained relatively constant during the course of the study. The number of infected trees infected each year on the treated plots fluctuated erratically but no consistent improvement in the disease situation occurred.'

The Missouri tests were conducted on a number of 40-acre plots over a 5-year period and are perhaps the most exhaustive ones made. The author is of the opinion that several factors beyond man's control will determine eventual success or failure in overcoming wilt. Among these are the number and efficiency of agents which spread the fungus. The fact that no highly efficient means of spread has been discovered after 10 years of intensive research, and that the fungus has apparently been infecting trees for nearly 75 years, even though its causal nature was discovered only 15 years ago, suggest that the disease may not be so serious as chestnut blight, Dutch elm disease, or phloem necrosis. Nevertheless, it is to be hoped that a control will be found for this disease.

TWIG BLIGHT. Chestnut oak and, at times, red and white oaks in ornamental plantings are attacked by the twig blight disease.

Symptoms. A sudden blighting of the leaves on twigs and branches scattered over the tree is the most striking symptom. The dead twigs and small branches, with their light brown, dead leaves still attached, are readily visible at some distance from the tree because of the sharp contrast with unaffected leaves near them.

Diseased bark becomes sunken and wrinkled, and the sapwood beneath is discolored.

*Cause.** Twig blight has been ascribed by some investigators to the fungus *Diplodia longispora*, and by others to the fungi *Sphaeropsis quercina* and *S. malorum*. In all cases, the causal fungi overwinter in dead twigs and in cankers on larger branches. During the rainy weather of the following spring, spores in large numbers ooze out from the dead areas and are splashed onto young shoots and into bark injuries, where they germinate and cause new infections.

Under usual conditions, the disease is confined to twigs and smaller branches. If the trees are weakened as a result of prolonged droughts and continuous defoliation by cankerworms, however, the fungus will kill large branches and even the entire tree.

Control. During the summer, prune infected twigs and branches to sound wood. As a rule, cutting to about 6 inches below the visibly infected area will ensure the removal of all fungus tissue in the wood. In severely weakened trees, considerably more tissue must be removed, inasmuch as the fungus may penetrate down the branch for a distance of 2 or more feet.

Valuable trees should be fertilized and watered and their foliage covered with a combination spray containing arsenate of lead or DDT and a copper fungicide to destroy leaf-chewing insects and prevent infections by leaf-spotting fungi. A phenyl mercury spray can be substituted for the copper fungicide.

ANTHRACNOSE. White oak and, at times, most other oaks are subject to the anthracnose disease. The causal fungus and the leaf symptoms are the same as those on planes. Unlike that on planes, however, the disease is confined primarily to the leaves and rarely extends to the twigs.

* A number of other fungi, particularly those belonging to the genera *Dothiorella*, *Coryneum*, *Cytospora*, *Phoma*, and *Diatrype*, are associated with many twig blights and cankers of oaks.

Control. Ordinarily, the destruction of all fallen leaves by burning is sufficient to keep severe outbreaks at a minimum. In very wet springs, as in 1958, sprays containing copper or phenyl mercury acetate are needed to prevent serious epidemics.

CANKER. Although primarily a disease of forest oaks, canker occasionally affects red and scarlet oaks in ornamental plantings. American beech, chestnut, red maple, tupelo, and pignut and shagbark hickories are also susceptible.

Symptoms. Several types of cankers, depending on the age of the tree and the rate of growth of the causal fungus, are produced on the trunks. Smooth-surfaced, diffuse, slightly sunken cankers are common on young trees with a diameter of 3 to 4 inches. Cankers on older trees have a rough surface ridged with callus tissue. Open wounds may be present in the center of large cankers as a result of secondary decay and shedding of the bark. Cankers extend more rapidly up and down the trunk than around. Death of the distal portions occurs when the canker completely girdles the trunk.

Cause. The fungus *Strumella coryneoidea* causes canker. It enters the trunk mainly through dead branch stubs.

Control. Prune dying and dead branches to eliminate an important possible source of inoculum. Remove small cankers on the trunk by surgical means, and fertilize and water the trees to improve their vigor.

BASAL CANKER. *Phaeobulgaria inquinans.* This fungus frequently forms in crevices of the sunken bark that overlies basal cankers. It enters through open wounds but then invades the surrounding bark and sapwood and eventually girdles and kills the tree. The mature fruit bodies are cup- or saucer-shaped and grow from short stems that extend into the bark. When moist, the fruiting bodies look and feel like rubber.

Control. Same as for the canker disease described above.

LEAF BLISTER. During cool, wet springs almost all species of oak are subject to the leaf blister disease. Circular, raised areas ranging up to ½ inch in diameter are scattered over the upper leaf surface, causing a depression of the same size on the lower surface (Fig. 125). The upper surface of the bulge is yellowish-white, and the lower, yellowish-brown. The leaves remain attached to the tree, and there is rarely any noticeable impairment in their functions.

Cause. Leaf blister of oaks is caused by the fungus *Taphrina coerulescens.* New infections in spring are caused by spores from overwintered leaves and possibly by spores lodged in bud scales and on twigs.

Control. Other than the destruction of infected leaves, preventive measures are rarely adopted. Bordeaux mixture or any other copper spray applied before rainy periods in spring will prevent infections on valuable trees.

LEAF SPOT. Like all other trees, oaks are subject to a number of leaf spot diseases. These rarely cause much damage to the trees, inasmuch as they become numerous rather late in the growing season.

FIG. 125. Leaf blister of oak.

The fungus *Marssonina ochraleuca* produces spots commonly 1 inch in diameter. The spots are circular and contain yellow to brown concentric markings. The fungus *Monochaetia desmazierii* produces even larger spots, with yellow centers surrounded by a reddish-brown border. A number of other fungi belonging to the genera *Gloeosporium*, *Phyllosticta*, and *Septoria* may also produce spotting of the leaves.

Control. Gathering and burning all fallen leaves is usually sufficient to keep down most outbreaks in seasons of normal rainfall. Valuable oaks may be protected by several applications of copper sprays at 2-week intervals starting in early spring when the leaves unfold. A delayed dormant spray of phenyl mercury acetate might also give good control.

WOOD DECAY. Oaks are subject to attacks by a number of fungi that produce decays beneath the bark, usually near the soil line. Two types of decay, known as white heart rot and brown heart rot, generally result. Fungi belonging to the genera *Stereum, Polyporus, Fomes,* and *Fistulina* are most commonly associated with these rots. The same tree may show infection by more than one of these fungi, although in most cases only one is involved.

The fungus *Stereum gausapatum* has been isolated from most basal decays of oaks. In the early stages of decay, it forms white lines through the sound wood, producing a mottled effect when viewed longitudinally. These white lines or channels usually follow the spring wood vertically, but they branch frequently and at times penetrate through the annual growth rings. Later much of the summer wood decays, and in the final stages, all the wood becomes light colored and brittle.

Control. Once wood decays have become extensive, little can be done to check their advance. Trees live for a long time, however, despite their presence. Except to eliminate a breeding place for vermin, cavity work is rarely done where extensive decay occurs.

SHOESTRING ROOT ROT. Oaks are especially subject to this disease, a complete discussion of which is found on page 243.

INSECTS

DATANA CATERPILLAR. *Datana ministra.* The leaves are chewed by the Datana caterpillar, a black and yellowish-white striped larva, 2 inches long. The adult female moth is cinnamon-brown with dark lines across the wings, which have a spread of 1½ inches. Eggs are deposited in batches of 25 to 100 on the lower leaf surface. The insect hibernates in the pupal stage in the soil.

Control. Spray the leaves with arsenate of lead, DDT, or malathion when the caterpillars are young.

SADDLE-BACK CATERPILLAR. *Sibine stimulea.* Broad, spine-bearing, red caterpillars with large green center spots edged in white, and attaining an inch in length, occasionally chew the leaves of ornamental and streetside oaks. The adult is a small moth with a wingspread of 1½ inches. The upper wings are dark reddish-brown, the lower, a light grayish-brown.

Control. The same as for Datana caterpillar.

WHITE BLOTCH OAK LEAF MINER. *Lithocolletis hamadryadella.* White, blotched patches on the leaves are produced as a result of feeding between the leaf surfaces by this oak leaf miner, a larva ⅓ inch long. The adult female is a small moth with white fore wings and silvery hind wings, which have a spread of ¼ inch. The larva hibernates in dry leaves on the ground.

Control. Gather and burn fallen leaves and, just before mines appear, spray with malathion, lindane, or chlordane.

OBSCURE SCALE. *Chrysomphalus obscurus.* Tiny, circular, dark gray scales, 1/10 inch in diameter, may occasionally cover the bark of twigs and branches. The pest overwinters in the partly grown adult stage.

Control. Spray with oil or winter-strength lime sulfur before growth starts in spring or with malathion in late spring.

LECANIUM SCALE. *Lecanium corni* and *L. quercifex.* Branches and twigs infested with these pests are covered with brown, downy-covered, hemispherical scales. The adults are ⅙ inch in diameter. The winter is passed in the partly grown stage.

Control. Spray with winter-strength oils or lime sulfur before the buds open in spring, or with malathion in early summer when the young scales are moving about.

KERMES SCALE. *Kermes pubescens.* The leaves are puckered and some twigs are killed as a result of infestation along leaf veins and terminal twigs by the Kermes pest, a globular, light brown scale, ⅛ inch in diameter. Bur, pin, red, and white oaks appear to be most susceptible. Winter is passed in the immature stage. The young scales are covered with a white down.

Control. A dormant oil spray applied in March or early April is very effective. Malathion sprays in late spring are effective against the crawling stage.

GOLDEN OAK SCALE. *Asterolecanium variolosum.* Trees infested by the golden oak scale have a ragged, untidy appearance. Young trees may be killed outright, the lower branches dying first. Shallow pits are formed in the bark by these circular, greenish-gold scales, which attain a diameter of only 1/16 inch.

Control. A miscible oil spray applied in early spring will destroy most of the pests. Malathion sprays in late spring will control the crawlers.

OAK GALLS. *Cynipoidea.* Swellings of various shapes commonly occur on the leaves and stems of oaks (Figs. 126 and 127). A tiny maggot lives for some time inside these galls. The adult stages are usually wasps.

Control. No effective method of controlling these pests has been developed, principally because they are not particularly destructive. Where control is desired, a dormant oil or lime sulfur spray, followed by several malathion sprays after leaves emerge, should provide control.

TWO-LINED CHESTNUT BORER. *Agrilus bilineatus.* Large branches are killed on some trees as a result of the formation of tortuous galleries underneath the bark by the two-lined chestnut borer, a white, flat-headed larva, ½ inch long. The adult, a slender greenish-black beetle, ⅜ inch long, appears in late June. The larvae pass the winter beneath the bark.

FIG. 126. Oak apple gall caused by a species of gall wasps.

Control. Cut down and burn all badly infested trees. Increase the vigor of the remaining trees by fertilizing and watering. DDT-malathion sprays applied to bark of trunk and branches early in July will also help.

OAK MITE. *Oligonychus bicolor.* Mottled yellow foliage results from the sucking of the leaf juices by the oak mite, a yellow, brown, or red, eight-legged pest. The pest usually overwinters in the egg stage.

Control. Spray with a dormant contact insecticide during early spring or with one of the following during summer: 1 pound of wettable sulfur in 30 gallons of water to which a good spreader has been added, 2 per cent summer oil, or 1 part lime sulfur in 49 parts of water.

LACE BUG. *Corythucha arcuata.* The leaves of many species of oaks, particularly the white oak, may turn whitish-gray when heavily infested with this lace bug.

Control. Spray with malathion or lindane when the young begin to feed. Direct the spray to the lower leaf surface.

FIG. 127. Gouty oak gall caused by the gall wasp *Andricus punctatus.*

PIN OAK SAWFLY. *Caliroa lineata.* The lower surface of pin oak leaves are chewed by ⅜ to ½ inch long greenish larvae or slugs, leaving the upper epidermal layer of cells and a fine network of veins. Injured leaves turn a golden brown, and when the feeding by the larvae is extensive, the injury can be readily distinguished at a considerable distance. The adult stage is a small, shiny black, four-winged insect ¼ inch long.

Control. Spray the leaves with arsenate of lead, DDT, or malathion when the larvae begin to feed in early summer, directing the spray primarily to the lower surfaces of the leaves. The upper parts of the tree appear to be preferred by this pest.

ASIATIC OAK WEEVIL. *Cyrtepistomus castaneus.* Severe damage to the leaves of oaks and chestnuts can be caused by a deep red or blackish weevil, ¼ inch long with scattered green scales that have a metallic lustre. Their long antennae are characteristic of only one other species in this country, the long-horned weevil, *Calomycteris setarius.* Like the elm leaf beetle and the box-elder bug, they move into homes in the fall to hibernate.

Control. DDT sprays on foliage at the time the weevils begin to feed provide good control.

CECROPIA MOTH. See under Linden.

CARPENTER WORM. See under Ash (Chapter 15).

JUNE BUGS. See under Hickory.

LEOPARD MOTH. See under Maple (Chapter 15).

TWIG PRUNER. See under Maple.

PLANE (*Platanus*)

Diseases

BLIGHT. This, also known as anthracnose and scorch, is the most common disease of planes. It is locally severe in some sections of the United States every year, and was especially prevalent in the northeastern United States in late spring of 1958. The American plane, *Platanus occidentalis,* is by far the most susceptible species. The London plane, *P. acerifolia,* though much more resistant, shows some injury during epidemic years. The destructiveness of this disease is usually underestimated. While a single attack seldom results in serious harm, repeated annual outbreaks will eventually so weaken the tree that it becomes susceptible to borer attack and winter injury.

Symptoms. The first symptom, often confused with late frost injury, is the sudden browning and death of single leaves or of clusters of leaves as they are expanding in the spring. Later, brown, dead areas along and between the veins appear in other leaves. The dead tissues assume a triangular form as a result of the death of the veinal tissues. The leaf falls prematurely when several lesions develop or when the leaf stem is infected. Many trees are completely defoliated and remain bare until late summer, when a new crop of leaves is formed. Small twigs are also attacked and killed. Cankers appear on the leaf spurs and on twigs as sunken areas with slightly raised margins. When these completely girdle the infected parts, the distal portions, including the leaves, are killed. Moderately infected trees show dead clusters of leaves scattered over the entire tree, which contrast sharply with the unaffected foliage.

Cause. Blight is caused by a fungus known as *Gnomonia veneta.* The fungus has several distinctly different types of spores. It may overwinter in the vegetative stage (mycelium) in fallen infected leaves and in twig cankers. In the South it passes the winter in the spore stage on dormant buds. Initial infections of the young leaves in the spring may originate from spores formed in overwintered leaves on the ground, from spores developed on infected twigs, and from spores on the dormant buds. Additional crops of spores for late infections are formed on the lower surfaces of the newly infected leaves in early spring.

The prevalence and severity of attack are governed mainly by weather conditions, frequent rains and cool temperatures favoring rapid spread. The cool, wet weather during the spring of 1958 was largely responsible for the severe epidemic on American planes in the eastern United States.

Control. Control measures are rarely attempted on American planes growing in open fields or in woodlands. They are justified, however, on particularly valuable specimens and on those used for shade and ornament.

All fallen leaves and twigs should be gathered and burned in autumn to destroy the overwintering mycelium, which produces spores for the following spring's infections. Infected spurs and dead twigs should be pruned, whenever feasible, and burned.

Organic mercury sprays such as Puratized Agricultural Spray and Tag 331 provide outstanding control. In seasons of mild infection a single application made just as the buds begin to swell will give good control. During very rainy springs two additional applications are needed, one when the leaves are half grown and the other about 2 weeks later. In some parts of the country, arborists still prefer to use three or more periodically applied bordeaux mixture or other copper sprays.

Trees suffering from repeated attacks should be heavily fertilized in the fall or the following spring to increase their vigor.

FIG. 128. Canker at base of large plane branch caused by the cankerstain disease fungus *Ceratocystis fimbriata*.

CANKERSTAIN. This disease has destroyed thousands of London planes (*P. acerifolia*) in the last 25 years. While it is most destructive along the Atlantic seaboard from Philadelphia to Washington, scattered cases have been found in New York City and as far south as Vicksburg, Mississippi. An outbreak involving dozens of trees occurred in St. Louis, Missouri, in 1947. According to Mr. E. G. Rex, Chief of the Bureau of Plant Pathology, of the New Jersey Department of Agriculture, the American plane is not susceptible to this disease.

Symptoms. The conspicuous reduction in both amount and size of foliage is the most obvious symptom. This is followed within a year or two by complete death of the tree. A less obvious but more important symptom is the presence of elongated cankers or sunken areas in the bark of the trunk or larger branches (Fig. 128). The formation of numerous small or a few large cankers in such areas results in girdling, which is primarily responsible for the poor top growth and eventual death of the tree. The wood beneath the cankers is marked by reddish-brown to bluish-black streaks. A cross section of the trunk or large branches in the vicinity of the cankers reveals numerous black streaks, usually radiating as wedge-shaped sectors (Fig. 129).

Cause. Canker is caused by a fungus closely resembling *Ceratocystis fimbriata.* Several closely related species cause blue stain in lumber, and one, *C. ulmi,* is a stage of the organism causing the Dutch elm disease.

The fungus enters the trunk or branches through bark injuries and grows toward the center. From this point it grows outward radially along various lines, forming cankers wherever it reaches the bark. Large numbers of spores are formed beneath recently killed bark or on freshly exposed wood surfaces.

Many cankers center around injuries made by saw cuts, pole pruner cuts, rope burns, accidental saw scratches, and injuries by climbers' boots. Observations indicate that cankers may also start from pruning cuts that have been painted with a wound dressing containing fungus-infested sawdust. The fungus can apparently survive in some of the commonly used asphaltum wound dressings. The disease has been experimentally transmitted from infected to healthy trees by means of contaminated pruning saws. It is highly probable that the use of contaminated tools and tree paints constitutes the most important means of spread on streetside and shade trees. Injuries to the roots and trunk during curb and sidewalk installation and those made by boys with sharp instruments are also important foci for fungus invasion. Natural bark fissures and frost cracks may also constitute entrance points.

Control. Diseased trees should be removed and burned as soon as the diagnosis is confirmed. All injuries to sound trees should be avoided. Saws and other implements used in pruning planes should be thoroughly disinfested by washing in denatured alcohol or some other strong disinfestant after use on each tree in the more heavily infested zones. Because of the possibility of spreading the fungus by means of infested wound dressings, the addition to

FIG. 129. Cross section of London plane trunk showing discoloration of wood produced by the cankerstain fungus.

the dressing of some mild disinfestant that is toxic to the fungus bodies but harmless to the tree tissues is suggested. Where such a disinfestant is not used, in localities where the disease is prevalent it is best not to apply a dressing on fresh wounds.

Until recently plant pathologists believed the disease could not be spread by way of contaminated saws if pruning was done in winter, from December 1 to February 15. But newly uncovered evidence suggests that this belief was only wishful thinking and that all pruning tools should be sterilized with 70 per cent denatured alcohol after use on each tree regardless of season.

BOTRYOSPHAERIA CANKER. *Botryosphaeria ribis.* This disease, like Verticillium wilt, and shoestring root rot, is far more widespread than most arborists realize. It has long been known to cause cankers and dieback on redbud, but the author was among the first to show that its asexual stage, *Dothiorella,* caused a highly destructive disease of London planes in New York City. By cross-inoculation tests he also proved that the same fungus can infect other important shade trees including sweetgum. The fungus also has been isolated from naturally infected sweetgums showing branch dieback and discolored wood.

Among other trees known to be susceptible to this fungus are apple, avocado, Japanese persimmon, hickory, pecan, poplar, quince, and willow.

In the case of natural infections of London planes growing along city streets, the author, working with Mr. E. C. Rundlett, Arboriculturist for the New York City Department of Parks, was first to record that leaf fires made beneath streetside trees predisposed them to infection by the Dothiorella stage of the fungus. He also pointed out that the same situation exists with infections by the cankerstain fungus discussed earlier.

Control. No control is possible once the fungus has invaded the main trunk. Where infections occur on the branches, pruning well below the cankered area may remove all the infected material. Wounds and damage to the bark by fires should be avoided. The trees should be kept in good vigor by feeding, watering during dry spells, and sprays to control leaf-chewing and leaf-sucking insects.

OTHER DISEASES. Several diseases of minor importance are occasionally found on planes. The powdery mildew disease caused by the fungus *Microsphaera alni* results in defoliation of young trees. Small brown spots may be produced on the leaves by the fungi *Mycosphorella platanifolia, Phyllosticta platani,* and *Septoria platanifolia.* Most of these can easily be kept under control by adopting the spray schedule suggested for the control of blight. The powdery mildew is more effectively controlled with either sulfur or Mildex (Karathane) sprays.

INSECTS

LACE BUG. *Corythucha ciliata.* The leaves turn pale yellow and many drop prematurely as a result of the sucking by the lace bug. The adult, ⅛ inch long, white and with lace-like wings, deposits eggs on the lower leaf surface. There are several generations a year. The adult passes the winter in bark crevices and other sheltered places in the vicinity of the trees.

Control. Spray with malathion as soon as the eggs hatch on the leaves, or about June 1. Occasionally additional sprays are necessary in late July or early August to control the second generation. Lindane or DDT, applied as above, will also control this pest. Because lace bugs overwinter as adults, some control is possible with a dormant oil spray applied just before growth starts in spring.

SYCAMORE PLANT BUG. *Plagiognathus albatus.* This bug is ⅛ inch long, tan or brown in color, with dark eyes and brown spots on the wings. The young bugs are yellow-green, with conspicuous reddish-brown eyes. These pests suck out the plant juices on the upper sides of sycamore leaves, presumably leaving a poisonous material which results in yellowish or reddish spots. As the leaves grow, the injured areas drop out leaving holes.

Control. Malathion sprays applied in early May and again 2 weeks later will control this insect.

SCALES. Several species of scale insects including the black, *Saissetia oleae;* cottony maple, *Pulvinaria innumerabilis;* grape, *Aspidiotus uvae;* oyster shell, *Lepidosaphes ulmi;* sycamore, *Stomacoccus platani;* and terrapin, *Lecanium nigrofasciatum* occasionally infest the several species of planes.

Control. Malathion sprays in late spring when the young scales are crawling about are effective. A dormant spray containing a miscible oil should be applied in early spring where the infestations are unusually heavy.

BAGWORM. See under Arborvitae (Chapter 17).

WHITE FLY. See under Dogwood (Chapter 15).

POPLAR (*Populus*)

Diseases

Several destructive diseases affect the trunks and limbs of poplars. The generic name of the causal fungus is used to differentiate the various types of cankers produced. A die-back, which is not clearly understood, and a number of leaf diseases, which rarely cause much damage, also occur on poplars.

DOTHICHIZA CANKER. The dying and dead Lombardy poplars (*Populus nigra italica*) in the eastern United States are ample evidence of this destructive disease. Black and eastern cottonwoods and balsam, black, and Norway poplars also may be affected, although these species are much less susceptible than Lombardy poplar.

FIG. 130. Trunk of Lombardy poplar showing elongated sunken cankers.

Symptoms. Elongated, dark, sunken cankers occur on the trunk, limbs, and twigs (Fig. 130). The bark and cambium in the cankers are destroyed, and the sapwood is invaded and discolored. When the cankers completely girdle the trunks or branches, the distal portions die. Trees less than 4 years old are usually killed outright in a relatively short while. Older trees do not succumb so rapidly, but are readily disfigured and become virtually worthless as ornamentals or windbreaks.

Cause. The fungus *Dothichiza populea* causes this disease. It forms small raised pustules on diseased bark, from which cream-colored masses of spores ooze out in April and May. The spores are splashed onto leaves and into bark injuries, where they germinate and infect these parts. Invasion often advances down the leaf petiole into the twig, causing death of the latter.

Control. No effective control measures are known. Wounds of all sorts should be avoided. Pruning of diseased parts, as suggested for most canker diseases, does not appear to help control this disease; in fact, it often spreads it.

Inasmuch as the leaves are attacked and the fungus enters the twigs through infected leaves, some investigators have recommended repeated applications of copper sprays to reduce leaf infections. As a rule, individual trees are not sufficiently valuable to justify four or more applications of a fungicide each year. Such a practice might be warranted, however, in nurseries where the disease threatens young stock.

The only hope lies in the development of resistant varieties. The Japan poplar, *P. maximowiczii,* appears to show some resistance. Conflicting reports exist concerning the susceptibility of the Simon poplar, *P. simonii;* some persons assert that it is very resistant, others, that it is very susceptible.

CYTOSPORA CANKER. Another destructive canker disease of poplar, Cytospora canker, primarily affects trees poor in vigor. Among the factors that most commonly contribute to the weakened condition are drastic pruning and the unfavorable environment of city soils. Carolina and silver-leaf appear to be the most susceptible of the poplars. Maples, mountain ash, and willows are also susceptible to the disease. The Rio Grande cottonwood in the West is resistant to this disease.

Symptoms. Brown, sunken areas covered with numerous red pustules first appear on young twigs. The fungus moves down the stem and invades larger branches or even the trunk. Here, circular cankers are formed, which continue to expand until the entire member is girdled and the parts above die.

Cause. The fungus *Cytospora chrysosperma,* a stage of the fungus *Valsa sordida,* is the causal agent. It lives on dead branches, but can attack live branches and trunks when the tree is weakened. Spores are developed in the black pin-point fruiting structures on dead wood. They ooze from these bodies and are splashed by rains or carried by birds and insects to bark wounds or weakened and dead branches, where they germinate and infect.

Control. Because Cytospora canker is primarily a disease of weak trees, the most effective preventive is the maintenance of the trees in high vigor by fertilization, watering, and the control of insect and fungus parasites of the leaves. In addition, dead and dying branches should be removed and all unnecessary injuries avoided.

HYPOXYLON CANKER. A highly destructive disease, Hypoxylon canker occurs less commonly than Dothichiza and Cytospora cankers. In the northeastern states it affects quaking and large-tooth aspen and balsam poplar. It is most harmful to young trees and is more common on forest trees than on ornamentals.

Symptoms. Gray cankers of varying sizes appear along the trunk but never on the branches. The color changes to black as the outer bark falls away from the surface of the canker. When the bark is peeled off, blackened sapwood is visible around the edge of the canker. Wefts of fungus tissue, resembling the chestnut blight fungus but different in color, are also visible beneath the peeled bark. Many cankers attain a length of several feet and cause considerable distortion of the trunk before complete girdling of the trunk and death occur.

Cause. This disease is caused by the fungus *Hypoxylon pruinatum.* Spores are believed to initiate infections through bark injuries.

Control. This disease is so highly contagious and destructive that infected trees should be cut down and burned as soon as the diagnosis is confirmed. Injuries should be avoided as much as possible.

SEPTORIA CANKER. A serious stem canker disease of hybrid poplars, Septoria canker may also infect leaves of all species of native poplars.

Symptoms. Leaves on young shoots or lowermost branches of native poplars are spotted early in the season. Cankers appear later in the season on the stems of hybrid poplars with black, balsam, and cottonwood parentage. When cankers girdle the stem, the distal portion dies.

Cause. Mycosphorella populorum causes leaf spot and stem canker. The asexual stage of this fungus is *Septoria musiva* and hence the common name for the disease. Fruit bodies of the Septoria stage are frequently found in the cankered areas.

Control. The most effective way of combating the canker stage of this disease is to use hybrid poplar clones that have proved to be naturally resistant. The leaf spot stage on ornamental poplars can probably be prevented by periodic applications of a copper fungicide.

BRANCH GALL. Small globose galls up to 1½ inches in diameter occasionally occur at the base of poplar twigs. The bark is hypertrophied primarily, although some swelling of the woody tissues also occurs. Twigs and some branches may be killed, but the disease rarely becomes serious.

Cause. Small black pin-point fruiting bodies of the fungus *Macrophoma tumefaciens* are embedded in the bark of the gall, especially along the fissures. Fungus penetration is initiated in the developing buds in early spring, and the swelling of the tissue probably results from some stimulant secreted by the fungus.

Control. Pruning the galled branches and dead twigs to sound wood and burning the removed wood are usually sufficient to hold the fungus in check.

LEAF RUST. Several species of rust fungi attack poplars in the eastern United States. These produce yellowish-orange pustules, usually on the lower leaf surface. The two species most common in the East are *Melampsora medusae* and *M. abietis-canadensis,* which require alternate hosts to complete their life cycles. The alternate host of the former is larch, and that of the latter, hemlock.

Control. Leaf rusts rarely cause enough damage to necessitate special control measures. Where continual and heavy infections occur, larch and hemlock should be separated at least 100 yards from the poplars.

LEAF BLISTER. Brilliant-yellow to brown blisters of varying sizes occasionally appear on poplar leaves following extended periods of cool, wet weather. The blisters, which may be more than an inch in diameter, result from stimulation of the leaf cells by the fungus *Taphrina aurea.*

Control. Spraying the leaves in early spring with bordeaux mixture or any ready-made copper fungicide will control leaf blister. Such a practice is suggested only where particularly valuable specimens are involved.

LEAF SPOT. A number of leaf spots common on poplar are caused by fungi belonging to the genera *Marssonina, Phyllosticta,* and *Septoria.* Of these, the fungus *Marssonina populi,* which produces brown spots with a darker brown margin and premature defoliation, is by far the most common. It also invades and kills the twigs.

Control. The first requisite in the control of leaf spot diseases is the gathering and burning of all fallen leaves to remove an important source of inoculum. Poplars are rarely sprayed with fungicides, but bordeaux or any ready-made copper fungicide or an organic mercury may be used early in the spring on valuable trees.

DIE-BACK. A die-back of the top and complete death of Lombardy poplar, cottonwood (*P. balsamifera*), and the goat willow (*Salix caprea*) have been reported on trees in the vicinity of Washington, D. C., and in western Tennessee. On Lombardy poplar, the wood appears first water-soaked, then red, and finally brown. Death of the tree occurs when the entire cross section of the trunk is stained brown.

Cause. The cause of this trouble is not known. Bacteria have been isolated from the margins of the discolored area. There is a strong possibility that the water-soaked appearance and the subsequent staining actually result from extremely low winter temperatures and that the bacteria enter after the injury occurs.

Control. No control measures are known.

POPLAR INKSPOT. *Sclerotinia bifrons.* Thick, round or elongated black spots form on leaves and then drop out, leaving holes. Aspen, Lombardy, and Carolina poplars are most subject to inkspot, whereas the large-toothed aspen (*P. grandidentata*) appears to be immune.

Control. Where the fungus attacks valuable specimen poplars, or nursery stock, the use of bordeaux mixture or some fixed copper is suggested.

CANKER. See under Plane.

Insects

RED-HUMPED CATERPILLAR. *Schizura concinna.* The leaves are chewed by clusters of red-humped caterpillars, yellow and black striped forms with red heads and red humps. The adult, a grayish-brown moth with a wingspread of 1¼ inches, deposits masses of eggs on the lower leaf surface in July. Winter is passed in a cocoon on the ground.

Control. Spray with arsenate of lead, DDT, or malathion when the caterpillars are small.

TENTMAKER. *Ichthyura inclusa.* The leaves are chewed and silken nests appear on twigs as a result of infestations of the tentmaker, a black larva with pale yellow stripes, which attains a length of 1¼ inches at maturity. The adult female is a moth with white-striped gray wings. The pupae overwinter in fallen leaves.

Control. Remove and destroy nests or spray with arsenate of lead, DDT, or malathion when the larvae are small.

POPLAR BORER. *Saperda calcarata.* Blackened and swollen scars on the limbs and on the trunk and sawdust at the base of the tree usually result from attack by the poplar borer, a white larva, 1¼ to 1½ inches long when full grown. The upper and lower parts of the larva's body have horny points. The adult female is a bluish-gray beetle, with black spots and yellow patches, and is slightly over an inch in length.

Control. Remove and burn badly infested trees. Spraying the trees with DDT at 2-week intervals, starting in late July, may be justified on valuable trees.

WILLOW LEAF BEETLE. *Plagiodera versicolora.* Holes are chewed in the leaves by the willow leaf beetle, the metallic blue adult form, ⅛ inch long; and the leaves are skeletonized by the black larvae, which are ¼ inch long. The adult lives through the winter and deposits eggs on the lower leaf surface in the spring.

Control. Spray with lead arsenate, DDT, or malathion after the eggs hatch in spring.

BRONZE BIRCH BORER. *Agrilus anxius.* Weakened trees are always more susceptible than vigorous trees to attack by the bronze birch borer. Branches are girdled and the upper parts of the tree die back as a result of feeding by the borer, a white, legless larva, ¾ inch long. The adult beetle, which appears in June and feeds on the leaves for a short while, is ½ inch long and bronze in color. The female deposits eggs in bark crevices in June.

Control. Spray the leaves, trunk, and branches with DDT in early June, and repeat in 3 weeks. Remove and burn badly infested parts. Water important trees during drought periods.

FALL WEBWORM. See under Ash (Chapter 15).

SWEETGUM (*Liquidambar*)

Because of their increased use in recent years, the diseases and insect pests of sweetgums have received more than the usual amount of attention.

Diseases

BLEEDING NECROSIS. The author was first to discover the bleeding necrosis disease of sweetgum in New Jersey and on Staten Island, New York, as far back as 1941.

Symptoms. The most striking symptom, readily visible at some distance, is the profuse bleeding of the bark, usually at the soil line or a few feet above, but occasionally observed as high as 20 feet on the main trunk and lower branches. The exudation looks like heavy motor oil poured over the bark.

The condition of the inner bark and sapwood beneath the oozing area, however, presents a more positive symptom of the disease. Such bark is dark reddish-brown with an occasional pocket containing a white crystalline solid. The outer layer of sapwood may be brown or olive-green in color.

Infected trees may exhibit undernourished foliage and more or less extensive dying-back of terminal branches by the time profuse bleeding is visible on the trunk.

Cause. The fungus *Botryosphaeria ribis* causes bleeding necrosis. When first reported by the author in the early 1940's the fungus was listed under the asexual stage Dothiorella. There is some evidence that raising the grade around the tree makes it more subject to this fungus.

Control. No control measures are known.

OTHER DISEASES. Several other important diseases of sweetgum have been reported in recent years. For most, the causal agent is still unknown. One, Leader Die-back, is caused by the fungus *Diplodia theobromae.* It was found capable of causing die-back of the leader branch under certain predisposing environmental conditions.

The disease known as Sweetgum Blight has caused death of many trees in Maryland and adjacent states. At this writing no causal organism has been associated constantly with the disease.

Insects

LEAF TIER. *Salebria afflictella.* The leaves are tied and matted together by small larvae.

Control. Spray with arsenate of lead, DDT, or malathion before the leaves are matted together.

CATERPILLARS. The leaves of sweetgum are occasionally chewed by the caterpillars listed below, as well as by those of several species of moths including the luna, *Actias luna;* polyphemus, *Antheraea polyphemus;* and the promethea, *Callosamia promethea.*

BAGWORM. See under Arborvitae (Chapter 17).

FALL WEBWORM. See under Ash (Chapter 15).

FOREST TENT CATERPILLAR. See under Maple (Chapter 15).

TULIP TREE (*Liriodendron*)

Diseases

LEAF YELLOWING. Starting in midsummer, during dry, hot periods, many tulip tree leaves turn yellow and drop prematurely. The yellowing is due to climatic conditions and not to any destructive organism. The leaves of recently transplanted trees exhibit yellowing more frequently than those of well-established trees. Very often, small, angular, brownish specks appear on the leaves between the leaf veins as a preliminary stage of yellowing and defoliation. This trouble closely resembles leaf scorch, discussed on page 194.

LEAF SPOT. Several leaf spot fungi belonging to the genera *Cercospora*, *Cylindrosporium*, *Gloeosporium*, *Phyllosticta*, and *Ectostroma* attack tulip trees. One of the most common species, *Gloeosporium liriodendri*, produces large brown blotches (Fig. 131); the other species cause smaller, more sharply defined spots.

FIG. 131. Anthracnose of tulip leaves caused by the fungus *Gloeosporium liriodendri*.

Control. Leaf spots rarely become sufficiently destructive to warrant control measures aside from the gathering and burning of infected leaves. The use of copper or organic mercury sprays may be justified on valuable trees.

POWDERY MILDEW. The powdery mildew fungi, *Phyllactinia corylea* and *Erysiphe polygoni*, may parasitize the leaves.

Control. Where particularly valuable specimens are affected, the use of Mildex (Karathane) may be justified.

DIE-BACK. A branch die-back preceded by yellowed and undersized leaves, has been reported on tulip trees in the South.

Cause. The fungus *Myxosporium* has been constantly associated with die-back, but its exact role is not clearly understood.

Control. No control measures are known. It is likely that severe pruning of affected branches, spraying with a copper fungicide, and feeding the trees might be justifiable on importantly placed ornamental tulip trees.

NECTRIA CANKER. See under Magnolia (Chapter 15).

INSECTS

TULIP SCALE. *Toumeyella liriodendri.* Trees may be killed by heavy infestations of oval, turtle-shaped, often wrinkled, brown scales, ⅓ inch in diameter (Fig. 132). The lower branches, which are usually first to die, may be com-

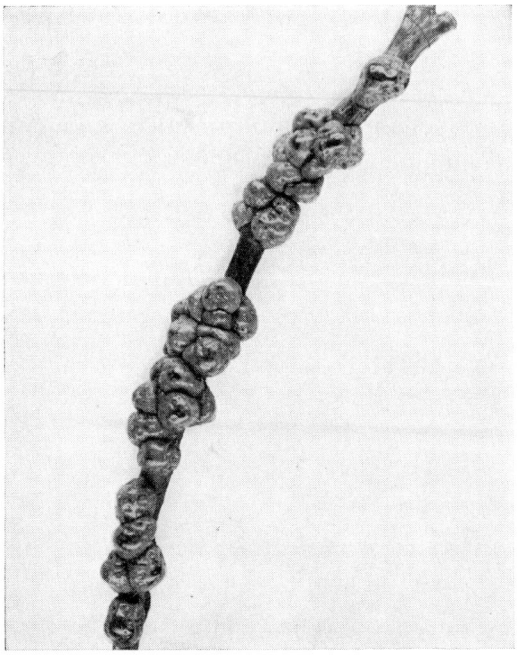

FIG. 132. LEFT: Adult stage of the tulip-tree scale.

FIG. 133. RIGHT: Crawling stage of tulip scales which appear in late summer.

pletely covered with scales. The secretion from the scales drops on leaves and is soon covered by sooty black mold. Tulip scales overwinter as partly grown young. They grow rapidly until they mature in August. Young scales are produced in late August, at which time the old females dry up and fall from the tree (Fig. 133).

Control. Spray the trees with a dormant miscible oil on a relatively warm day in late March or early April before the buds open. Or spray with malathion in late August or early September when the young are crawling about, and repeat the spray 2 weeks later.

In addition to the very common and destructive tulip tree scale, this host is also subject to oyster shell and willow scale.

TULIP TREE APHID. *Macrosiphum liriodendri.* This small green aphid, like the tulip scale insect, secretes copious quantities of honeydew. Hence leaves of other plants growing beneath the tree are coated with honeydew, which is then overrun with the sooty-mold fungus.

Control. Spring and summer applications of malathion or lindane will control this aphid.

TUPELO (*Nyssa*)

Tupelo, also known as sourgum, makes a fine lawn or park tree. Because it is difficult to transplant, it is not commonly used.

Diseases

TUPELO LEAF SPOT. *Mycosphorella nyssaecola.* Irregular purplish blotches, which later enlarge to an inch or more in width, are commonly scattered over the upper surfaces of the leaves of young tupelo trees in the southeastern states. Minute, black fruiting bodies of the fungus are visible in the infected areas.

Two other fungi, *Cercospora nyssae* and *Pirostoma nyssae*, also cause leaf spots.

Control. Copper fungicides are suggested as preventive sprays. Raking and burning infected leaves in the fall should also help to reduce infections.

RUST. *Aplospora nyssae.* This rust fungus occasionally infects tupelos from Maine to Texas.

Control. Control practices are not warranted.

Insects

TUPELO LEAF MINER. *Antispila nyssaefoliella.* Leaves are first mined by the larval stage. When mature, the larvae cut oval sections out of the leaves and fall to the ground with the severed pieces.

Control. Spray with malathion in May when the adult moths emerge. Repeat within 10 days.

WALNUT (*Juglans*)

Black walnut and butternut are rarely used as street trees. They require exacting soil conditions, are subject to wind and ice damage, are untidy, and their nuts are sought by children.

In ornamental plantings, the black walnut may produce detrimental effects on other types of plants growing near by. The author has observed extensive damage to native and hybrid rhododendrons and to mountain laurel, when these plants were grown in close proximity to black walnut roots. Other observers have reported similar damage to apple trees, alfalfa, tomato, potato, and a number of perennial ornamental plants growing near black walnut and butternut as well.* In most instances, the removal of the nut trees resulted in the disappearance of the harmful effects.

A number of important diseases and insect pests occur on the different species of walnut.

Diseases

CANKER. This disease is also common on apple; aspen; beech; birch; mountain, red, and sugar maple; dogwood; butternut; pignut hickory; and red, white, and black oaks.

Symptoms. Scattered, often numerous, rough, sunken, or flattened cankers with a number of prominent ridges of callus wood arranged more or less concentrically on the trunks or branches are characteristic symptoms. The size of the cankers varies according to age; on larger branches and on the trunk some attain a length of 4 feet and a width of 2½ feet, and have as many as 24 concentric ridges. Complete girdling and death of the distal portions may occur within a few years on smaller branches and in 20 or more

* A most exhaustive study of this problem is presented in West Virginia Agri. Exp. Sta. Circular 347 by M. G. Brooks.

years on large trunks. The concentric pattern of the canker is less evident on twigs and smaller branches.

On the dead bark and wood in late fall and winter, small, globose, red masses, the fruiting structures of the causal fungus, are barely visible to the unaided eye.

Cause. Canker is caused by the fungus *Nectria galligena.* In late winter and in the spring, the red fruiting bodies produce spores, which are forcibly ejected into the air. These are then carried by wind and rain to bark injuries made by insects, fungi, ice, low temperatures, wind, rubbing plant parts, or dead branch stubs. The spores germinate and produce mycelium, which penetrates the bark and wood. Penetration progresses rather slowly, usually about ½ inch of tissue being killed around the developing canker annually.

Control. All badly diseased trees should be felled, and the cankered tissues cut out and burned.

Valuable trees only mildly infected can be saved by removing the cankers by surgical methods, followed by shellacking the edges of the wound and thoroughly coating the open area with a good wound dressing. In addition, the trees should be fertilized, watered, and sprayed to control leaf pests and ensure rapid callusing.

Sites exposed to full sunshine should be used for new walnut plantings, inasmuch as the fungus does not appear to thrive under such conditions.

BACTERIAL BLIGHT. English and Persian walnuts grown for ornament and for nut production in the eastern United States are susceptible to this destructive blight.

Symptoms. Small, water-soaked spots, which turn reddish-brown, appear on the young, unfolding leaves in the spring. The spots are usually isolated, but they may coalesce, causing considerable distortion of the leaves. Black, sunken lesions also appear on the twigs, which may be completely girdled, causing death of the parts above. Infection of the husks usually results in premature dropping of the fruit, or in browning and decay of any fruit that remains attached.

Cause. The bacterium *Xanthomonas juglandis* causes blight. It is believed to overwinter mainly in diseased buds and is splashed by rain to various parts of the tree in the spring.

Control. Three well-timed applications of bordeaux mixture 2-2-50 will prevent blight. The first should be made when the terminal leaves of the new shoots have parted to barely expose the female or nut-bearing flowers; the second, just as the male flowers come into bloom but before the blossom ends of most of the female blossoms open; and the third, immediately after the female flowers are pollinated and the nuts are about the size of peas. The second application is the most important of the three. The addition of 1 pint of summer white oil in 50 gallons of the bordeaux mixture for the first two applications will reduce the severity of foliage injury.

Recent research has shown that sprays containing the antibiotic streptomycin or phenyl mercury acetate will also control the bacterial blight disease.

DIE-BACK. Though black walnut occasionally dies back, this disease is most prevalent and destructive on butternut and the Japanese walnut (*Juglans ailantifolia*).

Symptoms. On butternut the smaller branches first die back; the dying back progressing slowly until the main branches are involved. The bark on the affected branches changes from the normal greenish-brown, to reddish-brown, and finally gray. Small, black, pimple-like pustules soon cover the dead bark. These disappear within a year or two, leaving small, irregular holes in the loose outer bark.

Cause. The fungus *Melanconis juglandis* causes the slow dying back and eventual death of the tree. Olive-gray elliptical or kidney-shaped spores are exuded in a black inky mass from the black pustules on the bark to initiate new infections.

Control. Severely affected trees should be removed and destroyed. Where the disease is still confined to the upper ends of the branches, pruning to sound wood will help to check further spread. In addition, fertilization, watering, sanitation, and insect and leaf-disease control measures should be adopted to help the tree regain its vigor.

LEAF SPOT. Among the numerous leaf spots that occur on black walnuts and butternuts used for ornament, two are so destructive and appear so frequently that preventive sprays must be applied annually to curb them. One, *Microstroma juglandis*, also causes the witches'-broom disease of shagbark hickory.

Symptoms. Irregularly circular, brown spots are produced early in the summer by the fungus *Marssonina juglandis*. The spots result from infection by spores originating from diseased leaves that have overwintered on the ground. Additional infections develop during the summer from spores produced in these spots. When several spots form on a leaflet, premature defoliation results.

Other fungi known to spot leaves are *Cercospora juglandis*, *Phleospora multimaculans*, and *Marssonina californica*.

Control. Because the leaves harbor the causal fungi, all fallen leaves should be gathered and burned to eliminate this important source of inoculum. Black walnuts and butternuts of ornamental value should be sprayed periodically with ziram or a copper fungicide. The first application should be made when the leaves are nearly half-grown; the second, about 10 days later; and the third, when the leaves are full-grown.

TRUNK DECAY. The trunks of black walnut and butternut often show a white or brown decay of the heartwood. The former is caused by the so-called false tinder fungus, *Fomes igniarius*, which forms hard, gray, hoof-shaped fruiting bodies up to 8 inches in width, along the trunk in the vicinity of the decay. The latter is caused by the fungus *Polyporus sulphureus*, which forms soft, fleshy, shelf-like fruiting structures, which are orange-red above and brilliant yellow below. The fruiting bodies become hard, brittle, and dirty white as they age. These fungi usually enter the trunk through bark injuries and dead branch stubs.

Control. Once these decays become extensive, little can be done to eradicate them. Cleaning out the badly decayed portions and applying a wound dressing are suggested. Initial infections can be kept at a minimum by maintaining the vigor of the trees and by protecting the wounds as soon as they are formed.

WITCHES'-BROOM. VIRUS. This disease has been discovered in the eastern United States. Besides affecting the black walnut (*J. nigra*), it also occurs on butternut (*J. cinerea*) and Japanese walnut (*J. ailantifolia*).

Symptoms. The disease is characterized primarily by the appearance of brooms or sucker growths on main stems and branches, tufting of terminals, profuse development of branchlets from axillary buds, leaf dwarfing, and, at times, death of the entire tree. These symptoms vary from mild to severe and are particularly pronounced on Japanese walnut.

Cause. Witches'-broom is probably caused by a virus, inasmuch as transmission of the disease has been effected by grafting pieces of bark from infected trees to healthy ones.

Control. No controls have been developed. Because it is highly likely that the virus is spread by insects, valuable trees should be sprayed regularly with a mixture of DDT and malathion.

Insects

WALNUT DATANA. *Datana integerrima.* Trees are defoliated by the black Datana caterpillar, which is covered with long white hairs and which grows to 2 inches in length. The adult female moth has a wingspread of 1½ inches. Its dark buff wings are crossed by 4 brown lines. Eggs are deposited in masses on the lower sides of the leaves in July. Winter is passed in the pupal stage in the soil.

The following caterpillars also attack walnut: hickory horned devil, orange tortrix, omnivorous looper, red-humped, and yellow-necked.

Control. Spray with lead arsenate, DDT, or malathion when the larvae are small.

WALNUT SCALE. *Aspidiotus juglans-regiae.* Part or all of a tree may be severely weakened by masses of round brown scales, ⅛ inch in diameter. The adult female is frequently encircled by young scales. It overwinters on the bark.

Nut trees are also susceptible to many other scales including black, California red, calico, cottony-cushion, greedy, oyster shell, Putnam, scurfy, and white peach scale.

Control. Spray with lime sulfur when the trees are dormant, or with malathion in late spring and again in early summer.

WILLOW (*Salix*)

Willow are rarely used as street trees or in locations where the breaking off of their weak-wooded branches is likely to cause injury. Some species, however, are used extensively in landscape plantings, especially around ponds and streams.

Diseases

LEAF BLIGHT. Blight, or scab, is the most destructive disease of willows in the United States. It has already killed hundreds of trees in the northeastern states and appears to be spreading westward and southward.

Symptoms. Soon after growth starts in the spring, a few small leaves turn black and die. Later, all the remaining leaves on the tree suddenly wilt and blacken, as if they had been burned by fire. Cankers on twigs may result from the leaf infections. Following rainy periods, dense olive-brown fruiting structures of the causal fungus appear on the undersides of blighted leaves, principally along the veins and midribs.

Cause. Leaf blight is caused by the fungus *Venturia chlorospora.* It passes the winter in infected twigs, on which it produces spores during early spring. The spores are washed by rain onto the newly developing leaves and initiate early infections. Such leaves produce large numbers of spores of the asexual stage, *Fusicladium saliciperdum,* which are responsible for the severe blighting of the remaining leaves later in the season.

Control. Pruning of dead twigs and branches will eliminate the most important source of inoculum for early-season infections. Spraying with bordeaux mixture or any ready-made copper spray 3 or 4 times at 10-day intervals starting when the leaves begin to emerge in the spring will usually give control. These practices are recommended only for valuable specimens.

A number of willows, including the weeping, the bay-leaved, the osier, the purple, and the pussy willow, appear resistant to blight. These should be planted in place of the more susceptible species, such as the crack and the heart-leaved willow. Golden willow has been reported by some investigators to be susceptible, and by others to be resistant.

BLACK CANKER. Closely associated with blight is the disease known as black canker.

Symptoms. The symptoms of black canker resemble those of blight but usually appear later in the summer. Dark brown spots, many of which show concentric markings, appear on the upper leaf surface. Whitish-gray or gray elliptical lesions, with black borders, appear subsequently on the twigs and stems. Clusters of minute black fruiting bodies develop in the stem lesions. Successive attacks over a 2- or 3-year period usually result in death of the entire tree.

Cause. Black canker is caused by the fungus *Physalospora miyabeana.* Two types of spores are produced on the stem lesions: one, exuding from fruiting bodies as small pink masses in early summer; and the other, released from tiny, black, spherical bodies in early spring. The fungus lives over the winter on diseased twigs in the latter type of fruiting bodies.

Control. Pruning and spraying as suggested for willow blight are recommended. The use of resistant varieties also offers a means of avoiding this destructive disease. The bay-leaved, the osier, and the weeping willow appear to be resistant, whereas the crack, the heart-leaved, the white, the purple, and the almond willow appear to be susceptible. As in the case of blight, golden willow has been reported by some observers to be extremely susceptible, and by others to be resistant.

CYTOSPORA CANKER. Willows are subject to this canker disease, caused by the fungus *Cytospora chrysosperma*, the perfect stage of which is *Valsa sordida*, already discussed under Poplar on page 362. The control measures listed under that host hold true also for willows. In addition, the use of resistant varieties offers a means of avoiding serious trouble. One investigator has observed that the disease occurs rarely on black willow and on peach-leaf willow, whereas crack and golden willows appear to be extremely susceptible.

Other cankers on willow may be caused by *Botryosphaeria ribis* (p. 359), *Cryptodiaporthe salicina*, and *Discella carbonacea*.

BACTERIAL TWIG BLIGHT. Leaves turn brown and branches blight back for some distance as a result of a bacterial infection. Brown streaks are produced in the wood of infected branches.

Cause. The bacterium *Pseudomonas saliciperda* causes this disease. It overwinters in cankers on branches and twigs and infects leaves as soon as they unfurl.

Control. Pruning as many infested twigs as possible, and spraying as suggested for leaf blight and black canker, are the recommended control measures.

LEAF SPOT. Willows are subject to a number of leaf spot diseases, which rarely become serious. The fungus *Rhytisma salicinum* produces thick tar-like spots, ¼ inch in diameter, on the upper leaf surface. The fungus *Cylindrosporium salicinum* produces numerous small brown spots, resulting in premature defoliation. Other fungi known to produce leaf spots on willow are: *Ascochyta salicis*, *Cercospora salicina*, *Marssonina kriegeriana*, *Myrioconium comitatum*, *Phyllosticta apicelis*, *Ramularia rosea*, and *Septogloeum salicinum*.

Control. Because most leaf spot fungi overwinter on diseased fallen leaves, gathering and burning the leaves is a recommended practice. Valuable specimens should be sprayed with ferbam, zineb, phenyl mercury, or copper fungicides when leaf spots cause considerable defoliation annually.

GRAY SCAB. A disease affecting a number of species of willow, including *Salix fragilis*, *S. lasiandra*, and *S. lasiolepis* in the United States.

Symptoms. Round, irregular, somewhat raised, grayish-white spots with narrow, dark brown margins form on the leaves. Affected portions of the leaves frequently drop away.

Cause. The fungus *Sphaceloma murrayae* is known to cause gray scab. Remove and burn old leaves to destroy inoculum for later infections. Valuable trees can be protected with periodic applications of copper fungicides.

LEAF RUST. Several species of rust fungi attack willows. The two most common ones are *Melampsora abieti-capraearum* and *M. bigelowii*. The alternate host of the former is the fir, and that of the latter, the larch.

Small orange-yellow pustules appear on the lower leaf surface about the first week in June. The number of pustules increase rapidly until the entire surface is covered with the spore masses. After August first, a second type of spore is produced in small, raised, brown blisters on both leaf surfaces. These spores germinate the following spring, producing wind-blown secondary spores that can infect the alternate hosts.

Control. Although rust infections are not considered serious, they may result in heavy defoliation of young trees. Gathering and burning fallen leaves will help to prevent serious outbreaks. The use of copper or sulfur fungicides as preventives is suggested only in rare instances.

INSECTS

APPLE TREE TENT CATERPILLAR. *Malacosoma americanum*. The leaves are chewed and small silken nests are formed in branch crotches by the tent caterpillar, a yellow-haired, black form with white lines down the back (Fig. 134). It attains a length of nearly 2 inches, prefers to feed on wild cherry, but will feed on apple, peach, and plum in addition to willow. When these trees are scarce, the insects will also defoliate ash, beech, birch, elm, maple, oak, poplar, and many shrubs. The adult female, a fawn-colored moth, has a wing expanse of 2 inches. Eggs are laid in cylindrical clusters on twigs in July (Fig. 135).

Control. Remove and burn egg clusters and nests when they appear, and spray with arsenate of lead, DDT, or malathion when new twig growth is 2 to 4 inches long in spring.

CONE GALL. *Rhabdophaga strobiliodes*. Cone-shaped galls at the branch tips that hinder bud development are produced by small maggots. The adult, a small fly, deposits eggs in the opening buds. The larvae hibernate in cocoons inside the galls.

Control. Remove and burn galls in the fall, or spray thoroughly with malathion when the buds are swelling in the spring.

WILLOW SHOOT SAWFLY. *Janus integer*. Terminal shoots wilt in early spring from the feeding in the pith by a borer, the larva of the sawfly. The adult, a wasp-like insect, lays eggs in the shoots in early spring and then girdles the shoots below the point of deposition.

FIG. 134. Apple tree tent caterpillar massed above their characteristic tent.
FIG. 135. INSET: Egg masses of apple tree tent caterpillar on hawthorn twig.

Control. Prune and burn infested twigs. Spray with arsenate of lead, DDT, or malathion when the larvae are small.

EUROPEAN WILLOW GALL MIDGE. *Rhabdophaga salicis.* Swollen, distorted twigs may be produced by yellowish, jumping maggots of the European willow gall midge. The adults appear in early spring.

Control. Prune and burn infested twigs.

MOTTLED WILLOW BORER. *Cryptorhynchus lapathi.* Swollen and knotty limbs may be produced by mottled willow borers, white legless larvae ½ inch long, which eat through the cambium, sapwood, and heartwood. The adult beetle, ⅓ inch long, has a long snout and grayish-black, mottled wing covers. Eggs are laid in the fall, and the larvae overwinter in tunnels beneath the bark.

Control. Spray the leaves, branches, and trunk thoroughly during the latter part of August with 3 gallons of 18.5 per cent dieldrin emulsifiable solution in 100 gallons of water (4 liquid ounces in 3 gallons).

WILLOW SCURFY SCALE. *Chionaspis salicis-nigrae.* Branches and even small trees may be killed by heavy infestations of willow scale, a pear-shaped white scale, ⅛ inch long. The pest passes the winter in the egg stage underneath the scale of the female.

Control. In spring, just after the eggs hatch, spray with malathion.

WILLOW FLEA WEEVIL. *Rhynchaenus rufipes.* Tiny jet-black elliptical weevils feed on opening buds and new leaves in May. The larval stage, a 1/12 inch long, dirty white grub, mines the leaves of willow and poplar.

Control. To control the adult weevil spray early in May with arsenate of lead, DDT, or malathion. The grub stage can be controlled in early June with lindane or malathion sprays.

WILLOW LACE BUG. *Corythucha mollicula.* The leaves may be severely mottled and yellowed by this sucking insect.

Control. Spray with lindane or malathion in late spring.

BAGWORM. See under Arborvitae (Chapter 17).

CARPENTER WORM. See under Ash (Chapter 15).

WILLOW LEAF BEETLE. See under Poplar.

RED-HUMPED CATERPILLAR. See under Poplar.

YELLOWWOOD (*Cladrastis*)

The yellowwood is remarkably free of insect pests and fungus parasites. It is subject to the fungus *Polyporus spraguei,* which causes a brown friable rot of the heartwood of the butt and roots of living trees. No control is known. Avoiding bark wounds on the trunk may help to prevent infections.

An occasional tree has been reported infected with Botryosphaeria canker (p. 359) and Verticillium (p. 247).

The powdery mildew fungus *Phyllactinia corylea* has been reported from Oklahoma.

17

DISEASES AND INSECT PESTS OF EVERGREEN TREES

The treatment of evergreens in this chapter parallels closely that in the previous chapter. Evergreen trees are used primarily for ornamental purposes. In a few areas in the country some of them are occasionally used as street trees.

ARBORVITAE (*Thuja*)

In the wild, the American arborvitae, also called white cedar, inhabits moist, sometimes swampy areas along banks of streams. When under cultivation, it also prefers a moist soil, but such soil must be well drained. The unthrifty condition of many trees in ornamental plantings is often associated with waterlogged soils.

Many horticultural varieties of American arborvitae are now available from nurserymen. Most of these are subject to the same troubles that affect the parent tree.

Diseases

LEAF BLIGHT. Although most destructive on the giant arborvitae of the Northwest, leaf blight occasionally appears on the American arborvitae. The disease is most prevalent on trees under 4 years old, but may attack trees of all ages.

Symptoms. From 1 to more than 3 irregularly circular, brown to black cushions appear on the tiny leaves in late spring. The leaves then turn brown, and the affected areas appear as though scorched by fire. Toward fall the leaves drop, leaving the branches bare.

Cause. The fungus *Didymascella thujina* causes leaf blight. The brown to black cushions on the leaves are the fruiting bodies of this fungus. Spores are discharged into the air from these bodies during rainy periods in the summer, to initiate new infections.

Control. Small trees or nursery stock can be protected by several applications of bordeaux mixture or other copper sprays in midsummer and early autumn.

TIP BLIGHT. This disease resembles leaf blight but occurs primarily on varieties of *Thuja orientalis,* the golden biota or golden arborvitae being most susceptible. The leaves near the branch tips turn brown in late spring or early summer. Tiny black bodies, which are visible through a magnifying lens, occur on infected leaves.

Cause. The fungus *Coryneum berckmanii* causes tip blight. Several other fungi including *Cercospora thujina, Pestalotia funerea,* and *Phacidium infestans* are also associated with tip and twig blights.

Control. Spray with a copper fungicide before rainy periods.

JUNIPER BLIGHT. The fungus *Phomopsis juniperovora,* which is the cause of the blight of the red cedar and other species of *Juniperus,* also attacks arborvitae. A complete discussion of this disease is presented on page 402.

BROWNING AND SHEDDING OF LEAVES. The older, inner leaves of arborvitae turn brown and drop in the fall. When this condition develops within a few days or a week, as it does in some seasons, many persons feel that a destructive disease is involved. Actually it is a natural phenomenon similar to the dropping of leaves of deciduous trees. When the previous growing season has been favorable for growth, or when pests, such as red spider, have not been abundant, the shedding occurs over a relatively long period through the fall and consequently is not so noticeable.

INSECTS

BAGWORM. *Thyridopteryx ephemeraeformis.* Spindle-shaped bags suspended from twigs and each inhabited by a worm are the most characteristic signs of the bagworm (Fig. 136). The young worms appear in late May and feed on the leaves. The adult stage is a moth. The pests overwinter in the egg stage in the old female bags.

Control. Hand-pick and burn the bags in winter. Spray with arsenate of lead or malathion when the young worms first appear. The time of this appearance varies from May 1 to June 1, depending on the locality. Spray 7 to 10 days later if live bagworms are still present. DDT sprays are not effective.

SCALE. *Lecanium fletcheri.* This dark brown, flat, more or less hemispherical scale belongs to the so-called naked or unarmored scales. Newly hatched young have an amber color, which changes to pale orange-yellow. The insect overwinters as a partly grown scale on the stems, branches, and leaves. The pest is relatively unimportant on this host but is rather serious on yews. It may also infest pachysandra. Another species *L. coryli* occasionally attacks arborvitae and yews.

FIG. 136. Spindle-shaped bag harboring a bagworm on fir twig.

Control. Malathion sprays are very effective if applied during midsummer when the young are crawling about.

LEAF MINER. *Argyresthia thuiella.* The leaf tips turn brown as a result of feeding within the leaves by a small maggot. The adult stage is a tiny gray moth with a wingspread of ⅓ inch. The maggots overwinter in the leaves.

Control. Trim and burn infested leaves. Spray with malathion in late June or early July, when eggs are hatching. A mixture of DDT and a miticide is also effective.

CEDAR TREE BORER. *Semanotus ligneus.* This pest attacks the bark and wood of cedars, redwood, Douglas fir, and Monterey pine. The adult beetle is ½ inch long, black with orange or red markings on the wing cover. The larval stage bores into the inner bark and wood, frequently girdling the trees.

Control. Spray with DDT or malathion when the beetles emerge from the tree.

SPIDER MITE. See under Juniper.

BALD CYPRESS (*Taxodium*)

This genus has both evergreen and deciduous members. It is relatively free of fungus parasites and insect pests.

Diseases

TWIG BLIGHT. This disease, also present on a number of other conifers, including arborvitae, junipers, cypress, and yews, is rarely serious. It causes a spotting of leaves, cones, and bark and, in very wet seasons, a twig blight.

Cause. The fungus *Pestalotia funerea* is frequently associated with twig blight. It is not considered a vigorous parasite but becomes mildly pathogenic on trees weakened by mites, dry weather, sun scald, or low temperatures.

Control. Control measures are rarely applied, although a copper fungicide would probably be effective.

WOOD DECAY. A number of fungi belonging to the genera *Fomes, Lenzites, Poria,* and *Polyporus* are associated with wood decay of bald cypress. The last is sometimes found on living trees.

Control. Avoid wounding the trunk base of valuable trees. Keep trees in good vigor by watering and, if needed, by feeding.

Insects

CYPRESS MOTH. *Recurvaria apicitripunctella.* The larval stage of this moth mines the leaves of bald cypress and hemlock, then webs them together in late summer. The female adult is a small yellow moth with black markings and fringed wings.

Control. Spray with malathion when the larvae begin to mine the leaves in late spring and before they web the leaves together.

BOXWOOD (*Buxus*)

Boxwood requires a well-drained, neutral soil with an occasional light application of ground limestone. In poorly drained soil, it is more susceptible to winter injury and fungus diseases. A few kinds, particularly the tree box, grow into small trees and hence are discussed here.

Diseases

CANKER. One of the most destructive diseases of boxwood is canker.

Symptoms. The first noticeable symptom is that certain branches or certain plants in a group do not start new growth so early in the spring as do others, nor is the new growth so vigorous as that on healthy specimens. The leaves turn from normal to light green and then to various shades of tan. Infected leaves turn upward and lie close to the stem instead of spreading out like the leaves on healthy stems. The diseased leaves and branches show small, rose-colored, waxy pustules, the fruiting bodies of the fungus. The bark at the base of an infected branch is loose and peels off readily from the gray to black discolored wood beneath. Infection is frequently found to take place at the bases of small dead shoots or in crotches where leaves have been allowed to accumulate.

Cause. Canker is caused by the fungus *Pseudonectria rousselliana,* which is found on the stem cankers. Another stage of the same fungus, known as *Volutella buxi,* attacks the leaves and twigs, producing pale rose-colored spore masses on yellowed leaves. A species of Verticillium is occasionally associated with the die-back of twigs. The fungus *Nectria desmazierii,* whose imperfect stage is known as *Fusarium buxicola,* is also capable of causing canker and die-back of boxwood.

Control. Dead branches should be removed as soon as they are noticeable, and cankers on the larger limbs should be treated by surgical methods. The annual removal and burning of all leaves that have lodged in crotches is recommended. Four applications of bordeaux mixture 3-3-50 or lime sulfur 1-50 have been shown to be very effective in preventing canker. The first should be made after the dead leaves and dying branches have been removed and before growth starts in the spring; the second, when the new growth is half completed; the third, after spring growth has been completed; and the fourth, after the fall growth has been completed. The boxwood should be fed with occasional applications to the soil of well-rotted cow manure, or commercial fertilizers, and ground limestone.

BLIGHT. Several fungi are associated with the blighting of boxwood leaves. The most common are *Phoma conidiogena* and *Hyponectria buxi*. The exact role of these fungi in this disease complex is not clear.

Control. The fungicides recommended for canker control will also control blight.

LEAF SPOT. Leaf spot, common on boxwood, may be caused by one of several fungi. *Macrophoma candollei* is the fungus usually found associated with leaf spotting. Infected leaves are straw-yellow, and the under surface is thickly dotted with small black bodies, the fruiting structures of the organism. The fungus *Phyllosticta auerswaldii* is also frequently found in spotted leaves, as is a species of *Colletotrichum*. All of these fungi are apparently limited in their attacks to foliage weakened by various causes, for inoculations into healthy foliage on vigorous plants have failed to produce infection.

Control. Leaf spots may be controlled by shaking out all fallen and diseased leaves from the center of the bush and destroying them by burning. All dead branches in the center of specimen plants or hedges should be removed to allow better aeration. An application of bordeaux mixture 2-2-50 or any good ready-made copper fungicide before growth starts in the spring is beneficial. This spray will discolor the foliage, but the unsightly effect is soon hidden by new growth.

WINTER INJURY AND SUN SCALD. Most boxwood troubles in the northeastern United States are due to freezing and sun scalding, which primarily injure the cambium of unripened wood. Several distinct types of symptoms are exhibited by winter injury. Young leaves and twigs may be injured when growth extends far into the fall or begins too early in the spring. Leaves may turn rusty brown to red as a result of exposure to cold, dry winds during the winter. A die-back of leaves, twigs, and even the entire plant may occur on warm winter days when the above-ground tissues thaw rapidly and lose more water than can be replaced through the frozen soil and roots. Another type of winter injury is expressed by the splitting and peeling of the bark. The bark becomes loosened and the stems are entirely girdled, resulting in death of the distal portions.

Control. Fertilizers should be applied in late fall, preferably, or very early in the spring. Adequate windbreaks should be provided during the winter, especially in the more northern latitudes. In the latitude of New York City spraying with Wilt-Pruf on a mild day in December will provide as much protection from winter winds as do burlap windbreaks. A heavy mulch consisting of equal parts of leaf mold and cow manure should be applied to prevent deep freezing and to aid in supplying water continuously.

Insects

LEAF MINER. *Monarthropalpus buxi.* Oval, water-soaked swellings on the lower leaf surface result from the feeding inside the leaves by the leaf miner, a yellowish-white maggot, ⅛ inch long. The adult is a tiny midge, 1/10 inch long, which appears in May.

Control. When the adult midges are seen in mid-May (about the time the weigelas are in full bloom) spray the leaves with a mixture of malathion and DDT. Or spray malathion after all the eggs are hatched in mid-June. The latter will control young miners already inside the leaves.

MEALY BUGS. *Pseudococcus comstocki* and *Rhizoecus falcifer.* The former, known as the Comstock mealy bug, attacks many hosts including catalpa, holly, horsechestnut, maple, pine, and poplar. It is one of the few species capable of wintering outdoors. The latter, the ground mealy bug, feeds on the roots of many shrubs and trees.

Control. Malathion sprays applied to the leaves and stems will control the Comstock mealy bug. A dilute solution of lindane (1 teaspoon of 15 per cent lindane powder per gallon of water) applied to the soil around the base of the boxwood should control the ground mealy bug.

SCALES. Five species of scales, California red, cottony-cushion, cottony maple, lesser snow, and the oyster shell, may infest boxwood.

Control. Spray with a dormant oil to control overwintering scales. Where infestations are heavy, follow with a malathion spray when the young are crawling about in May and June.

BOXWOOD PSYLLID. *Psylla buxi.* Terminal leaves are cupped and young twig growth is checked by the boxwood psyllid, a small gray, sucking insect covered with a cottony or white waxy material. The adult is a small green fly with transparent wings having a spread of ⅛ inch.

Control. Spray in mid-May and again 2 weeks later with either lindane or malathion.

GIANT HORNET. *Vespa crabro* var. *germana.* The bark is occasionally stripped from the branches by the giant hornet, which has a 1-inch wing expanse and dark orange markings. The pest nests in trees, in buildings, or underground.

Control. Locate the nests and blow chlordane or DDT dust into the openings. Apply a DDT spray to the trunk and branches of trees subject to bark-tearing by this pest.

BOXWOOD MITE. *Eurytetranychus buxi.* A light mottling followed by brownish discoloration of the leaves is caused by infestations of 8-legged mites, about 1/64 inch long when full grown. Winter is passed in the egg stage. The eggs hatch in April and the young mites begin to suck out the leaf juices. By June or July, considerable injury may occur on infested plants. As many as six generations of mites may develop in a single season.

Control. The dormant oil spray recommended for scales will destroy many overwintering mites. Malathion, Aramite, Dimite, or Ovex applied during May and June will also control mites.

Other Pests

NEMATODE. Leaf bronzing, stunted growth, and general decline of boxwood may result from invasion of the roots by meadow nematodes (*Pratylenchus* sp.). These are tiny eel-like worms, 3/10 to 7/10 of a millimeter long, which are visible only through a microscope. They enter the roots, usually near the tips, and move through the cortical tissue. Invaded portions soon die and the plant forms lateral roots above the invaded area. These laterals in turn are infected. Repeated infections and lateral root production result in a stunted root system resembling a witches'-broom. Even heavy rains may fail to wet such densely woven root bundles.

Control. In recent years many chemicals have been tested in attempts to control this pest. The most promising one is the chemical 1, 2-dibromo-3-chloropropane sold under the trade name Nemagon. Old, heavily infested boxwoods are said to have been cured by inserting this chemical into the soil around the roots in the spring and fall for several years. This treatment should be supplemented by shearing and fertilization to stimulate root activity. Details on how to use Nemagon can be obtained from the manufacturer, the Shell Chemical Company, New York City.

The life of infected but untreated plants may be prolonged by providing good care and by soaking the soil thoroughly during dry spells. Before boxwood is replaced in infested soil, the planting site should be fumigated with any one of several materials available for that purpose. Plant pathologists at state experiment stations will advise on selection and use of fumigants.

CEDARS (*Cedrus*)

This genus contains several very beautiful evergreen trees like the Atlas cedar, *C. atlantica*, and the cedar of Lebanon, *C. libani*. Cedars are relatively free of pests and diseases.

Diseases

TIP BLIGHT. The fungus *Diplodia pinea*, formerly called *Sphaeropsis ellisii*, occasionally causes canker and die-back of branch tips in the South.

Control. The same as for tip blight of pine (p. 406).

ROOT ROT. Several fungi including *Armillaria mellea*, *Clitocybe tabescens*, and *Phymatotrichum omnivorum* are associated with root and trunk decay. The last attacks a great variety of trees, shrubs, ornamental, and food plants in the South.

Control. No effective, practicable control measures are known.

Insects

BLACK SCALE. *Saissetia oleae*. This dark brown to black scale is primarily a pest of citrus on the west coast. It attacks a wide variety of trees and shrubs in the South and West, including the Deodar cedar, *C. deodara*. Besides extracting juice from the plant, it secretes on the leaves and stems a substance on which the sooty mold fungus grows.

Control. Malathion sprays when the young are crawling about in May and June are effective.

DEODAR WEEVIL. *Pissodes nemorensis*. This brownish snouted weevil feeds on the cambium of leader and side branches of Deodar, Atlas, and Lebanon cedars. It deposits eggs in the bark, and the ⅓ inch long, white grubs which hatch from the eggs burrow into the wood. Eventually the leaders and terminal twigs turn brown and die. Small trees may be killed by this pest.

Control. A DDT-malathion spray applied in April, when the beetles are feeding, will control this pest.

CRYPTOMERIA (*Cryptomeria*)

This tree is practically free of fungus parasites and has no important insect pests.

Diseases

LEAF BLIGHT. Leaves and twigs may be blighted by a species of *Phomopsis* during rainy seasons.

Control. Pruning out affected leaves and twigs is usually sufficient. Copper fungicides can be used on valuable specimens. Dormant oil sprays which may cause serious damage, should never be used on this host.

LEAF SPOT. Two fungi, *Pestalotia cryptomeriae* and *P. funerea,* are frequently associated with a leaf spotting of this host. Infection probably follows winter injury or some other agent.

Control. Same as for leaf blight.

CYPRESS (*Cupressus*)

Diseases

CANKER. Twigs and branches may be girdled by cankers, and the entire tree may be killed. Cypress, particularly the Monterey cypress, junipers, and oriental arborvitae are all subject to this disease.

Cause. The fungus *Coryneum cardinale* forms spores in branch and twig cankers. These are spread by wind-splashed rain, by pruning tools, and perhaps by insects and birds.

Control. Control is difficult. Remove and burn severely infected trees. Drastically prune mildly infected ones and spray periodically with a copper fungicide starting with the beginning of the rainy period.

CYTOSPORA CANKER. The columnar form of Italian cypress, *C. sempervirens,* along the California coast is most subject to this disease. Occasionally it also affects the horizontal form of Italian cypress and the smooth cypress.

Symptoms. Smooth reddish brown cankers, from which resin flows, develop on young branches. Diseased bark on older branches becomes cracked and distorted with a more abundant flow of resin.

Cause. The fungus *Cytospora cenisia* var. *littoralis* causes this canker.

Control. Prune and burn dead or dying branches. Trees with cankers on the trunks should be removed and burned.

Cypress is also subject to several other diseases: CROWN GALL (p. 240); BLIGHT caused by *Phomopsis juniperovora* (p. 402); and TWIG BLIGHT caused by *Coryneum berckmanii* (p. 383).

Insects

CYPRESS APHID. *Siphonatrophia cupressi.* This large green aphid infests blue and Monterey cypress.

Control. Spray with malathion when the young begin to feed.

CYPRESS MEALY BUG. *Pseudococcus ryani.* Primarily a pest of Monterey cypress in California, this species also infests arborvitae, redwood, and other species of cypress.

CATERPILLARS. The caterpillars of the following moths feed on this host: tip moth, *Argyresthia cupressella;* Webber, *Epinotia subviridis;* imperial, *Eacles imperialis;* and the white-marked tussock moth, *Hemerocampa leucostigma.*

Control. Spray with arsenate of lead, DDT, or malathion when the caterpillars are small.

BARK SCALE. *Ehrhornia cupressi.* This pink scale covered with loose white wax infests Monterey cypress primarily. It occasionally attacks Guadalupe and Arizona cypress and incense cedar. Leaves of heavily infested trees turn yellow then red or brown.

Control. Malathion sprays applied to the trunk and branches are said to be effective in combating this scale. Several other scales including the cottony-cushion and the juniper scale occasionally infest cypress. These are also controlled with malathion sprays.

FIRS (*Abies* and *Pseudotsuga*)

Several parasitic micro-organisms and insects attack firs, but rarely produce extensive damage in ornamental plantings. Unthriftiness in many trees is more likely caused by an unfavorable environment.

Firs thrive best in light, porous, acid soils, which are well drained and yet are continually moist. They also require full sunlight but appear to grow more vigorously in valleys or protected places than near hilltops or exposed situations.

DISEASES

NEEDLE AND TWIG BLIGHT. In the northeastern United States, needle and twig blight occurs quite commonly on balsam fir and to a lesser extent on Colorado and Alpine firs. Noble and Fraser firs are also susceptible, but are attacked infrequently.

Symptoms. The needles of the current season's growth turn red and shrivel, and the new twigs are blackened and stunted. Severely infected trees appear as though scorched by fire or damaged by frost. The lower branches are most heavily infected. Needles infected in previous years remain attached to the twigs.

Cause. The fungus *Rehmiellopsis balsameae* causes needle and twig blight. It overwinters on diseased needles and twigs. In the spring, fruiting bodies mature on these parts and release spores, which infect the newly developing needles. The fungus *Cenangium abietis* occasionally attacks firs but is more common on pine (p. 410). Another fungus, *Botrytis cinerea*, may kill leaves at the tips of the branches of Douglas fir during wet springs. The true firs, larches, and other conifers are occasionally infected by the same organism.

Control. In ornamental plantings, control is possible by pruning and burning infected twigs and by applying a copper fungicide 3 times at 12-day intervals, starting when the new growth begins to emerge from the buds.

LEAF CAST. A number of closely related species of fungi cause the so-called leaf cast disease of firs. Affected needles turn yellow, then brown, and drop prematurely. Some young trees are completely defoliated.

Among the most common fungi responsible for leaf casts on the firs are *Bifusella abietis*, *B. faullii*, *Hypodermella nervata*, *H. mirabilis*, *Lophodermium autumnale*, and *L. lacerum*. These produce elongated black bodies along the middle vein of the lower surface. In July, spores are shot from small black fruiting bodies, on the upper leaf surface, to young needles. These germinate and penetrate the new growth but do not produce striking symptoms until the following spring.

The Douglas fir is susceptible to a disease resembling leaf cast in which the older needles are blighted by the fungus *Rhabdocline pseudotsugae*. Yellow spots first appear near the needle tips in the fall. These spots enlarge in the spring, then turn reddish-brown, contrasting sharply with the green tissues adjacent to them. With continued moist weather, the discoloration spreads until the entire needle turns brown. When many groups of needles are so affected, the trees have a brown, scorched aspect when viewed from a distance.

Two other fungi produce leaf cast of Douglas fir: *Adelopus gaumanni* and *Rhabdogloeum hypophyllum*. The former produces symptoms closely resembling those of *Rhabdocline pseudotsugae*, and both may be present in the same tree.

Control. No control measures for large trees have been reported. Severe outbreaks in nurseries or on small trees can probably be prevented by spraying with copper fungicides at the time the spores are being discharged.

NEEDLE RUST. Firs are susceptible to several rust fungi that primarily attack the needles and in a few cases produce distortion of twigs.

Symptoms. White, bladdery pustules appear on the lower surfaces of the needles during the summer.

Cause. Fungi belonging to the genera *Milesia* and *Uredinopsis* are mainly responsible for rust on firs. The white pustules they form break open and release yellow or white powdery spores. These are blown to the fronds of ferns or other alternate hosts (depending on the species of rust), where they infect and produce red or yellow spots. The following spring another type of spore is produced in these spots and is blown to fir needles. Here infection occurs and the life cycle is completed.

The rust fungus *Melampsora albertensis* affects Douglas fir as well as its alternate host, the poplars.

Control. As with many rust diseases where two different hosts are necessary for the completion of the life cycle, the elimination of one host will result in nearly complete disappearance of the fungus. The rusts on firs resemble one another so closely that only a specialist can identify them. Because these have different alternate hosts, they should be correctly identified so that the proper alternate host can be eradicated.

Although sulfur sprays and dusts are commonly used for rust control, they are not recommended for large trees, because of the difficulties involved in their application. Moreover, these diseases cause little damage to large trees.

CANKER. A number of canker diseases, some of which are occasionally serious, occur on firs in ornamental plantings. On balsam fir, the fungus *Aleurodiscus amorphus* forms narrowly elliptical cankers with a raised border on the main trunk of young trees and centering around a dead branch.

A few other fungi which cause cankers on the true firs are *Cytospora pinastri*, *Cryptosporium macrospermum*, *Scoleconectria balsamea*, and *S. scolecospora*.

The fungi *Dasyscypha pseudotsugae*, *D. ellisiana*, and *Phacidiopycnis pseudotsugae* are associated with cankers on Douglas fir.

Control. As with other canker diseases, sanitation, avoidance of bark and branch injuries, and fertilization to maintain the tree in good vigor are recommended.

SOOTY MOLD. Fir needles are covered more frequently than most other species of evergreens by black soot-like material. This substance consists of fungus tissues that exist on the secretions of aphids and other insects. The black mold will usually disappear if the insects are kept under control.

WITCHES'-BROOM. In the western states Douglas firs are subject to the mistletoe *Arceuthobium douglasii*.

SHOESTRING ROOT ROT. Firs are subject to the fungus *Armillaria mellea* (p. 243).

WOOD DECAY. Most of the wood-decay fungi affecting other evergreen trees attack firs also. Fungi belonging to the genera *Fomes, Polyporus, Lenzites,* and *Stereum* are associated with most of the decays in fir.

Control. The wood decays are most prevalent in old, neglected trees. The best preventives are avoidance of bark injuries and maintenance of the tree in good vigor.

INSECTS

BALSAM TWIG APHID. *Mindarus abietinus.* The leaves and shoots of white and balsam firs, as well as spruce, are attacked by this green aphid, which is covered with white waxy secretions. Affected shoots are roughened and curled.

Control. Spray with malathion when the aphids are seen in spring.

BARK BEETLE. *Pityokteines sparsus.* Seepage of balsam from the trunk, reddening of the needles, and death of the upper parts of the tree result from infestations of the balsam bark beetle, a pest $\frac{1}{10}$ inch long. Vigorous trees are attacked.

Control. Prune and burn infested parts.

CATERPILLARS. Several caterpillars feed on the needles of firs. These are the larval stage of the hemlock looper moth, the spotted tussock moth, the balsam fir sawfly, the Zimmerman pine moth, and the pine butterfly.

Control. Arsenate of lead, DDT, or malathion sprays will control these pests.

BAGWORM. See under Arborvitae.

CEDAR TREE BORER. See under Arborvitae.

COOLEY GALL APHID. See under Spruce.

SPIDER MITE. See under Juniper.

HEMLOCK (*Tsuga*)

Hemlocks are among our most graceful and highly prized evergreens. They grow best in a fairly damp soil where their roots may be cool. The soil must be well drained, however, and moderately acid. Hemlocks are among the few evergreens that thrive near the trunks of large deciduous trees. Like many other conifers, they are not adapted to city conditions and rarely prosper in small front yards of suburban homes.

Hemlocks are less susceptible to diseases and most insect pests than are other conifers, such as firs, pines, and spruces.

Diseases

LEAF BLIGHT. In late summer, leaves of Eastern hemlock turn brown and drop prematurely when attacked by the fungus *Didymascella tsugae*.

Small, black fruiting bodies of the fungus occur on the fallen leaves. These produce spores the following spring, which initiate new infections.

Control. Leaf blight rarely damages the trees sufficiently to necessitate measures other than gathering and burning fallen infected leaves in the autumn.

NEEDLE RUST. Of the six rusts that attack the needles of hemlock, only one, *Melampsora farlowii*, produces extensive damage. Eastern hemlock and, to a lesser extent, Carolina hemlock are attacked by this fungus.

Symptoms. In late May or early June some of the new leaves turn yellow. Within two weeks, the shoots to which these leaves are attached turn yellow, become flaccid, and droop. Most of the needles then drop from the affected shoots. Severely rusted trees appear as though their branch tips had been scorched by fire. Red, waxy, linear fungus bodies occur on the lower leaf surface, on the shoots, and on the cones.

Melampsora farlowii does not require an alternate host to complete its life cycle. Spores developing on the overwintered twigs and cones, infected the preceding spring, can infect the new growth on hemlock directly.

Control. A spray consisting of 4 pounds of dry lime sulfur in 50 gallons of water applied at weekly intervals in May has given good control of the disease on small trees in nursery plantings. The use of this material for large trees may be justified in unusual circumstances.

CANKERS. Several fungi including *Dermatea balsamea*, *Hymenochaete agglutinans*, and a species of *Cytospora* cause cankers on this host.

Control. Prune affected branches and spray with a copper fungicide where affected trees are particularly valuable.

SAPWOOD ROT. The fungi *Ganoderma lucidum* (p. 288) and *Coniophora puteana* frequently cause decay of the tissues immediately beneath the bark at the base of the trunk, resulting in death of the tree.

Control. No effective control measures are known. Avoid wounding the bark, and feed and water trees to keep them in good vigor.

Insects

HEMLOCK LOOPER. *Lambdina fiscellaria*. Hemlocks may be completely defoliated by a pale yellow caterpillar with a double row of small black dots along the body, which is more than an inch long at maturity. The adult moth has tan to gray wings, which expand to more than an inch. Another species *L. athasaria*, occurs less frequently but also defoliates hemlocks.

Control. Spray with arsenate of lead, DDT, or malathion when the larvae are small.

SPOTTED HEMLOCK BORER. *Melanophila fulvoguttata.* Wide, shallow galleries in the inner bark and sapwood result from boring by a white larva, ½ inch long. The adult, a flat metallic-colored beetle with three circular, reddish-yellow spots on each wing cover, deposits eggs in bark crevices.

Control. Prune and burn severely infested branches. Keep the tree in good vigor by fertilizing and watering.

FIORINIA HEMLOCK SCALE. *Fiorinia externa.* This scale may infest hemlock needles, causing them to turn yellow and drop prematurely. Both male and female scales are elongated. The females are pale yellow to brown and are almost completely covered with their own cast skins. There are two generations a year in the New England states.

Control. Spray with malathion just after the crawlers leave the protective coating of the adult females in late June and again in September.

GRAPE SCALE. *Aspidiotus uvae.* Hemlock hedges can be destroyed by infestations of this small, dingy white scale which has yellowish nipples or exuviae.

Control. Malathion sprays applied in mid- to late-June provide good control.

SPIDER MITES. Hemlocks in ornamental plantings are extremely susceptible to several species of mites. One, the spruce mite, *Oligonychus ununguis*, is most prevalent on spruces and junipers in addition to hemlock. The other, the so-called two-spotted mite, is discussed under Juniper on page 405.

BAGWORM. See under Arborvitae.

CYPRESS MOTH. See under Bald Cypress.

HOLLY (*Ilex*)

The American holly, *Ilex opaca*, is relatively free of insect and fungus pests. Most of the troubles experienced with trees in ornamental plantings result from improper transplanting practices or unfavorable soil conditions. In localities where it can withstand the winters, holly will thrive in almost any type of soil that is well drained and contains considerable amounts of humus. Incorporating several bushels of well-rotted oak leaf mold into the soil at transplanting time will help to provide the conditions favored by this tree. When the tree is established in its new site, cottonseed meal should be occasionally worked into the soil to supply nitrogen.

Many other species of holly both of native and Asiatic origin are also grown. Unless specifically noted, the diseases and insects described here affect primarily the American holly.

Diseases

TAR SPOT. Native holly and the English holly (*I. aquifolium*) are subject to the tar spot disease.

Symptoms. Yellow spots appear on the leaves during late May. These turn reddish-brown and finally black by fall. A narrow border of yellow tissue remains around the darkened spots. The efficiency of the leaf is reduced when a large number of spots develop. Premature defoliation seldom occurs.

Cause. Tar spot is caused by the fungus *Phacidium curtisii.* Spores produced in the blackened areas initiate new infections in early spring.

Control. Although no experimental work on control has been attempted, preventive sprays containing copper applied at 2-week intervals beginning in late spring should keep this disease in check.

OTHER LEAF SPOTS. Many other species of fungi are known to cause leaf spots of holly. Among these are *Cercospora ilicis, C. pulvinula, Englerulaster orbicularis, Gloeosporium aquifolii, Macrophoma phacidiella, Microthyriella cuticulosa, Phyllosticta concomitans, P. terminalis, Rhytisma ilicinicolum, R. velatum,* and *Septoria ilicifolia.*

Control. Infected leaves should be picked off and burned, and the vigor of the tree improved by incorporating oak leaf mold or cottonseed meal into the soil. In addition, water should be provided during dry spells. Applications of bordeaux mixture or any other copper spray in late summer and in early fall will largely prevent the formation of the spots. More lasting results are obtained, however, by improving the soil conditions. Moreover, copper fungicides may cause some injury, especially to leaves that have been punctured by the holly leaf miner. They also leave an unsightly residue which persists for some time. Ferbam sprays should be used where copper sprays are apt to cause injury.

CANKER. Stem cankers are caused by several species of fungi including *Diaporthe eres, Nectria coccinea, Physalospora ilicis, Phomopsis crustosa,* and a species of *Diplodia.*

Control. Prune diseased branches and spray with a copper fungicide several times in the spring.

DIE-BACK. A die-back and wilt of branches of American holly along the east coast has been prevalent for the last 20 years.

Cause. The fungus *Fusarium solani* var. *martii* causes this disease.

Control. Prune out and burn diseased branches. Spray the trees 3 times at 2-week intervals with ferbam. Copper sprays also control this disease, but in cool, wet springs they may injure the leaves.

SPOT ANTHRACNOSE. Chinese holly leaves (*Ilex cornuta*) in the South are occasionally affected by this disease.

Symptoms. Two types of lesions occur on the leaves: numerous tiny black spots, and a large leaf-distorting spot more than an inch in length, which is confined to half of the leaf blade. Lesions on the shoots and berries may also occur.

Cause. The fungus *Elsinoe ilicis* causes this disease.

Control. Periodic applications of copper fungicides will control this disease.

TWIG DIE-BACK. Black stem cankers and black spots on the leaves of English holly grown in the Pacific Northwest are typical of this relatively new fungus disease. During cool rainy weather, complete defoliation and severe twig blighting of the lowermost branches is common.

Cause. The fungus *Phytophthora ilicis* causes this disease. Earlier it was thought that the fungi *Boydia insculpta* and *Phomopsis crustosa* were causes.

Control. Although no control measures have been developed, it is highly probable that periodic applications of copper fungicides will provide control.

SPINE SPOT. Small gray spots with purple halos are caused by puncturing of the leaves by the spines of adjacent holly leaves. A careful examination with a hand lens will reveal tiny, circular holes or an irregular tear at the center of each spot.

Spine spot is often confused with slits made by the holly leaf miner. The latter have neither a gray center nor a purple halo.

LEAF SCORCH. A browning or scorching of the leaves, common on holly in late winter or early spring, is of nonparasitic origin. Occasionally it is caused by the presence of water or ice on the leaves at the time the sun is shining brightly. This causes a scalding, followed by invasion by secondary organisms and finally by scorching.

Hollies planted in wind-swept areas are also more susceptible to so-called winter-drying. The leaves in late winter or early spring lose water faster than it can be replaced via the roots. As a result the leaf edges wilt and turn brown. In exposed situations, newly transplanted holly should be protected with some sort of windbreak, or sprayed with an antidesiccant such as Wilt-Pruf.

Insects

LEAF MINER. *Phytomyza ilicis.* Yellow or brown serpentine mines or blotches in leaves are produced by the leaf miner, a small yellowish-white maggot, ⅙ inch long, that feeds between the leaf surfaces (Fig. 137). The adult is a small black fly that emerges about May 1 and makes slits in the lower leaf surfaces, where it deposits eggs.

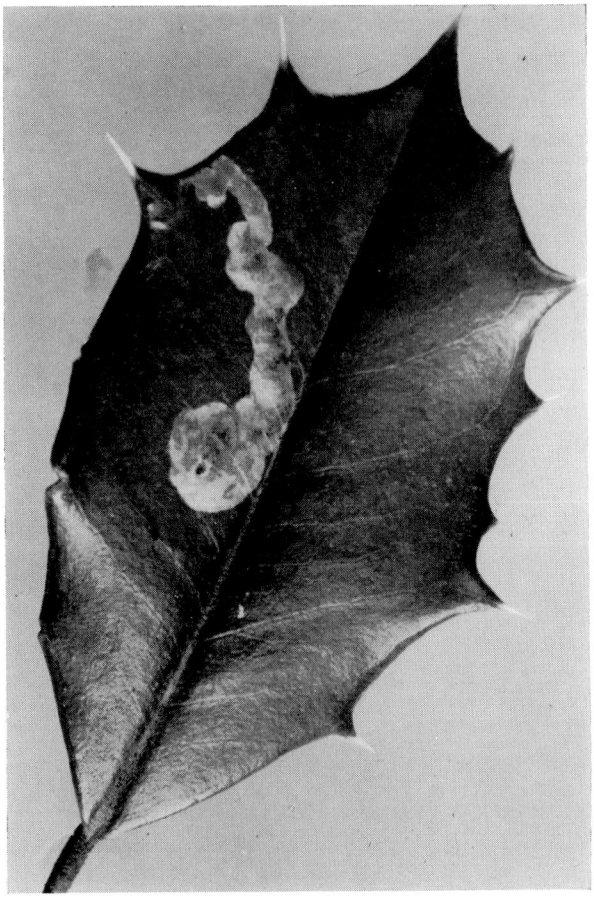

FIG. 137. Blotch produced by the holly leaf miner.

Another species of leaf miner, *P. iliciola*, produces very slender mines and may occur on the same tree.

Control. Spray with a mixture of DDT and malathion about mid-May and again 10 days later. The malathion will help to keep down mites which flourish when DDT alone is used. These sprays are directed against the adults. Where such sprays have not been used and the young larvae have already entered the leaves, it is still possible to achieve control by spraying with either lindane or dieldrin in July.

Picking off and burning blotched and mined leaves in winter also helps to keep down these pests.

Some recent research at the University of Maryland suggests that systemic fungicides are effective in combating leaf miner infestations. The organic phosphate sold as Systox, for example, painted in a wide band around the trunk of the tree just below the branches was as effective in controlling leaf miner as were foliage sprays. Apparently enough of the poison is absorbed through the bark and translocated to the leaves to kill the leaf miners there. Organic phosphates are deadly poisons and must be handled with special care.

BUD MOTH. *Phopobota naevana* var. *ilicifolia*. Holly in the Pacific Northwest is subject to this pest, the larval stage of which feeds on the buds and terminal growth inside a web.

Control. Spray with arsenate of lead, DDT, or malathion when the larvae begin to feed.

HOLLY SCALE. *Aspidiotus britannicus*. Circular, flat, 1/16-inch scales infest the berries, leaves, and twigs of holly in the West.

Control. Malathion sprays applied to the young crawling stage give excellent control.

PIT-MAKING SCALE. *Asterolecanium puteanum*. This scale is becoming more prevalent on American holly and Yaupon (*I. vomitoria*) in the southern states. Oval, 1/16 inch in diameter when mature, and pale yellow in color, this insect imbeds itself in the bark and causes a pitted and swollen condition of the stems somewhat resembling that produced by the golden oak scale (p. 351). Branches are distorted, the leaves take on an abnormal color, and at times there is considerable die-back from the branch tips.

Control. A toxaphene spray applied in late April in the latitude of the state of Delaware has given excellent control of this pest. Malathion sprays should also be effective.

SOUTHERN RED MITE. *Oligonychus ilicis.* This mite can become a serious pest of holly, particularly where DDT is used frequently.

Control. Malathion or miticides such as Aramite, Dimite, and Ovex give excellent control. The last should not be used early in the season because it may cause some injury to the leaves. The first of two applications at 10-day intervals should be made about mid-May.

WHITE FLY. See under Dogwood (Chapter 15).

JUNIPER (*Juniperus*)

Because of their evergreen habit, diversity of form and other desirable qualities, the species and varieties in the genus *Juniperus*, known commonly as cedars and junipers, are extensively used in ornamental plantings.

DISEASES

TWIG BLIGHT. Although primarily a disease of seedlings and nursery stock, twig blight may appear on 8- to 10-foot trees in ornamental plantings, and on larger native red cedars in some parts of the country. The disease becomes progressively less serious, however, as the trees become older, and little damage occurs on trees over 5 years old.

Though primarily a disease of the common red cedar (*Juniperus virginiana*) and its horticultural varieties, twig blight has been found on more than a dozen other groups, among which are arborvitae, cypresses, retinosporas, white cedar, and other species of Juniperus.

Symptoms. The tips of branches first turn brown, followed by progressive dying back until an entire branch or even the entire young tree is killed.

Cause. Twig blight is caused by the fungus *Phomopsis juniperovora.* Older trees in the vicinity of an evergreen nursery may harbor the fungus on diseased twigs. The fungus may also be carried on small, diseased branch tips that accompany purchased seed.

In the spring and summer, large numbers of spores are produced in fruiting bodies on infected twigs and branches. During rainy periods the spores ooze out from the bodies and are spread by wind, rain, and laborers, to other plants.

Control. The first essential in any control program is the elimination of the source of inoculum. Infected twigs and branches should be pruned and burned. Seed collected in sections of the country where native red cedars are infected should be carefully cleaned to eliminate debris that might harbor the fungus.

Extremely susceptible types such as *J. virginiana* and its varieties should be sprayed every 10 days with a copper fungicide. Because such sprays do not adhere readily to the waxy juniper foliage, a good sticker and spreader should be added according to the manufacturer's directions. During rainy seasons as many as 10 sprayings may be necessary to give good control.

Recently Kromad and Merbam sprays have been found to give as good control as the copper fungicides. They are not applied so frequently as the copper sprays.

The use of resistant varieties offers considerable promise as a means of avoiding twig blight. The spiny Greek juniper, *J. excelsa* var. *stricta;* the Keteleer red cedar, *J. virginiana* var. *Keteleeri;* and the Hill Juniper, *J. virginiana* var. *pyramidiformis hillii,* are reported resistant to attacks.

CEDAR RUSTS. The several species of rust fungi that attack various species of *Juniperus* occasionally produce material damage. These rusts require hosts other than cedar for the completion of their life cycle, inasmuch as the spores produced on one are unable to reinfect the plant upon which they are formed. The fungi cannot survive, therefore, if either of the alternate hosts is removed. For this reason, several important apple-growing states prohibit the planting of susceptible cedars in the vicinity of apple orchards.

The rust fungi all belong to the genus *Gymnosporangium.* The cedar hosts are *Juniperus virginiana* and many of its varieties, *J. sabina, J. scopulorum, J. communis,* and *J. sibirica.* The alternate hosts for these rusts belong to the order *Rosales,* including the cultivated apple, pear, and ornamental crab; *Crataegus; Amelanchier; Cydonia; Sorbus; Aronia;* and others of less importance.

Although at least seven different species of rust fungi attack *Juniperus,* only three are important in the East. These are known as the cedar-apple rust, the cedar-quince rust, and the cedar-hawthorn rust.

The cedar-apple rust, caused by the fungus *Gymnosporangium juniperi-virginianae,* attacks junipers, apple, and hawthorn. On juniper twigs it produces large, globose galls, ranging up to 1½ inches in diameter. Long gelatinous orange tendrils or spore horns protrude from these galls during rainy weather in the spring. The spores formed on the orange tendrils are carried by wind and insects to developing apple and hawthorn leaves and fruits. Yellow or orange lesions are then produced on the leaves or fruits as a result of infection by these spores. This stage of the disease is most destructive, inasmuch as severe defoliation may occur, especially on apple.

Control of Cedar-Apple Rust. On apple, hawthorn, and other rosaceous hosts, as many as 6 applications, at 10-day intervals, of a wettable sulfur or a mixture of wettable sulfur and ferbam may be needed for good control. The initial application should be made just before an expected rainy spell and as soon as the leaves emerge in the spring. On junipers, a single application of 100 parts of cycloheximide in one million parts of water when the galls have formed will prevent spore formation and thus prevent infections on nearby apples and crabapples. This concentration is prepared by dissolving one 380-milligram Actispray tablet in each gallon of water.

The cedar-quince rust, caused by the fungus *G. clavipes*, attacks junipers, quince, and the fruit of the hawthorn, the juneberry, and the apple. It appears later in the season and differs somewhat in appearance from the cedar-apple rust. On juniper, it produces slight fusiform swellings on the twigs, branches, and occasionally the trunk. During wet weather, reddish orange spore masses break through the bark. Of the three rusts discussed, quince rust is the most destructive on the evergreen hosts. On the fruits of the rosaceous hosts, quince rust produces slightly swollen areas covered with tiny deep-red dots. Long, white tube-like structures eventually emerge, open, and expose orange-brown spore masses. These spores then reinfect the *Juniperus* hosts.

Control of the Cedar-Quince Rust. On quince, hawthorn, and other rosaceous hosts, 3 applications, at 10-day intervals, of the sprays recommended for cedar-apple rust will give control. The first application should be made when the blossom buds are opening. On *Juniperus* hosts, the cycloheximide spray should be applied, as suggested above, when the galls have formed but before the gelatinous horns emerge.

The cedar-hawthorn rust, caused by the fungus *G. globosum*, attacks junipers, hawthorn, and to a lesser extent, apple. On junipers, it produces small irregularly shaped galls, less than an inch in diameter, with wedge-shaped, gelatinous, orange spore masses. The symptoms on hawthorn and apple are similar to those produced by the cedar-apple rust.

Control of Cedar-Hawthorn Rust. The spray schedule suggested for the control of cedar-apple rust is also recommended for the control of cedar-hawthorn rust, except that the spray on junipers should be applied about August 1 rather than in mid-July.

WOOD DECAY. Several species of wood-decay fungi are occasionally found on junipers but rarely become destructive enough to warrant special treatments. The fungi usually enter through injuries at the base of the tree. Prevention and treatment of wounds, and fertilization to maintain the tree in good vigor are recommended.

INSECTS

JUNIPER SCALE. *Diaspis carueli.* The needles, particularly of the Pfitzer juniper, turn yellow as a result of sucking by tiny circular scales, which are at first snow-white, then turn gray or black. The pest overwinters in the female adult stage.

Control. Spray with 1 part of concentrated lime sulfur solution in 10 parts of water before growth starts in spring, or with 1 pound of 25 per cent wettable malathion powder in 25 gallons of water during warm weather in May. A second malathion spray should be applied about mid-June for best results.

JUNIPER MIDGE. *Contarinia juniperina.* Blisters at the bases of the needles and death of leaf tips are produced by small yellow maggots, the adult stage of which is a small fly.

Control. Spray the leaves in mid- to late-April with malathion. Cultivating the soil beneath infested trees or treating it with lindane will destroy the overwintering stage.

JUNIPER WEBWORM. *Dichomeris marginella.* The twigs and needles are webbed together, and some turn brown and die, when infested by the juniper webworm, a small, ½ inch long, brown larva, with longitudinal reddish-brown stripes. The adult female, a moth with a wingspread of ⅗ inch, appears in June and deposits eggs that hatch in 2 weeks. The winter is passed in the immature larval stage.

Control. Spray with arsenate of lead or malathion in early spring or in July.

RED CEDAR APHID. *Cinara sabinae.* Twig growth may be checked and the entire tree weakened by heavy infestations of the red cedar aphid, a reddish-brown insect, ⅛ inch long. The so-called honeydew it secretes is a good medium for the sooty blotch fungus, which may completely coat the leaves and further weaken the tree.

Control. Spray with lindane or malathion when the aphids are visible.

SPIDER MITE. *Tetranychus telarius.* The leaves assume a gray or yellow cast when severely infested by tiny green, yellow, or red mites, 1/50 inch long. Infested leaves and twigs are occasionally covered with fine silken webs. Winter is passed primarily in the egg stage.

Control. Spray with 1 part concentrated lime sulfur in 10 parts of water before growth begins in the spring, and with a good miticide such as Aramite, Dimite, or Ovex during the growing season. Malathion is also a fairly effective mite killer.

BAGWORM. See under Arborvitae.

TAXUS MEALY BUG. See under Yews.

PINE (*Pinus*)

Although pines fare better than hemlock, fir, or spruce in the environment of larger cities, they grow best in small towns and in the country. Among the species most commonly used in ornamental plantings are Austrian pine, mugho pine, Scotch pine, and white pine. Each is subject to a number of fungus diseases and insect pests.

Diseases

DIPLODIA TIP BLIGHT. Within the last 25 years, Diplodia tip blight has become a disease of major importance on several of the most desirable pines for ornamental plantings. Austrian pine is, without question, most susceptible to the disease. Scotch, red, mugho, Western yellow, and white pines are decreasingly susceptible in the order named. Douglas fir and blue spruces are occasionally attacked.

Symptoms. The most prominent symptoms are the stunting of the new growth and browning of the needles. The lower branches are most heavily infected, although in wet springs, the branches over the entire tree may show browned tips.

Tip blight is often confused with troubles caused by low temperatures, drought, winter-drying, and pine-shoot moth injury. It can be differentiated from these by the presence of small, black, pin-point fungus bodies, which break through the epidermis at the base of the needle (Fig. 138). These bodies also occur on affected twigs and cone scales, but are less readily discernible than on needles, especially if the examination is made soon after the needles turn yellow and wilt. Placing such material in a moist atmosphere for a day or so will induce the black bodies to break through the epidermis and thus make them visible.

Cause. The fungus *Diplodia pinea,* formerly called *Sphaeropsis ellisii,* causes tip blight. The generic name is used to distinguish this disease from twig blights caused by other micro-organisms. It overwinters in infected needles, twigs, and cones. In the spring, the small fruiting bodies release egg-shaped, light brown spores, which are splashed by rain and wind to the newly developing needles. The fungus grows down through the needles and into the twigs, where it destroys tissues as far as the first node.

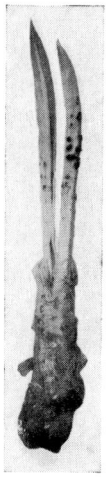

FIG. 138. Single needle cluster of Austrian pine greatly magnified to show fruiting bodies of causal fungus, *Diplodia pinea*.

Diplodia pinea has also been associated with a root and root-collar rot of red pine in nurseries, where such trees were completely destroyed. By inserting bits of the fungus into wounded bark near the root collar, the author has induced the same disease on Austrian, mugho, Scotch, and white pines, on Douglas fir, and on Norway spruce. Canadian scientists recently reported that the fungus may follow an infestation of the spittle bug (p. 415).

Control. As soon as noticed, the infected needles and twigs should be pruned to sound tissues and burned. Pruning should be done when the branches are dry, because there is less danger of spreading the spores by contact with the operator and with tools. Where infection has been particularly severe, preventive fungicides are also recommended. Bordeaux mixture 2-2-50 or any ready-made copper spray should be applied at least three times in the spring; the first, as soon as the new growth or candles appear; the second, when the candles are half grown; and the third, about 10 days or 2 weeks later. In rainy springs, a fourth application, about 10 days after the third, may be necessary. As a supplementary treatment, fertilization of the trees in the fall or early spring is also suggested.

WHITE PINE BLISTER RUST. This disease is probably the most important disease of white pine in the northeastern United States. Other pines, such as Western white pine and sugar pine, having five needles in the leaf cluster, are also susceptible. Blister rust also attacks both wild and cultivated forms of currants and gooseberries, but produces less damage on these than on pines.

Symptoms. On pines, reddish-brown, drooping needles appear on one or more dead branches. Rough, swollen cankers are found at the bases of affected branches or on the trunk. In late spring, orange-yellow blisters filled with powdery spores break through the bark of the cankered areas. These blisters are most conspicuous from April to June and are the most positive sign of the disease. After that time they may be eaten by squirrels or covered with resin. The disease is fatal when trunk cankers or branch cankers completely girdle the affected parts.

The symptoms are less conspicuous on currants and gooseberries, where minute, orange-yellow pustules or brown, curved tendrils appear on the lower surface. Most of the leaves are stunted when the pustules become numerous.

Cause. White pine blister rust is caused by the fungus *Cronartium ribicola.* Winter is passed in cankered bark of living pines. The orange-yellow spores produced in these cankers in late spring are blown to susceptible currant and gooseberry leaves, where they produce infection. A new type of spore is produced on these leaves in the fall. This type is blown to pines where they infect the needles to complete the life cycle. After the fungus enters the needle it penetrates slowly downward into the twig and eventually the trunk. The whole process takes place slowly: several years may elapse from the needle infection stage to the appearance of the orange-yellow blisters on the branch or trunk.

Control. Because currants and gooseberries are absolutely essential for the survival of the fungus, blister rust can be controlled by eradication of these undesirable alternate hosts. Local control of the disease can be accomplished by destroying all currants and gooseberries within 900 feet of the pines. Furthermore, the European or cultivated black currant should not be grown within a mile of pines. This species, which is more susceptible to rust than are other currants and wild gooseberries, has been found to be the chief agent in the extensive spread and establishment of the disease in previously uninfected areas.

Many valuable pines showing moderate infection can be saved by prompt surgical treatments. Small branches with cankers should be removed, and cankers on large branches or on the trunk cut out. The latter practice is justified only where the cankers extend around a relatively small part of the trunk or branch, and is not practicable where they nearly completely girdle the affected member.

Cankers can be inactivated by the antibiotic cycloheximide (Actidione). Concentrations ranging from 150 to 600 parts of Actidione in a million parts of water applied to blister rust cankers are effective. The outer bark is first removed, cuts are made in the remaining bark, and then the whole area is painted with the antibiotic.

The removal of all alternate hosts within the distances already suggested, is, of course, essential to prevent new infections on pines.

Investigators of the United States Department of Agriculture and some state experiment stations, Wisconsin in particular, have selected and bred white pines resistant to blister rust. This work looks very promising, inasmuch as only one strain of the blister rust fungus is believed to exist. In other rust diseases a number of strains of the same fungus occur, which often makes breeding for resistance to such diseases a very unfruitful undertaking.

NEEDLE RUST. Several species of rust fungi, all belonging to the genus *Coleosporium*, cause blister rust of pine needles. As with white pine blister rust, these also require certain alternate hosts for completion of their life cycle. One of the most frequently encountered needle rusts, known as *Coleosporium solidaginis*, attacks many species of two- and three-needled pines, especially the red pine (Fig. 139). This fungus requires certain species of cultivated and wild asters and goldenrod as alternate hosts.

Control. Needle rusts rarely cause appreciable damage, although some defoliation may occur. Consequently, control measures are not generally attempted. Small valuable pines can be protected by sulfur sprays. Eradication of asters and goldenrod in the vicinity of valuable trees or near nurseries might be warranted under unusual conditions.

OTHER RUSTS. Many other rusts, too numerous to mention in this book, also occur on pines. These appear on all the above-ground parts of the tree, producing many and varied types of symptoms. They rarely become important parasites, however, on pines in ornamental plantings.

LEAF CASTS. Like other narrow-leaved evergreens, pines are subject to the so-called leaf-cast diseases. Reddish-brown spots or elongated brown areas appear along the needles. These spots are often confused with discolorations resulting from unfavorable weather or soil conditions, and can be distinguished only by microscopical examination.

The fungus *Lophodermium pinastri* is most commonly associated with leaf cast of red pine. Other species of *Lophodermium* are associated with the same disease on other pines.

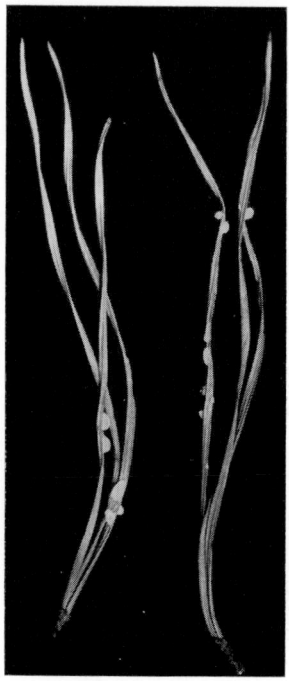

FIG. 139. Leaf blister rust of pine.

Control. Leaf-cast diseases rarely cause sufficient damage to large trees to warrant preventive practices. They can be controlled on small trees and on nursery stock by applying bordeaux mixture 4-4-50, or any other copper spray, when the needles are half-grown and again about 3 weeks later. Because such sprays do not readily adhere to pine foliage, casein soap or some other material should be added as a spreading agent.

CENANGIUM TWIG BLIGHT. The twigs of exotic species of pine and of white pine may be killed by the fungus *Cenangium abietis*. Early symptoms are the dying of terminal buds and reddening of the needles. Infection rarely spreads beyond the current season's growth.

Control. Affected twigs should be pruned to sound wood and burned. Because twig blight is usually severe only on weakened trees, fertilization and watering are suggested for valuable pines.

CANKERS. A large number of canker diseases occur on all pines. These are more common on trees in forests than in ornamental plantings.

The fungus *Tympanis pinastri* attacks red pine and, to a lesser extent, white pine. It produces elongated stem cankers with or without definite margins, and with depressed centers that become roughened and open after two or three years. Each canker centers at a node, indicating that the fungus enters the branch at the base of the lateral branch. Trees in weakened condition are most susceptible.

Other cankers on pines are associated with the fungi *Dasyscypha pini, Atropellis pinicola, A. tingens, Caliciopsis pinea, Sphaeropsis* sp., and *Phomopsis* sp.

Control. Removal of dead and weak branches, avoidance of bark injuries, and fertilization to increase the vigor of the trees are suggested. Surgical treatments of cankers should be attempted only on valuable specimens.

NEEDLE BLIGHT. *Dothistroma pini.* This fungus causes slightly swollen, dark spots or bands on 1-year-old Austrian pine needles in late summer. The distal portion of the swollen needle turns light brown and dies. Severely affected trees show sparse foliage as a result of the premature dropping of needles.

Control. An organic mercury spray, such as Puratized Agricultural Spray, or bordeaux mixture applied first in late April or early May and for a second time 3 weeks later will control this disease.

BLIGHT OF WHITE PINE. A needle blight of white pine (*P. strobus*) which appears as a browning of entire needles or tips of needles of the current season's growth has been prevalent throughout the northeastern United States for the last 15 years. The primary cause of the trouble is unknown. Some experts have attributed it to a deficiency or an excess of soil moisture, others to a nutrient deficiency and to fungi, and still others to a virus. The disease is not always fatal. Some trees recover without special treatment.

WOOD DECAY. Because pine wood contains a high percentage of resins, wood decays are not so frequent as in hardwoods and in some other evergreens. Some of the fungi commonly associated with wood decay are *Stereum sanguinolentum, Fomes pini, F. officinalis, F. roseus, Polyporus circinatus,* and *P. schweinitzii.* The first mentioned is commonly associated with improperly treated pruning wounds.

FIG. 140. Pine bark aphid secretions on trunk and branches of white pine.

Control. No control is feasible where extensive decay occurs. Wounds should be avoided and the trees kept in good vigor.

INSECTS

PINE BARK APHID. *Pineus strobi.* White woolly patches on trunks and limbs (Fig. 140), especially of white pine, are signs of infestations by the bark aphid. The adult females overwinter under the woolly material.

Control. Spray with either malathion or lindane in late fall or early spring.

PINE LEAF APHID. *Pineus pinifoliae.* This insect infests species of pines and spruces. It winters over on pine, and then moves over to spruces in spring where it causes terminal galls. In summer the aphids move to white pine where they give birth to nymphs. These then suck out the sap from the new shoots, resulting in either undersized leaves or in the death of the new shoots.

Control. Spray the white pines with malathion just after the first galls open on the spruces in June.

SAWFLIES. *Neodiprion sertifer* and *N. lecontei.* The needles of pines are voraciously chewed by these species of sawfly larvae. The former are green with shiny black heads that feed in groups and are difficult to detect because they resemble the pine needles (Fig. 141). The latter larvae are whitish with a brownish head when young, and become yellowish with 6 rows of black spots as they age. When full grown, this larva is almost 1 inch long and has a red head. The adult stages look like small bees or fuzzy flies.

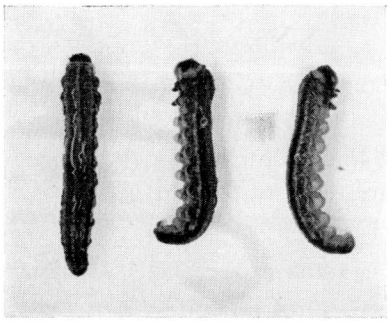

FIG. 141. Pine sawfly larvae.

In the East, another sawfly, also known as pine false webworm, *Acantholyda erythrocephala,* chews the needles of red and white pines, pulling them into a web.

Control. Spray with arsenate of lead, DDT, or malathion as soon as the larvae hatch in spring and begin to feed. The larvae of the red-headed sawfly appear about June 7 in the latitude of New York City.

WEBWORM. *Tetralopha robustella.* Masses of brown frass at the ends of terminal twigs result from infestations of the webworm, a yellowish-brown larva, with a black stripe on each side of the body, which is ⅘ inch long.

Control. Spray with arsenate of lead, DDT, or malathion when the larvae are young and before the needles are webbed.

EUROPEAN PINE SHOOT MOTH. *Rhyacionia buoliana.* Wilting of lateral shoots of pine usually results from feeding by the European pine shoot moth larvae, a brown-bodied, black-headed caterpillar, ⅔ inch long. The adult, a moth with silvery-banded, reddish-orange frontal wings and brown hind wings, which have a spread of ¾ inch, deposits eggs in late June or early July on buds or terminal twigs. Austrian and mugho pines are most susceptible to this pest.

Another species, Nantucket pine moth, *R. frustrana,* attacks 2- and 3-needled pines in the eastern United States.

Control. Remove and burn infested shoots in fall, winter, or early spring. Spray with DDT in April when the buds begin to swell. Where this application is skipped, spray in late June before the larvae enter the branch tips.

PINE LEAF SCALE. *Phenacaspis pinifoliae.* Pine needles may appear nearly white when heavily infested with pine leaf scale, an elongated insect, $\frac{1}{10}$ inch long, which is white with a yellow spot at one end (Fig. 142). The pest overwinters in the egg stage under female scales. The eggs hatch in May. The black pine needle scale, *Aspidiotus californicus,* occasionally infests some species of pine. Austrian pines may be attacked by the scale *Poliaspis pini,* recently introduced from the Orient.

Control. Spray with lime sulfur during the dormant period, or with malathion during late May or early June, when the young scales are crawling about.

PINE NEEDLE MINER. *Exoteleia pinifoliella.* Ornamental pine needles turn yellow and dry up as a result of mining by a ⅕-inch long brown larva.

Control. Spray with malathion on the first warm day in late March or early April. A second application in summer when the yellow-brown moths are flying about is suggested where this pest is particularly prevalent.

BARK BEETLES. *Scolytidae.* Trees in weakened or dying condition are subject to infestation by bark beetle larvae, small worms that mine the bark and engrave the sapwood. Tiny shot holes are present on the bark surface where beetles, varying from 1/20 to 1/3 inch in size (depending on the species), emerge.

Control. No effective control measure is known. Trees should be kept in vigorous condition by feeding and watering. Severely infested trees should be cut down and burned, or, if left standing, should be debarked.

FIG. 142. A heavy infestation of pine-leaf scale.

SPITTLE BUG. *Aphrophora parallela.* A spittle-like substance covering the needles and twigs (Fig. 143) is produced by the young of the spittle bug, a frog-shaped brown insect, 3/8 inch long. Eggs are deposited in the fall, and the young hatching from these the following spring cover themselves with the frothy substance. Another species, the Saratoga spittle bug, *A. saratogensis,* seriously damages jack and red pine.

Control. Spray with DDT or malathion in June.

WHITE PINE WEEVIL. *Pissodes strobi.* The leader shoot wilts and dies as a result of girdling of the inner bark by the white pine weevil larva, a white legless form, ⅓ inch long. The adult is a reddish-brown beetle, about ¼ inch long, mottled with brown and white scales. It makes incisions and deposits eggs in the bark of the leader shoot in June. The adult hibernates in ground litter.

FIG. 143. Spittle bug on Austrian pine.

Control. Remove and burn infested leader shoots * in the spring. Spray with DDT in late April or early May when the buds begin to swell.

RED PINE SCALE. *Matsucoccus resinosae.* This species now infests red pines in the area of southeastern New York and southwestern Connecticut. The

* The next lower lateral should be tied in an erect position with a small stick and soft rope to ensure straight development of the stem.

current season's growth on infested trees consists of yellowed needles which turn brick-red; then the tree dies. The young and adult stages, very difficult to detect, are yellow to brown in color and are hidden in the bark or inside the needle clusters.

Control. Severely infested trees should be cut down and burned. Spray less severely infested trees with 1 pound of 25 per cent chlorthion wettable powder in 25 gallons of water 3 or 4 times at 2 to 3-week intervals starting in June.

PALES WEEVIL. *Hylobius pales.* The bark of young white, red, and Scotch pines may be chewed by a night-feeding reddish-brown to black weevil ⅓ inch long. Young trees may be girdled completely.

Control. Spray the trunk, particularly the lower portions, with DDT in late May.

BAGWORM. See under Arborvitae.

REDWOOD (*Sequoia*)

Redwoods are unusually free of diseases and insect pests.

Diseases

CANKER. A bark canker is caused by the fungus *Dermatea livida.* Small brown to black fruiting bodies break through the bark surface at maturity to release one-celled colorless spores.

Control. No satisfactory control measures have been developed.

NEEDLE BLIGHT. Two fungi, *Chloroscypha chloromela* and *Mycosphorella sequoiae,* can blight redwood leaves. The giant sequoia is susceptible to two other leaf-blighting fungi, *Cercospora sequoiae* and *Pestalotia funerea.*

Control. Small trees infected with these fungi should be sprayed with a copper fungicide.

SHOESTRING ROOT ROT. See page 243.

Insects

CEDAR TREE BORER. *Semanotus ligneus.* The larvae of this beetle make winding burrows in the inner bark and sapwood of redwood, cedars, arborvitae, Douglas fir, and Monterey pine, occasionally girdling and killing the trees. The adult beetle is ½ inch long, black with orange and red markings on its wing covers.

Control. Young trees should be kept in good vigor by feeding and watering when necessary. Sprays containing DDT or malathion on the trunk and branches, when the adult beetles emerge, should provide control.

SEQUOIA PITCH MOTH. *Vespamima sequoiae.* The cambial region of redwoods and many other conifers in the West is mined by an opaque, dirty white larva. The adult moth resembles a yellow-jacket wasp because it is black and has a bright yellow segment at its lower abdomen.

Control. This pest is rarely abundant enough to warrant control measures.

MEALY BUGS. Three species of mealy bugs may infest redwoods: citrus, *Planococcus citri;* cypress, *Pseudococcus ryani;* and yucca, *Puto yuccae.*

Control. Spray young, valuable trees with malathion. Repeat in a few weeks where the first spray did not provide good control.

SCALES. Two scale insects have been reported on this host: greedy, *Aspidiotus camelliae,* and oleander, *A. hederae.*

Control. Spray with malathion when the young scales are crawling about.

SPRUCE (*Picea*)

Besides being valuable forest trees, the spruces are highly prized as ornamentals. They thrive best in moist, sandy loam soil and can tolerate more shade than most other conifers. The varieties most commonly used in ornamental plantings are Norway spruce, Colorado blue spruce, and Koster's blue spruce. They are subject to a number of destructive fungus diseases and insect pests.

DISEASES

CYTOSPORA CANKER. The most prevalent disease on Norway spruce and Colorado blue spruce is known as Cytospora canker. Koster's blue spruce and Douglas-fir are also subject to it, but to a lesser extent. Although the disease occurs on young trees, those over 15 years old appear to be most susceptible.

Symptoms. The most striking symptoms is the browning and death of the branches, usually starting with those nearest the ground and slowly progressing upward. Occasionally branches high in the tree are attacked, even though the lower ones are healthy.

The needles may drop immediately from infected branches or may persist for nearly a year, eventually leaving dry brittle twigs that contrast sharply with unaffected branches (Fig. 144). White patches of pitch or resin may appear along the bark of dead or dying branches (Fig. 145). Cankers occur in the vicinity of these exudations. They are not readily discernible, but can be found by cutting back the bark in the area that separates diseased from healthy tissues. In the cankered area, tiny, black, pin-point fruiting bodies of the causal fungus are a positive sign of the disease.

FIG. 144. Blue spruce affected by Cytospora canker disease. Note the dead, leafless branches which contrast with uninfected branches.

FIG. 145. Close-up of Cytospora-infected branch showing white, resinous exudation.

Cause. The fungus *Cytospora kunzei* causes this disease. The fruiting structures protrude slightly during wet weather, but are concealed beneath the bark scales during dry spells (Fig. 146). Under the former conditions they release long, curled, yellow tendrils, which contain millions of spores. Wind and rain splash the spores to bark wounds on branches, where infection takes place. When the infection spreads completely around the branch, the distal portions die as a result of the girdling.

Many other trees weakened by adverse growing conditions such as drought, winter injury, insects, fire, and mechanical injuries are subject to different species of Cytospora canker fungi. Following are the most common, in addition to *C. kunzei* mentioned above and *C. chrysosperma* discussed on page 362: *C. ascerina* on Japanese maple; *C. ambiens* on American elm; *C. annularis* on black and red ash; *C. corni* on dogwood; *C. leucostoma* on Norway spruce; *C. microspora* on American mountain ash; *C. mollissima* on Chinese chestnut; *C. pinastri* on balsam fir; *C. rubescens* on American and European mountain ash, and an undetermined species on walnut and white mulberry.

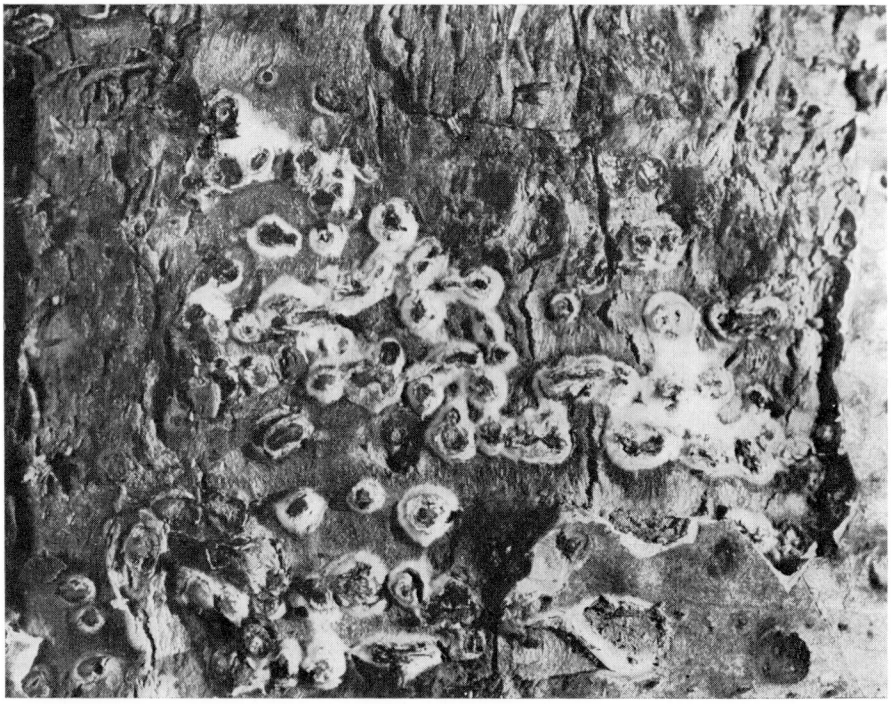

FIG. 146. Fruiting bodies of the Cytospora canker fungus on spruce.

Control. Infected branches cannot be saved. They should be cut off a few inches below the dead or infected parts, or at the point of attachment to the main stem. Because of the danger of spreading spores to uninfected branches, pruning should never be undertaken while the branches are wet. Inasmuch as the available evidence indicates that the fungus enters the branches only through wounds, bark injuries by lawn mowers and other tools should be avoided. Some plant pathologists believe, however, that there is no fool-proof way of preventing infections. Others report that spraying the lower branches and the trunk several times in the spring with bordeaux mixture 4-4-50, or any other copper fungicide, will help to prevent further spread. The applications should be spaced at 2 to 3-week intervals.

The disease appears to be most prevalent on trees in poor vigor. Consequently, trees should be fertilized at least every few years, watered during dry spells, and the soil improved as suggested in Chapter 2, to increase or maintain their vigor.

NEEDLE CAST. As with many other conifers, spruces are subject to the leaf-cast disease. The symptoms resemble those on pines, but the causal fungus is *Lophodermium filiforme*. The lower branches of red and black spruce may be completely defoliated. Another species, *L. piceae*, is also common. The fungus *Rhizosphaera kalkhoffi* causes leaf yellowing, starting with the lowermost needles and progressing upward.

Control. Spray valuable ornamental specimens several times in the spring with bordeaux mixture or any other copper spray.

NEEDLE RUST. Several species of rust fungi occur on spruce needles. Each requires an alternate host to complete its life cycle. On the spruce, the fungi appear as whitish blisters on the lower leaf surface. Affected needles turn yellow and may drop prematurely.

The fungus *Chrysomyxa cassandrae* attacks black, blue, and red spruces and requires leatherleaf (*Chamadaphne calyculata*) as its alternate host.

Chrysomyxa ledi occurs on black, Norway, and red spruces, and its alternate stage is found on various species of Labrador tea, *Ledum glandulosum*, *L. groenlandicum*, and *L. decumbens*.

Chrysomyxa ledicola affects blue, black, red, white, Englemann, and Sitka spruces and requires several species of Labrador tea as the alternate hosts.

Control. Removal of the alternate hosts in the vicinity of valuable spruces is suggested. Protective sprays containing sulfur or ferbam are rarely justified.

WOOD DECAY. Spruces are subject to a number of wood decays caused by various fungi. Among fungi most commonly associated with various types of decay are *Trametes pini*, *Polyporus schweinitzii*, *Fomes pinicola*, and *Polyporus sulphureus*.

Control. Little can be done once these decays become extensive. Avoidance of wounds and fertilization to increase vigor are suggested preventive measures.

Insects

SPRUCE GALL APHID. *Chermes abietis.* Elongated, many-celled, cone-shaped galls (Fig. 147), less than an inch long, result from feeding and irritation by the spruce gall aphid. Trees are weakened and distorted when large numbers of these galls are formed. Norway spruce is most seriously infested; white, black, and red spruces are less so. In the spring the adult wingless females deposit eggs near where the galls later develop. Immature females hibernate in the bud scales.

FIG. 147. Galls of spruce gall aphid *Chermes abietis* on Norway spruce.

Control. Spray with malathion plus a spreader-sticker when the trees are still dormant in late March or early April to kill overwintering females at the base of the buds.

COOLEY GALL APHID. *Chermes cooleyi.* Galls ranging from ½ to 2½ inches in length on the terminal shoots of blue, Englemann, and Sitka spruces and Douglas fir are produced by feeding and irritation of an aphid closely related to that which infests the Norway spruce. The adult female overwinters on the bark near the twig terminals and deposits eggs in that vicinity in the spring.

Control. Because Douglas fir is an alternate host for this pest, avoid planting that species near spruces. Spray with malathion in spring just before new growth emerges.

SPIDER MITE. *Oligonychus ununguis.* Yellow, sickly needles, many of which are covered with a fine silken webbing, indicate a severe infestation of the spider mite, a tiny pest, only 1/64 inch long. The young are pale green; the adult female is greenish-black. Winter is passed in the egg stage on the twigs and the needles. In addition to spruces, arborvitae, junipers, and hemlock are also attacked.

Control. Spray with 1 part concentrated lime sulfur in 10 parts of water before growth begins in the spring, or with 1 part concentrated lime sulfur in 49 parts of water at any time during the growing season, especially when the weather is warm. Or spray in mid- or late-May with Aramite, Dimite, Ovex, or malathion to kill the young mites of the first generation. Repeat later in the season if necessary.

SPRUCE BUD SCALE. *Physokermes piceae.* Globular red scales, about ⅛ inch in diameter, may occasionally infest the twigs of Norway spruce. The young crawl about in late July, and the winter is passed in the partly grown adult stage.

Control. Before growth starts in spring, spray with malathion plus a spreader-sticker.

BUDWORM. *Choristoneura fumiferana.* One of the most destructive pests of forest and ornamental evergreens, this pest attacks spruce and balsam firs in forest plantations, and also infests ornamental spruce, balsam fir, Douglas fir, pine, larch, and hemlock. The opening buds and needles are chewed by the caterpillar stage, which is dark, reddish-brown with a yellow stripe along the side. The adult female moths, dull gray, marked with brown bands and spots, emerge in late June and early July.

Control. Forest areas can be protected with airplane applications of DDT. Ornamental spruces, pines, firs, hemlocks, and larches can be sprayed with arsenate of lead, DDT, or malathion just as the buds burst in spring.

SPRUCE EPIZEUXIS. *Epizeuxis aemula.* Ornamental spruces in the Northeast may be attacked by small brown larvae with black tubercles which web needles together and fill them with excrement.

Control. Spray with arsenate of lead or malathion when the larvae begin to feed in late summer.

SPRUCE NEEDLE MINERS. *Taniva albolineana, Epinotia nanana,* and *Recurvaria piceaella.* These insects mine the needles of many kinds of spruce.

Control. Spray with malathion or DDT in mid-May and again in mid-June.

BAGWORM. See under Arborvitae.

YEW (*Taxus*)

Yews are among the most useful evergreens for ornamental purposes. They withstand city conditions better than most other evergreens. The English yew, the Japanese yew, and a large number of other species and varieties are now planted extensively.

Diseases

DIE-BACK. The most prevalent trouble on yews is die-back.

Symptoms. The plants first turn yellow at the growing tips, followed by general yellowing, wilting, and death. Several months may elapse from the time the first symptoms appear to the complete wilting and death of the plant. The below-ground symptoms are a decay of the bark on the deeper roots, which sloughs off readily.

Cause. Die-back is not caused by parasitic fungi. Studies by the author show that it is associated with unfavorable soil conditions. In nearly every case investigated, the soil was very acid, pH 4.7 to 5.4, in addition to being heavy and poorly drained. Recent research at Rutgers University revealed that yews could be killed by immersing their roots in water for 32 to 64 hours and then drying out the soil.

Control. Improve drainage by embedding tile in the soil, or move plants to more favorable soil. Add ground limestone, according to the recommendation of a soil specialist, to increase the pH to about 6.5.

TWIG BLIGHT. Several fungi are associated with a twig blight of yew during rainy seasons. One of the most common is *Phyllostictina hysterella,* whose perfect stage is *Physalospora gregaria.* Others are *Pestalotia funerea* and a species of *Sphaeropsis.*

Control. Prune out diseased twigs and burn them. Spray several times at 2-week intervals during rainy springs.

TWIG BROWNING. This condition is not caused by fungus parasites but is due to snow and winter damage. The combined effect of heavy snow cover and low temperatures causes many small twigs at the ends of branches to turn brown in late winter and early spring. The condition was very prevalent in the spring of 1958 in the Northeast.

Control. Shake off heavy snow covers with a bamboo rake as soon as the snowfall stops.

NEEDLE BLIGHT. A needle blight of yew caused by the fungus *Sphaerulina taxi* appears occasionally.

Control. The disease is rarely serious enough to warrant applications of a protective fungicide containing copper.

CROWN GALL. This disease (p. 240) occurs on English yew.

INSECTS

BLACK VINE WEEVIL. *Brachyrhinus sulcatus.* This pest, by far the most serious one on yews, is so named because of its color and because it attacks grapes in Europe. It also attacks retinospores, hemlocks, and broad-leaved plants like rhododendrons and azaleas.

The leaves of yew turn yellow, and whole branches or even the entire plant may die when the roots are chewed by the larval stage, a white-bodied, brown-headed pest, ⅜ inch long. As few as 8 larvae are capable of killing a large-sized yew. The adult, a snout beetle, ⅜ inch long, feeds on the foliage of yews and other hosts at night. The edges of such leaves are scalloped.

Eggs deposited in the soil during July and August hatch into larvae which enter the soil and feed on the roots of susceptible hosts.

STRAWBERRY ROOT WEEVIL. *Brachyrhinus ovatus.* This insect is related to the black vine weevil, has similar habits, but is smaller, measuring ⅕ to ¼ inch in length. It feeds on hemlock, spruce, and arborvitae in addition to Taxus. It often wanders into homes in search of hibernating quarters and thus may become a nuisance.

Control. Spray the lower parts of susceptible plants and the soil surface beneath them in late June and again in early July with either dieldrin or chlordane. One gallon of either material properly diluted will treat an area of about 80 square feet. Following are the correct amounts for each material

FIG. 148. White, waxy coating on the twigs is typical of Taxus mealy bug infestations.

in 1 gallon of water: 2 level tablespoons of 25 per cent dieldrin wettable powder; or 1 level teaspoon of 18.5 per cent dieldrin emulsifiable solution; or 5 level tablespoons of 40 per cent chlordane wettable powder; or 1 level teaspoon of 72 per cent chlordane emulsifiable solution. For large-scale application, mix 2 pounds of 25 per cent dieldrin wettable powder or 5 pounds of chlordane wettable powder in 100 gallons of water.

The wettable powders are preferred even though they leave a whitish coating on the leaves. Dieldrin has a longer residual effect than the chlordane.

TAXUS MEALY BUG. *Pseudococcus cuspidatae*. First reported from a New Jersey nursery in 1915, this pest has become increasingly prevalent in the northern United States (Fig. 148). The ⅜ inch long bug, covered with white wax, may completely cover the trunk and branches of yews. Young mealy bugs winter over in bark crevices and mature in June. Although all species of Taxus are susceptible, those with dense foliage such as *Taxus cuspidata nana* and *T. wardii* are preferred. This pest has also been seen on apple, basswood, and rhododendron but probably does not breed on these plants.

Control. Good control is obtainable by spraying with malathion when the young stage is crawling about. In the area of Long Island and Connecticut, the latter part of June is the proper time; in cooler regions the application should be made later.

GRAPE MEALY BUG. *Pseudococcus maritimus* occasionally infests the Japanese yew.

Control. Where this pest infests above-ground parts of yews, spray with malathion.

COTTONY TAXUS SCALE. *Pulvinaria floccifera*. This small, light brown, flattened, hemispherical ⅛ inch long scale is primarily a greenhouse pest but it can infest yews outdoors. In the spring it produces long, narrow, fluted cottony egg masses.

Control. Apply a dormant oil spray while the plants are dormant in early spring when the temperature is well above freezing. In mid-May and again in mid-June spray with malathion.

OTHER SCALES. California red scale, *Aonidiella aurantii*; Fletcher, *Lecanium fletcheri*; oleander, *Aspidiotus hederae*; and purple, *Lepidosaphes beckii*, also attack yews.

Control. Spray with malathion in late spring when the young are crawling about.

ANTS AND TERMITES. The carpenter ant *Camponotus herculeanus pennsylvanicus*, and the subterranean termite, *Reticulitermes flavipes*, occasionally infest the trunks of older yews. Both are capable of excavating the trunk and making their nests therein.

Control. Apply 2 per cent chlordane dust or spray a 2 per cent chlordane liquid prepared from the wettable powder around the trunk base and soil surface.

Appendix

METHOD OF EVALUATING SHADE TREES *

In determining the monetary value of shade and ornamental trees, three basic factors must be considered. These are size, kind, and condition of the tree.

Tree size can be designated in several ways, such as the diameter of the trunk, the area of a cross section of the trunk, the height, the branch spread, or a combination of all or part of these measurements. It is the belief of the committee that the area of the cross section of the trunk at a point breast high (four and one-half feet above ground) is the best means of expressing the size of shade trees when a value related to size is to be determined. This cross-section trunk area can easily be calculated from the trunk diameter by using the formula $.7854\ d^2$. These calculations have been made for trees with trunk diameters ranging from 2 inches to 60 inches and are given on page 430.

It is the opinion of the committee that the conservative value of a perfect specimen shade tree in August, 1951, was $5.00 per square inch of trunk cross section. The value of our dollar varies with economic conditions and no price set upon a tree will remain constant for all time.† Should our money depreciate further, a proportionate increase in the value per square inch of trunk cross section will be necessary. Conversely, if money increases in value the unit price of $5.00 can be reduced.

Not all species and varieties of shade trees are of equal value. The same kind of tree may have different values in different areas of the country. Therefore, it is necessary that a list of trees growing in each area be compiled and that the trees be segregated into groups based on relative value. It is the committee's recommendation that the trees in Group 1 be valued at 100 per cent, in Group 2 at 80 per cent, in Group 3 at 60 per cent, in Group 4 at 40 per cent, and those in Group 5 at 20 per cent.

Very few shade trees are perfect specimens. As trees become large and old they

* Except for minor changes in style to conform with the remainder of the book, and for typographical errors, this material was taken verbatim from the booklet *Shade Tree Evaluation* published by The National Shade Tree Conference and the National Arborist Association in July 1957 at Wooster, Ohio.

† At the 1958 annual meeting of the National Shade Tree Conference, Mr. Norman Armstrong suggested that low-growing trees be given a higher value than the $5.00 per square inch of trunk cross section allowed for tall-growing or standard trees.

very often become defective in one or several ways. The person making an appraisal must consider the condition of the tree and judge how nearly each tree approaches a perfect specimen. With respect to condition of the tree, the committee suggests five percentage classes with values of 100 per cent of a perfect specimen, 80 per cent, 60 per cent, 40 per cent, and 20 per cent.

To illustrate the proposed method for determining the value of a shade tree, consider a tree with a trunk DBH (diameter breast high) of 20 inches. By referring to the table on this page, the trunk cross-section area of this tree is found to be 314 inches. The tree has a basic value of $5.00 x 314 or $1570. This would be the value of the tree, if it were in Group 1 or 100 per cent species group and also in Class 1 or 100 per cent condition class. But if the tree were in Group 3 with respect to species, the value would be 60 per cent of $1570 or $942. If this 20-inch tree also should be placed in the 60 per cent of perfect class with respect to condition, its value would be 60 per cent of $942 or $565. For the sake of simplicity, it is recommended that the cents be dropped and that tree values be expressed to the nearest dollar.

Any means or formula for evaluating shade trees must be flexible. No hard and fast rule can be set up that will hold in all cases. Instances will undoubtedly be encountered of such remarkable specimens that the basic value should be increased. This is a matter for the appraiser's judgment as is the Condition Class for any tree. Always the judgment and opinions of the appraiser will be a large factor in determining the value. If it is desirable to consider matters not included in the formula, such as effect of land value, or any other factor believed to influence the value of the particular tree that is left to the person making the appraisal.

Most small trees have a Replacement Value. It is suggested that if a definite replacement value can be established for a particular tree that this be considered the value of the tree. For medium and large trees where a replacement value is impractical the method which has been described will give fair values.

TRUNK DIAMETERS AND AREA OF TRUNK CROSS-SECTIONS

Trunk Diameter Inches	Cross-Section Area Sq. Inches	Trunk Diameter Inches	Cross-Section Area Sq. Inches
2	3.14	18	254.5
3	7.07	19	283.5
4	12.57	20	314.2
5	19.64	21	346.4
6	28.27	22	380.1
7	38.49	23	415.5
8	50.27	24	452.4
9	63.62	25	490.9
10	78.5	26	530.9
11	95.0	27	572.6
12	113.1	28	615.8
13	132.7	29	660.5
14	153.9	30	706.9
15	176.7	31	754.8
16	201.1	32	804.3
17	227.0	33	855.3

TRUNK DIAMETERS AND AREA OF TRUNK CROSS-SECTIONS (*Cont.*)

Trunk Diameter Inches	Cross-Section Area Sq. Inches	Trunk Diameter Inches	Cross-Section Area Sq. Inches
34	907.9	48	1809.6
35	962.1	49	1885.7
36	1017.9	50	1963.5
37	1075.2	51	2042.8
38	1134.1	52	2123.7
39	1194.6	53	2206.2
40	1256.6	54	2290.2
41	1320.3	55	2375.8
42	1385.4	56	2463.0
43	1452.2	57	2551.8
44	1520.5	58	2642.1
45	1590.4	59	2734.0
46	1661.9	60	2827.4
47	1734.9		

SUGGESTED FORM OF WORK SHEETS

Species and Variety	Diameter of Trunk in Inches	Sq. Inches in Cross Section	Basic Value in Dollars	Species and Variety Group	Condition Class	Value of Tree in Dollars
			ILLUSTRATIONS			
White Oak	20	314	$ 1570	1-100%	3-60%	$ 942
Black Locust	20	314	1570	5-20%	4-40%	125

CLASSIFICATION OF TREES BY REGIONS

Region I — New England: Connecticut, Maine, Massachusetts, New Hampshire, Rhode Island, Vermont.

Region II — Eastern: New Jersey, New York, Pennsylvania.

Region III — Southern: Alabama, Delaware, District of Columbia, Florida, Georgia, Louisiana, Maryland, Mississippi, North Carolina, South Carolina, Tennessee, Texas, Virginia, West Virginia.

Region IV — Central: Indiana, Kentucky, Michigan, Ohio.

Region V — Midwestern: Arkansas, Colorado, Illinois, Iowa, Kansas, Minnesota, Missouri, Nebraska, North Dakota, Oklahoma, South Dakota, Wisconsin.

Region VI — Western: Arizona, California, Idaho, Montana, Nevada, New Mexico, Oregon, Utah, Washington, Wyoming, British Columbia.

Region VII— Eastern Canada: Ontario and Quebec.

TREES OF REGION I—NEW ENGLAND

Class No. 1—100%

Acer saccharum—Sugar Maple
Amelanchier canadensis—Shadblow Serviceberry
Betula pendula—European White Birch
Fagus grandifolia—American Beech
Gleditsia triacanthos—Common Honey Locust
Liriodendron tulipifera—Tulip tree
Nyssa sylvatica—Black Tupelo
Pinus strobus—Eastern White Pine
Quercus alba—White Oak
Quercus bicolor—Swamp White Oak
Quercus borealis maxima—Eastern Red Oak
Quercus coccinea—Scarlet Oak
Quercus palustris—Pin Oak
Tsuga canadensis—Canada Hemlock

Class No. 2—80%

Acer rubrum—Red Maple
Betula lenta—Sweet Birch
Betula nigra—River Birch
Carya cordiformis—Bitternut Hickory
Carya glabra—Pignut Hickory
Carya ovata—Shagbark Hickory
Carya tomentosa—Mockernut Hickory
Fraxinus americana—White Ash
Fraxinus nigra—Black Ash
Fraxinus pennsylvanica—Red Ash
Fraxinus pennsylvanica lanceolata—Green Ash
Juglans nigra—Black Walnut
Malus pumila—Common Apple
Picea mariana—Black Spruce
Pinus resinosa—Red Pine
Pyrus communis—Common Pear
Quercus montana—Chestnut Oak
Sassafras albidum—Common Sassafras
Sorbus aucuparia—European Mountain Ash

Class No. 3—60%

Aesculus hippocastanum—Common Horse-chestnut
Betula lutea—Yellow Birch
Picea glauca—White Spruce
Platanus occidentalis—American Plane (Sycamore)
Populus candicans—Balm of Gilead Popular
Populus deltoides—Eastern Poplar
Populus tremuloides—Quaking Aspen
Quercus prinus—Swamp Chestnut Oak
Quercus velutina—Black Oak
Salix alba—White Willow
Salix babylonica—Babylon Weeping Willow
Salix fragilis—Brittle Willow
Salix lucida—Shining Willow
Salix nigra—Black Willow
Sorbus americana—American Mountain Ash
Tilia americana—American Linden (Basswood)

Class No. 4—40%

Abies balsamea—Balsam Fir
Acer saccharinum—Silver Maple
Ailanthus altissima—Tree-of-heaven (Ailanthus)
Betula populifolia—Gray Birch
Larix laricina—Eastern Larch (Tamarack)
Morus alba—White Mulberry
Morus rubra—Red Mulberry
Picea rubens—Red Spruce
Pinus banksiana—Jack Pine
Populus alba—White Poplar
Populus grandidentata—Big-tooth Aspen
Prunus serotina—Wild Black Cherry
Robinia pseudoacacia—Black Locust
Ulmus americana—American Elm
Ulmus thomasii—Rock Elm

Class No. 5—20%

Acer negundo—Boxelder
Catalpa speciosa—Northern Catalpa
Juglans cinerea—Butternut
Pinus rigida—Pitch Pine
Ulmus fulva—Slippery Elm

TREES OF REGION II—EASTERN

Class No. 1—100%

Acer platanoides erectum—Erect Norway Maple
Acer rubrum—Red Maple
Acer saccharum—Sugar Maple
Carpinus caroliniana—American Hornbeam
Cercidiphyllum japonicum—Katsura tree
Cornus florida—Flowering Dogwood
Crataegus oxyacantha paulii—Pauls Scarlet Hawthorn
Crataegus phaenopyrum—Washington Hawthorn
Fagus grandifolia—American Beech
Fagus sylvatica var.—European Beech
Ginkgo biloba—Ginkgo
Gleditsia triacanthos moraine—Moraine Thornless Honey Locust
Ilex opaca—American Holly
Magnolia acuminata—Cucumber-tree Magnolia
Malus sp. and var.—Flowering Crabapples
Oxydendrum arboreum—Sourwood
Quercus alba—White Oak
Quercus borealis maxima—Eastern Red Oak
Quercus imbricaria—Shingle Oak
Quercus palustris—Pin Oak
Quercus phellos—Willow Oak
Quercus shumardii—Shumard Oak
Taxodium distichum—Common Bald Cypress
Tsuga canadensis—Canada Hemlock
Ulmus americana—American Elm

Class No. 2—80%

Acer platanoides—Norway Maple
Acer platanoides schwedleri—Schwedler Norway Maple
Aesculus carnea—Red Horsechestnut
Aesculus hippocastanum—Common Horsechestnut
Amelanchier canadensis—Shadblow Serviceberry
Carya ovata—Shagbark Hickory
Cladrastis lutea—Yellowwood
Crataegus sp.—Hawthorns
Fraxinus americana—White Ash
Fraxinus pennsylvanica lanceolata—Green Ash
Juglans nigra—Black Walnut
Koelreuteria paniculata—Panicled Goldenrain tree
Liquidambar styraciflua—Sweetgum
Liriodendron tulipifera—Tulip tree
Nyssa sylvatica—Black Tupelo (Blackgum)
Phellodendron amurense—Amur Cork tree
Pinus strobus—Eastern White Pine
Platanus acerifolia—London Plane
Platanus occidentalis—American Plane (Sycamore)
Quercus bicolor—Swamp White Oak
Quercus coccinea—Scarlet Oak
Quercus lyrata—Overcup Oak
Quercus prinus—Swamp Chestnut Oak
Sophora japonica—Japanese Pagoda tree
Tilia cordata—Little-leaf Linden

Class No. 3—60%

Acer pseudoplatanus—Sycamore Maple
Aesculus glabra—Ohio Buckeye
Betula lutea—Yellow Birch
Betula populifolia—Gray Birch
Carya sp.—Hickories
Celtis occidentalis—Common Hackberry
Cercis canadensis—Eastern Redbud
Diospyros virginiana—Common Persimmon
Gleditsia triacanthos—Common Honey Locust
Gymnocladus dioicus—Kentucky Coffee tree
Ostrya virginiana—American Hophornbeam
Picea pungens glauca—Blue Colorado Spruce
Pinus nigra—Austrian Pine
Pinus resinosa—Red Pine
Quercus macrocarpa—Bur Oak
Quercus montana—Chestnut Oak
Quercus robur fastigiata—Upright English Oak
Quercus stellata—Post Oak
Sassafras albidum—Common Sassafras
Tilia americana—American Linden (Basswood)
Ulmus procera—English Elm

TREES OF REGION II—EASTERN (*Cont.*)

Class No. 4—40%

Acer saccharinum—Silver Maple
Asimina triloba—Common Pawpaw
Maclura pomifera—Osage-Orange
Morus sp.—Mulberry
Picea abies—Norway Spruce
Robinia pseudoacacia—Black Locust
Salix babylonica—Babylon Weeping Willow

Class No. 5—20%

Acer negundo—Boxelder
Catalpa speciosa—Northern Catalpa
Populus canadensis eugenei—Eugene Poplar
Populus nigra—Lombardy Poplar
Salix sp.—Willows
Ulmus pumila—Siberian Elm
Ailanthus altissima—Tree-of-heaven (Ailanthus)

TREES OF REGION III—SOUTHERN

Class No. 1—100%

Acer rubrum—Red Maple
Cornus florida—Flowering Dogwood
Fagus grandifolia—American Beech
Fagus sylvatica—European Beech
Ginkgo biloba—Ginkgo
Gleditsia triacanthos moraine—Moraine Thornless Honey Locust
Ilex opaca—American Holly
Magnolia grandiflora—Southern Magnolia
Pinus palustris—Long-leaf Pine
Pinus taeda—Loblolly Pine
Quercus alba—White Oak
Quercus coccinea—Scarlet Oak
Quercus imbricaria—Shingle Oak
Quercus laurifolia—Laurel Oak
Quercus lyrata—Overcup Oak
Quercus macrocarpa—Bur Oak
Quercus phellos—Willow Oak
Quercus shumardii—Shumard Oak
Quercus virginiana—Live Oak

Class No. 2—80%

Acer floridanum—Florida Maple
Acer saccharum—Sugar Maple
Carya illinoensis—Pecan
Gleditsia triacanthos inermis—Thornless Honey Locust
Juglans nigra—Black Walnut
Koelreuteria paniculata—Panicled Goldenrain tree
Liquidambar styraciflua—Sweetgum
Liriodendron tulipifera—Tulip tree
Nyssa sylvatica—Black Tupelo (Blackgum)
Pinus strobus—Eastern White Pine
Quercus prinus—Swamp Chestnut Oak
Taxodium distichum—Common Bald Cypress
Tsuga canadensis—Canada Hemlock
Tsuga caroliniana—Carolina Hemlock
Ulmus alata—Winged Elm
Ulmus americana—American Elm
Ulmus crassifolia—Cedar Elm
Ulmus thomasii—Rock Elm

Class No. 3—60%

Acer platanoides—Norway Maple
Aesculus glabra—Ohio Buckeye
Aesculus hippocastanum—Common Horsechestnut
Amelanchier canadensis—Shadblow Serviceberry
Carpinus caroliniana—American Hornbeam
Cedrus atlantica—Atlas Cedar
Cedrus libani—Cedar of Lebanon
Cladrastis lutea—American Yellowwood
Carya var.—Hickories
Fraxinus americana—White Ash
Magnolia virginiana—Sweetbay Magnolia
Ostrya virginiana—American Hophornbeam
Oxydendrum arboreum—Sourwood
Platanus occidentalis—American Plane (Sycamore)
Quercus falcata—Southern Red Oak
Salix babylonica—Babylon Weeping Willow
Sassafras albidum—Common Sassafras

Class No. 4—40%

Celtis occidentalis—Common Hackberry
Fraxinus pennsylvanica lanceolata—Green Ash
Maclura pomifera—Osage-Orange
Magnolia acuminata—Cucumber-tree Magnolia
Quercus marilandica—Blackjack Oak
Quercus nigra—Water Oak
Quercus stellata—Post Oak
Tilia cordata—Little-leaf Linden

Class No. 5—20%

Acer negundo—Boxelder
Acer saccharinum—Silver Maple
Ailanthus altissima—Tree-of-heaven (Ailanthus)
Catalpa bignonioides—Southern Catalpa
Cercis canadensis—Eastern Redbud
Diospyros virginiana—Persimmon
Melia azedarach—Chinaberry
Morus rubra—Red Mulberry
Pinus echinata—Short-leaf Pine
Populus canadensis eugenei—Eugene Poplar
Populus heterophylla—Swamp Poplar (Swamp Cottonwood)
Prunus caroliniana—Carolina Laurel Cherry
Quercus borealis maxima—Eastern Red Oak
Quercus palustris—Pin Oak
Robina pseudoacacia—Black Locust
Salix nigra—Black Willow
Ulmus pumila—Siberian Elm

TREES OF REGION IV—CENTRAL

Class No. 1—100%

LARGE TREES—Mostly 60 feet and above

Acer plataniodes—Norway Maple
Acer saccharum—Sugar Maple
Gleditsia triacanthos moraine—Moraine Thornless Honey Locust
Liquidambar styraciflua—American Sweetgum
Nyssa sylvatica—Black Tupelo (Sourgum)
Quercus alba—White Oak
Quercus borealis maxima—Eastern Red Oak
Quercus coccinea—Scarlet Oak
Quercus imbricaria—Shingle Oak
Quercus shumardii schneckii—Schneck Oak
Ulmus americana—American Elm
Ulmus carpinifolia sarniensis—(*Wheatleyi*) Sarnia Smooth-leaf Elm

MEDIUM TREE—Mostly 30 to 45 feet

Acer platanoides erectum—Erect Norway Maple
Cladrastis lutea—American Yellowwood
Ginkgo biloba fastigiata—Sentry Ginkgo
Ostrya virginiana—American Hophornbeam
Phellodendron amurense—Amur Cork tree
Phellodendron sachalinense—Sakhalin Cork tree
Quercus acutissima—Saw-tooth Oak
Quercus phellos—Willow Oak
Tilia cordata—Little-leaf Linden

SMALL TREES—Mostly 20 to 30 feet

Acer buergerianum—Trident Maple
Acer mono—Mono Maple
Acer tartaricum—Tartarian Maple
Carpinus betulus fastigiata—Pyramid European Hornbeam
Cornus florida—Flowering Dogwood
Cornus officinalis—Japanese Cornel Dogwood
Crataegus phaenopyrum (*cordata*)—Washington Hawthorn
Koelreuteria paniculata—Panicled Goldenrain tree
Magnolia soulangeana—Saucer Magnolia
Malus species and varieties—Flowering Crabapples

Class No. 2—80%

LARGE TREES—Mostly 60 feet and above

Acer platanoides, Crimson King—Crimson King Norway Maple
Acer rubrum—Red Maple
Acer rubrum schlesingerii—Schlesinger Red Maple
Fagus grandifolia—American Beech
Fagus sylvatica—European Beech
Fagus sylvatica heterophylla—Cut-leaf European Beech
Ginkgo biloba—Ginkgo
Gymnocladus dioicus—Kentucky Coffee tree

TREES OF REGION IV—CENTRAL (*Cont.*)

Liriodendron tulipifera—Tulip tree
Magnolia acuminata—Cucumber-tree Magnolia
Platanus acerifolia—London Plane
Quercus montana—Chestnut Oak
Quercus palustris—Pin Oak
Quercus velutina—Black Oak
Sophora japonica—Japanese Pagoda tree
Tilia euchlora—Crimean Linden
Tilia petiolaris—Silver-pendent Linden
Tilia tomentosa—Silver Linden
Ulmus procera (*campestris*)—English Elm

MEDIUM TREES—Mostly 30 to 40 feet

Acer campestre—Hedge Maple
Acer rubrum columnare—Column Red Maple
Acer saccharum columnare—Upright Sugar Maple
Aesculus carnea briotii—Ruby Red Horsechestnut
Betula nigra—River Birch
Carpinus caroliniana—American Hornbeam
Cercidiphyllum japonicum—Katsura tree
Eucommia ulmoides—Eucommia
Magnolia kobus borealis—Large Kobus Magnolia
Pseudolarix amabilis—Lovely Goldenlarch
Quercus muhlenbergii—Chinkapin Oak
Quercus robur—English Oak
Quercus robur fastigiata—Pyramidal English Oak
Sassafras albidum molle—Silky Sassafras
Ulmus parvifolia—Chinese Elm (Asiatic)
Zelkova serrata—Japanese Zelkova

SMALL TREES—Mostly 20 to 30 feet

Acer circinatum—Vine Maple
Acer diabolicum—Devil Maple
Acer ginnala—Amur Maple
Acer griseum—Paperbark Maple
Acer tschonoskii—Tschonoski Maple
Amelanchier laevis—Alleghany Serviceberry
Cercis canadensis—Eastern Redbud
Cercis canadensis alba—White Redbud
Cornus alternifolia—Pagoda Dogwood
Crataegus succulenta—Fleshy Hawthorn
Euonymus maackii—Maack Euonymus
Magnolia kobus—Kobus Magnolia
Magnolia liliflora nigra—Purple Lily Magnolia
Magnolia virginiana—Sweetbay Magnolia
Prunus species and varieties—Flowering Cherries
Syringa amurensis japonica—Japanese Tree Lilac

Class No. 3—60%

LARGE TREES—Mostly 60 feet and above

Acer pseudoplatanus—Sycamore Maple
Celtis laevigata (*mississippiensis*)—Sugar Hackberry
Celtis occidentalis—Common Hackberry
Fraxinus americana—White Ash
Fraxinus pennsylvanica lanceolata—Green Ash
Fraxinus quadrangulata—Blue Ash
Kalopanax pictus—Kalopanax
Larix leptolepis—Japanese Larch
Quercus laurifolia—Laurel Oak
Salix elegantissima—Thurlow Weeping Willow
Taxodium distichum—Common Bald Cypress
Tilia platyphyllos—Big-leaf Linden
Ulmus thomasii—Rock Elm

MEDIUM TREES—30 to 45 feet

Betula papyrifera—Paper Birch
Castanea mollissima—Chinese Chestnut
Evodia hupehensis—Hupeh Evodia
Fraxinus excelsior—European Ash
Halesia monticola—Mountain Silverball
Hemiptelea davidii—David Hemiptelea
Prunus sargentii—Sargent Cherry

SMALL TREES—20 to 30 feet

Euonymus yedoensis—Yeddo Euonymus
Evodia daniellii—Korean Evodia
Syringa pekinensis—Pekin Lilac

Class No. 4—40%

No large trees in this class.
Medium Trees—Mostly 30 to 45 feet

Betula pendula gracilis—Cut-leaf European White Birch

Class No. 5—20%

Large Trees—Mostly 60 feet and above

Acer saccharium—Silver Maple
Ulmus pumila—Siberian Elm

No medium or small trees in this class.

TREES OF REGION V—MIDWESTERN

Class No. 1—100%

Acer saccharum—Sugar Maple
Acer platanoides—Norway Maple
Acer rubrum—Red Maple
Fagus grandifolia—American Beech
Gleditsia triacanthos moraine—Moraine Thornless Honey Locust
Ostrya virginiana—American Hophornbeam
Quercus alba—White Oak
Quercus borealis maxima—Eastern Red Oak
Quercus macrocarpa—Bur Oak
Quercus palustris—Pin Oak

Class No. 2—80%

Carya ovata—Shagbark Hickory
Celtis occidentalis—Common Hackberry
Fraxinus americana—White Ash
Fraxinus pennsylvanica lanceolata—Green Ash
Ginkgo biloba—Ginkgo
Gleditsia triacanthos inermis—Thornless Honey Locust
Juglans nigra—Black Walnut
Tilia cordata—Little-leaf Linden

Class No. 3—60%

Aesculus glabra—Ohio Buckeye
Aesculus hippocastanum—Common Horsechestnut
Platanus occidentalis—American Plane (Sycamore)
Prunus serotina—Black Cherry
Ulmus fulva—Slippery Elm

Class No. 4—40%

Populus alba bolleana—Bolleana Poplar
Prunus pennsylvanica—Pin Cherry
Tilia americana—American Linden (Basswood)
Ulmus americana—American Elm
Ulmus americana moline—Moline Elm

Class No. 5—20%

Acer negundo—Boxelder
Acer saccharinum—Silver Maple
Ailanthus altissima—Tree-of-heaven (Ailanthus)
Catalpa speciosa—Northern Catalpa
Populus alba—White Poplar
Populus deltoides—Eastern Poplar (Cottonwood)
Robinia pseudoacacia—Black Locust
Salix sp.—Willows
Sorbus americana—American Mountain Ash
Ulmus pumila—Siberian Elm

Rocky Mountain Shade Trees for Irrigated Areas 4000 to 6000 Feet

Large Trees

Class No. 1—100%

Acer saccharinum—Silver Maple
Aesculus hippocastanum—Common Horsechestnut
Aesculus octandra—Yellow Buckeye
Celtis occidentalis—Common Hackberry
Gleditsia triacanthos inermis—Thornless Honey Locust
Gymnocladus dioicus—Kentucky Coffee tree
Quercus macrocarpa—Bur Oak
Tilia americana—American Linden
Tilia europaea—European Linden
Ulmus americana—American Elm

Class No. 2—80%

Acer platanoides—Norway Maple
Acer platanoides schwedleri—Schwedler Norway Maple

TREES OF REGION V—MIDWESTERN (*Cont.*)

Acer saccharinum laciniatum—Cut-leaf Silver Maple
Acer saccharum—Sugar Maple
Aesculus glabra—Ohio Buckeye
Betula papyrifera—Paper Birch
Betula pendula—European White Birch
Betula pendula gracilis—Cut-leaf European White Birch
Betula populifolia—Gray Birch
Catalpa speciosa—Northern Catalpa
Fraxinus pennsylvanica lanceolata—Green Ash
Quercus falcata—Southern Red Oak
Juglans nigra—Eastern Black Walnut
Quercus palustris—Pin Oak
Quercus robur—English Oak
Ulmus procera—English Elm

Class No. 3—60%

Acer pseudoplatanus—Sycamore Maple
Fraxinus americana—White Ash
Larix sp.—Larch
Morus alba tatarica—Russian Mulberry
Morus rubra—Red Mulberry
Platanus occidentalis—American Plane (Sycamore)
Populus sargentii—Plains Poplar
Prunus serotina—Black Cherry
Quercus alba—White Oak
Quercus coccinea—Scarlet Oak
Ulmus pumila—Siberian Elm

Class No. 4—40%

Acer negundo—Boxelder
Ginkgo biloba—Ginkgo
Juglans cinerea—Butternut
Liriodendron tulipifera—Tulip tree
Populus alba—White Poplar
Populus alba bolleana—Bolleana Poplar
Populus andrewsii—Andrews Poplar
Populus angustifolia—Narrow-leaf Poplar
Populus deltoides missouriensis—Southern Poplar (Balsam)
Populus canadensis eugenei—Eugene Poplar
Populus nigra italica—Lombardy Poplar
Populus simonii—Simon Poplar
Salix—Willows

Small Trees

Class No. 1—100%

Acer ginnala—Amur Maple
Catalpa ovata—Chinese Catalpa
Crataegus crusgalli—Cockspur Hawthorn
Crataegus phaenopyrum—Washington Hawthorn
Eleagnus angustifolia—Russian Olive
Koelreuteria paniculata—Panicled Goldenrain tree
Malus sp.—Dolgo Crabapple
Malus pumila—Hopa Crabapple
Malus pumila—Redsilver Crabapple
Prunus cerasus—Sour Cherry
Sophora japonica—Japanese Pagoda tree
Sorbus americana—American Mountain Ash
Sorbus aucuparia—European Mountain Ash
Sorbus hybrida—Oakleaf Mountain Ash
Tilia cordata—Little-leaf Linden
Xanthoceras sorbifolium—Shiny-leaf Yellowhorn

Class No. 2—80%

Alnus tenuifolia—Thin-leaf Alder
Crataegus coloradensis—Colorado Hawthorn
Crataegus oxyacantha paulii—Pauls Scarlet Hawthorn
Crataegus punctata—Dotted Hawthorn
Juglans rupestris—Texas Black Walnut
Malus baccata—Siberian Crabapple
Malus pumila—Niedzwetskyana Crabapple
Malus purpurea eleyi—Eley Purple Crabapple
Malus ioensis—Prairie Crabapple
Malus ioensis plena—Bechtel Crabapple
Morus alba pendula—Weeping Mulberry
Populus tremuloides—Quaking Aspen
Prunus sibirica—Siberian Apricot
Robinia neomexicana—New Mexico Locust
Salix pentandra—Laurel Willow
Syringa amurensis japonica—Japanese Tree Lilac

Class No. 3—60%

Acer campestre—Hedge Maple
Betula fontinalis—Water Birch
Cercis canadensis—Eastern Redbud
Cladrastis lutea—American Yellowwood
Crataegus soligna—Willow Hawthorn
Malus floribunda—Japanese Flowering Crabapple
Prunus americana—Newport American Plum

Class No. 4—40%

Ailanthus altissima—Tree-of-heaven (Ailanthus)
Alnus glutinosa—European Alder
Carpinus betulus—European Hornbeam
Carya sp.—Hickory
Catalpa bignonioides—Southern Catalpa
Fagus sylvatica—European Beech
Malus arnoldiana—Arnold Crabapple
Malus atrosanguinea—Carmine Crabapple
Malus halliana parkmani—Parkman Halls Crabapple
Malus sp. and vars.

For High Altitudes, Mountain Areas, 6000 to 9000 feet

In many very high towns, the native evergreens are the only street trees possible.

Large Trees

Class No. 1—100%

Populus acuminata—Lance-leaf Poplar
Populus angustifolia—Narrow-leaf Poplar

Class No. 2—80%

Abies lasiocarpa—Alpine Fir
Picea englemanni—Englemann Spruce
Picea pungens—Colorado Spruce
Pinus aristata—Bristlecone Pine
Pinus contorta latifolia—Lodgepole Pine
Pinus flexilis—Limber Pine
Populus canadensis eugenei—Eugene Poplar
Populus deltoides missouriensis—Southern Poplar (Balsam)
Salix alba vitellina—Yellow-stem White Willow
Salix acutifolia—Sharp-leaf Willow

Small Trees

Class No. 1—100%

Populus tremuloides—Quaking Aspen

Class No. 2—80%

Populus grandidentata—Big-tooth Aspen
Salix alba—White Willow

For Dry Plains Areas, 3500 to 5000 feet

Large Trees

Class No. 1—100%

Celtis occidentalis—Common Hackberry
Gleditsia triacanthos—Common Honey Locust
Populus sargentii—Plains Poplar
Ulmus pumila—Siberian Elm

Class No. 2—80%

Acer saccharinum—Silver Maple
Fraxinus pennsylvanica lanceolata—Green Ash
Ulmus americana—American Elm

Small Trees

Class No. 1—100%

Eleagnus angustifolia—Russian Olive

Class No. 2—80%

Morus alba tatarica—Russian Mulberry

TREES OF REGION VI—WESTERN

Far Northern Section (British Columbia)

Class No. 1—100%

Acer campestre—Hedge Maple
Acer circinatum—Vine Maple
Acer rubrum—Red Maple
Cornus nuttallii—Pacific Dogwood
Crataegus oxyacantha—English Hawthorn
Fagus sylvatica—European Beech
Liquidambar styraciflua—Sweetgum
Liriodendron tulipifera—Tulip tree
Quercus borealis—Northern Red Oak
Quercus coccinea—Scarlet Oak
Quercus palustris—Pin Oak
Ulmus americana—American Elm

Class No. 2—80%

Acer platanoides—Norway Maple
Acer saccharinum—Silver Maple
Acer saccharum—Sugar Maple
Aesculus carnea—Red Horsechestnut
Fraxinus ornus—Flowering Ash
Ilex aquifolium—English Holly
Juniperus virginiana—Eastern Red Cedar
Platanus acerifolia—London Plane
Prunus blireiana—Blireiana Plum
Prunus cerasifera pissardii—Purple-leaf Myrobalan Plum
Pseudotsuga taxifolia—Common Douglas Fir
Robinia pseudoacacia—Black Locust

Class No. 3—60%

Betula pendula—European White Birch
Chamaecyparis lawsoniana—Lawson False Cypress
Fraxinus sp.—Ash
Ginkgo biloba—Ginkgo
Laburnum anagyroides—Goldenchain Laburnum

Class No. 4—40%

Acer negundo—Boxelder
Populus sp.—Poplar
Salix babylonica—Babylon Weeping Willow
Tilia sp.—Linden

Northern Section (Oregon-Washington)

Class No. 1—100%

Acer monspessulanum—Montpelier Maple
Acer platanoides schwedleri—Schwedler Norway Maple
Liriodendron tulipifera—Tulip tree
Platanus racemosa—California Plane (Sycamore)

Class No. 2—80%

Acer campestre—Hedge Maple
Acer palmatum—Japanese Maple
Acer platanoides—Norway Maple
Acer pseudoplatanus—Sycamore Maple
Acer rubrum—Red Maple
Aesculus carnea—Red Horsechestnut
Betula pendula gracilis—Cut-leaf European Birch
Carpinus caroliniana—American Hornbeam
Castanea crenata—Japanese Chestnut
Cedrus atlantica glauca—Blue Atlas Cedar
Chamaecyparis obtusa—Hinoki Cypress
Crataegus phaenopyrum—Washington Hawthorn
Laburnum alpinum—Scotch Laburnum
Libocedrus decurrens—California Incense Cedar
Liquidambar styraciflua—Sweetgum
Magnolia kobus—Kobus Magnolia
Picea pungens glauca—Blue Colorado Spruce
Platanus acerifolia—London Plane
Prunus blireiana—Blireiana Plum
Prunus cerasifera pissardii—Purple-leaf Myrobalan Plum
Prunus serrulata—Oriental Cherries
Quercus robur—English Oak

Class No. 3—60%

Acer negundo variegatum—Silver-leaf Boxelder
Arbutus menziesii—Pacific Madrone
Betula lutea—Yellow Birch
Cedrus deodara—Deodar Cedar
Chamaecyparis lawsoniana—Lawson False Cypress
Chamaecyparis pisifera—Sawara False Cypress
Cornus nuttallii—Pacific Dogwood
Crataegus oxyacantha—English Hawthorn
Cryptomeria japonica—Cryptomeria
Fagus sylvatica atropunicea—Purple European Beech
Gleditsia triacanthos—Honey Locust
Laburnum anagyroides—Goldenchain Laburnum
Pinus nigra—Austrian Pine
Quercus coccinea—Scarlet Oak
Robinia pseudoacacia—Black Locust
Sassafras albidum—Sassafras
Sorbus aucuparia—European Mountain Ash

Class No. 4—40%

Aesculus hippocastanum—Horsechestnut
Araucaria araucana—Monkey-puzzle Araucaria
Catalpa bignonioides—Southern Catalpa
Ilex aquifolium—English Ivy
Pseudotsuga taxifolia—Common Douglas Fir
Salix babylonica—Babylon Weeping Willow
Thuja plicata—Giant Arborvitae

Class No. 5—20%

Acer macrophyllum—Big-leaf Maple
Acer saccharinum—Silver Maple
Fraxinum oregona—Oregon Ash

NORTHERN CALIFORNIA COASTAL SECTION

Class No. 1—100%

Aesculus carnea—Red Horsechestnut
Carpinus caroliniana—American Hornbeam
Crataegus phaenopyrum—Washington Hawthorn
Crataegus lavallei—Lavalle Hawthorn
Crataegus oxyacantha—English Hawthorn
Fagus sylvatica atropunicea—Purple Beech
Ginkgo biloba—Ginkgo
Libocedrus decurrens—California Incense Cedar
Liquidambar styraciflua—American Sweetgum
Liriodendron tulipifera—Tulip tree
Lyonothamnus asplenifolius—Fern-leaf Lyon tree
Magnolia grandiflora—Southern Magnolia
Prunus blireiana—Blireiana Plum
Quercus ilex—Holly Oak
Quercus palustris—Pin Oak
Taxus baccata fastigiata—Irish English Yew
Tilia cordata—Little-leaf Linden

Class No. 2—80%

Acer rubrum—Red Maple
Arbutus unedo—Strawberry Madrone
Betula pendula—European White Birch
Cedrus atlantica—Atlas Cedar
Cedrus deodara—Deodar Cedar
Ceratonia siliqua—Carob
Eucalyptus ficifolia—Scarlet Eucalyptus
Fraxinus velutina—Velvet Ash
Ilex aquifolium—English Holly
Koelreuteria paniculata—Panicled Goldenrain tree
Leitospermum laevigatum—Victoria Tea tree
Ligustrum lucidum—Glossy Privet
Melaleuca leucadendron—Cajeput tree
Phoenix canariensis—Canary Date
Pinus canariensis—Canary Pine
Pinus pinea—Italian Stone Pine
Platanus acerifolia—London Plane
Prunus cerasifera pissardii—Purple-leaf Myrobalan Plum
Sequoia gigantea—Giant Sequoia
Sequoia sempervirens—Redwood
Ulmus parvifolia—Chinese Elm

Class No. 3—60%

Acer macrophyllum—Big-leaf Maple
Arbutus menziesii—Pacific Madrone
Cinnamomum camphora—Camphor-tree
Eucalyptus globulus—Tasmanian Blue Eucalyptus

TREES OF REGION VI—WESTERN (*Cont.*)

Grevillea robusta—Silk-oak Grevillea
Hakea laurina—Sea-urchin Hakea
Picea pungens glauca—Blue Colorado Spruce
Pinus radiata—Monterey Pine
Pittosporum undulatum—Orangeberry Pittosporum
Prunus lyonii—Cataline Cherry
Salix babylonica—Babylon Weeping Willow
Schinus molle—California Pepper tree
Ulmus americana—American Elm
Washingtonia sp.—Washington Palm
Washingtonia filifera—California Washington Palm

Class No. 4—40%

Acacia baileyana—Bailey Acacia
Cordyline australis—Giant Draecena
Robinia pseudoacacia—Black Locust
Trachycarpus fortunei—Fortunes Windmill Palm

Class No. 5—20%

Acacia melanoxylon—Blackwood Acacia
Acer negundo—Boxelder
Populus nigra italica—Lombardy Poplar

Northern California Inland Section

Class No. 1—100%

Albizzia julibrissin—Silk-tree Albizzia
Carya illinoensis—Pecan
Cedrus atlantica glauca—Blue Atlas Cedar
Celtis australis—European Hackberry
Celtis sinensis—Chinese Hackberry
Ceratonia siliqua—Carob
Cercis canadensis—Eastern Redbud
Crataegus oxyacantha paulii—Pauls Scarlet Hawthorn
Crataegus phaenopyrum—Washington Hawthorn
Eriobotrya japonica—Loquat
Eucalyptus sideroxylon rosea—Pink Mulga Ironbark Eucalyptus
Ginkgo biloba—Ginkgo
Gleditsia triacanthos inermis—Thornless Honey Locust
Lagerstroemia indica—Common Crape Myrtle
Laurus nobilis—Grecian Laurel
Ligustrum lucidum—Glossy Privet
Liriodendron tulipifera—Tulip tree
Magnolia grandiflora—Southern Magnolia
Maytenus boaria—Chile Mayten
Olea europaea—Common Olive
Pinus halepensis—Aleppo Pine
Pinus pinea—Italian Stone Pine
Pistacia chinensis—Chinese Pistache
Pittosporum phillyraeoides—Willy Pittosporum
Podocarpus macrophyllus—Yew Podocarpus
Prunus ilicifolia—Holly-leaf Cherry
Quercus agrifolia—California Live Oak
Quercus cerris—European Turkey Oak
Quercus chrysolepis—Canyon Live Oak
Quercus ilex—Holly Oak
Quercus shumardii—Shumard Oak
Quercus wislizeni—Interior Live Oak
Sophora japonica—Japanese Pagoda tree

Class No. 2—80%

Acer buergerianum—Trident Maple
Aesculus carnea—Red Horsechestnut
Betula pendula—European White Birch
Cinnamomum camphora—Camphor tree
Diospyros kaki—Kaki Persimmon
Eucalyptus sideroxylon rosea—Pink Mulga Ironbark Eucalyptus
Eucalyptus rudius—Moitch Eucalyptus
Eucalyptus viminalis—Ribbon Eucalyptus
Fraxinus velutina—Velvet Ash
Juglans regia—Persian Walnut
Koelreuteria paniculata—Panicled Goldenrain tree
Laburnum watereri—Waterer Laburnum
Liquidambar styraciflua—American Sweetgum
Lithocarpus densiflorus—Tanoak
Malus species—Flowering Crabapples
Morus alba—White Mulberry (Fruitless)
Parkinsonia aculeata—Jerusalem Thorn
Pinus canariensis—Canary Pine
Pinus coulteri—Coulter Pine

Prunus lyonii—Catalina Cherry
Prunus mume—Japanese Apricot
Prunus persica—Flowering Peach
Prunus serrulata—Oriental Cherry
Prunus subhirtella—Higan Cherry
Prunus triloba—Flowering Plum
Prunus yedoensis—Yoshino Cherry
Quercus lobata—California White Oak
Quercus macrocarpa—Bur Oak
Quercus robur—English Oak
Robinia pseudoacacia decaisneana—Decaisne Black Locust
Salix alba vitellina—Yellow-stem Weeping Willow
Salix babylonica—Babylon Weeping Willow
Ulmus carpinifolia—Smooth-leaf Elm
Ulmus parvifolia—Chinese Elm
Ulmus procera—English Elm
Ulmus pumila—Siberian Elm
Zelkova serrata—Japanese Zelkova

Class No. 3—60%

Acer macrophyllum—Big-leaf Maple
Acer saccharinum—Silver Maple
Aesculus californica—California Buckeye
Aesculus hippocastanum—Common Horsechestnut
Brachychiton populneum—Kurrajong Bottle tree
Broussonetia papyrifera—Common Papermulberry
Castanea sativa—European Chestnut
Celtis occidentalis—Common Hackberry
Grevillea robusta—Silk-leaf Grevillea
Juglans hindsii—Hinds Black Walnut
Libocedrus decurrens—California Incense Cedar
Metasequoia glyptostroboides—Dawn Redwood
Morus nigra—Black Mulberry
Paulownia tomentosa—Royal Paulownia
Prunus amygdalus—Almond
Prunus cerasifera atropurpurea—Pissard Purple-leaf Plum
Pterocarya stenoptera—Chinese Wing-nut
Quercus palustris—Pin Oak
Taxodium distichum—Common Bald Cypress
Thuja orientalis—Oriental Arborvitae
Ulmus americana—American Elm
Washingtonia filifera—California Washington Palm
Zizyphus jujuba—Common Jujube

Class No. 4—40%

Catalpa speciosa—Northern Catalpa
Platanus acerifolia—London Plane
Platanus occidentalis—American Plane (Sycamore)
Platanus racemosa—California Plane
Populus nigra italica—Lombardy Poplar
Schinus molle—California Pepper tree
Sequoia sempervirens—Redwood
Umbellularia californica—California Laurel

Class No. 5—20%

Ailanthus altissima—Tree-of-heaven (Ailanthus)
Ficus carica—Common Fig
Maclura pomifera—Osage-Orange
Melia azedarach—Chinaberry
Phoenix canariensis—Canary Date
Pinus sabiniana—Digger Pine
Tamarix aphylla—Athel Tamarisk

SOUTHERN CALIFORNIA SECTION

Class No. 1—100%

Cedrus deodara—Deodar Cedar
Ceratonia siliqua—Carob
Fraxinus uhdei—Shamel Ash
Liquidambar styraciflua—American Sweetgum
Magnolia grandiflora—Southern Magnolia
Pittosporum undulatum—Orangeberry Pittosporum
Podocarpus macrophyllus—Yew Podocarpus
Quercus ilex—Holly Oak
Washingtonia robusta—Mexican Washington Palm

Class No. 2—80%

Cinnamomun camphora—Camphor tree
Eucalyptus sideroxylon rosea—Pink Mulga Ironback Eucalyptus
Fraxinus velutina—Velvet Ash

TREES OF REGION VI—WESTERN (*Cont.*)

Ginkgo biloba—Ginkgo
Lagerstroemia indica—Common Crape Myrtle
Ligustrum japonicum—Japanese Privet
Phoenix canariensis—Canary Date
Pittosporum rhombifolium—Diamond-leaf Pittosporum
Schinus terebinthifolia—Brazil Pepper tree
Ulmus parvifolia—Chinese Elm

Class No. 3—60%

Calodendrum capense—Cape Chestnut
Eucalyptus ficifolia—Scarlet Eucalyptus
Eucalyptus lehmanni—Lehmann Eucalyptus
Koelreuteria paniculata—Panicled Goldenrain tree
Liriodendron tulipifera—Tulip tree
Pinus canariensis—Canary Pine
Pinus pinea—Italian Stone Pine
Pistacia chinensis—Chinese Pistache
Schinus molle—California Pepper tree

Class No. 4—40%

Acacia baileyana—Cootamundra-wattle Acacia
Alnus rhombifolia—Sierra Alder
Arecastrum romanzoffianum—Queen Palm
Clethra arborea—Lily-of-the-valley Clethra
Libocedrus decurrens—California Incense Cedar
Prunus caroliniana—Carolina Laurel Cherry
Quercus agrifolia—California Live Oak
Trachycarpus fortunei—Fortunes Windmill Palm
Tristania conferta—Brisbanebox Tristania
Zelkova serrata—Japanese Zelkova

Class No. 5—20%

Acacia melanoxylon—Blackwood Acacia
Bauhinia purpurea—Purple Bauhinia
Betula pendula—European White Birch
Brachychiton populneum—Kurrajong Bottle tree
Casuarina stricta—Coast Beefwood
Grevillea robusta—Silk-oak Grevillea
Pinus halepensis—Aleppo Pine
Platanus acerifolia—London Plane
Quercus lobata—California White Oak
Washingtonia filifera—California Washington Palm

TREES OF REGION VII—CANADA

For Ontario, South of Ottawa

Class No. 1—100%

Acer platanoides—Norway Maple
Acer platanoides schwedleri—Schwedler Norway Maple
Acer saccharum—Sugar Maple
Fagus sylvatica atropunicea—Purple European Beech
Quercus borealis maxima—Eastern Red Oak
Tilia cordata—Little-leaf Linden

Class No. 2—80%

Carpinus caroliniana—American Hornbeam
Carya cordiformis—Bitternut Hickory
Fagus sylvatica—European Beech
Fraxinus americana—White Ash
Gleditsia triacanthos inermis—Thornless Honey Locust
Juglans nigra—Black Walnut
Platanus occidentalis—American Plane (Sycamore)
Quercus palustris—Pin Oak
Sorbus aucuparia—European Mountain Ash
Ulmus americana—American Elm

Class No. 3—60%

Acer saccharinum—Silver Maple
Aesculus hippocastanum—Common Horsechestnut
Betula pendula—European White Birch
Cladrastis lutea—American Yellowwood

Ginkgo biloba—Ginkgo
Liriodendron tulipifera—Tulip tree
Magnolia acuminata—Cucumber-tree Magnolia

Class No. 4—40%

Catalpa speciosa—Northern Catalpa
Gymnocladus dioicus—Kentucky Coffee tree
Ulmus pumila—Siberian Elm

Class No. 5—20%

Acer negundo—Boxelder
Ailanthus altissima—Tree-of-heaven (Ailanthus)
Populus sp.—Poplars

For Vicinity of Montreal

Class No. 1—100%

Acer platanoides—Norway Maple
Acer platanoides schwedleri—Schwedler Norway Maple
Acer saccharum—Sugar Maple
Quercus borealis maxima—Eastern Red Oak
Tilia cordata—Little-leaf Linden
Tilia euchlora—Crimean Linden
Ulmus americana—American Elm

Class No. 2—80%

Carya cordiformis—Bitternut Hickory
Celtis occidentalis—Common Hackberry
Crataegus phaenopyrum—Washington Hawthorn
Fraxinus americana—White Ash
Ginkgo biloba (male)—Ginkgo
Gleditsia triacanthos inermis—Thornless Honey Locust
Ostrya virginiana—American Hophornbeam
Picea pungens glauca—Blue Colorado Spruce
Pinus sylvestris—Scotch Pine
Pyrus calleryana—Callery Pear
Syringa amurensis japonica—Japanese Tree Lilac

Class No. 3—60%

Acer saccharinum—Silver Maple
Cladrastis lutea—American Yellowwood
Cornus controversa—Giant Dogwood
Fraxinus pennsylvanica lanceolata—Green Ash
Juglans nigra—Black Walnut
Malus baccata—Siberian Crabapple

Class No. 4—40%

Aesculus hippocastanum—Common Horsechestnut
Catalpa speciosa—Northern Catalpa
Gymnocladus dioicus—Kentucky Coffee tree
Phellodendron sachalinense—Sakhalin Cork tree
Salix alba—White Willow
Salix blanda—Wisconsin Weeping Willow
Salix elegantissima—Thurlow Weeping Willow
Salix pentandra—Laurel Willow

Class No. 5—20%

Betula ermani—Ermans Birch
Betula pendula—European White Birch
Populus sp.—Poplars
Robinia pseudoacacia decaisneana—Decaisne Black Locust

Selected Bibliography

Chapter 1 THE NORMAL TREE

1. Allen, S. W.: *An introduction to American forestry*. New York: McGraw-Hill. 1938. 402 p.
2. Armstrong, Norman: 'Shade tree evaluation formulas and their use.' *Proc. Twenty-third National Shade Tree Conference*: 38-48. 1947.
3. Gibbs, R. D.: 'Studies in tree physiology.' I. General Introduction. Water contents of certain Canadian trees. *Canad. Jour. Res.* **17**: 460-82. 1939.
4. Haller, J. M.: *Tree care*. New York: Macmillan. 1957. 224 p.
5. Kieran, John: *An introduction to trees*. Garden City, New York: Hanover House. 1954.
6. Miller, E. C.: *Plant physiology*. New York: McGraw-Hill. 1939.
7. Moon, Franklin and N. C. Brown: *Elements of forestry*. New York: Wiley. 1937.
8. Pack, C. L.: *School book of forestry*. Washington, D.C.: American Tree Assoc. 1922.
9. Pratt, Rutherford: *American trees*. New York: Dodd, Mead. 1952.
10. Priestley, J. H.: 'Radial growth and expansion growth in the tree.' *Forestry* (Gt. Brit.) **9**: No. 2, 84-95. 1935.
11. Purcell, C. R.: 'The reality value of trees.' *Proc. Thirty-second National Shade Tree Conference*: 128-36. 1956.
12. Rankin, W. H.: 'Looking into the heart of a tree.' *Proc. Sixth National Shade Tree Conference*: 107-11. 1930.
13. Steele, Fletcher: 'Tree values shift.' *Proc. Twenty-second National Shade Tree Conference*: 79-93. 1946.

Chapter 2 THE SOIL AND ITS RELATION TO TREES

1. Barker, B. T. P.: 'Studies on root development.' *Ann. Rept. Agr. and Hort. Res. Sta.* Long Ashton, Univ. Bristol 1921: 9-20. 1922.
2. Baver, L. D.: *Soil physics*. New York: Wiley. 1940. 370 p.

3. Boynton, D. and D. C. Compton: 'Normal seasonal changes of oxygen and carbon dioxide percentages in gas from the larger pores of three orchard soils.' *Soil Science* **57**: 107-17. 1944.
4. Burges, A.: 'On the significance of mycorrhiza.' *New Phytol.* **35**: 117-31. 1936.
5. Bushey, D. J.: 'Root extension of shade trees.' *Proc. Thirteenth National Shade Tree Conference*: 22-30. 1937.
6. Cannon, W. A.: *Physiological features of roots, with especial reference to the relation of roots to aeration of soil.* Washington: Carnegie Inst. Publ. 368. 1925.
7. Chadwick, L. C.: 'Soil acidity and plant growth.' *Proc. Tenth National Shade Tree Conference*: 14-25. 1934.
8. Gager, C. S.: 'Aeration of tree roots.' *Proc. Fifth National Shade Tree Conference*: 26, 27. 1929.
9. Hacskaylo, Edward and J. G. Palmer: 'Mycorrhizae in relation to tree nutrition.' *Arborist's News*, **20**: No. 11, 81-5. Nov. 1955.
10. Hatch, A. B.: 'The role of mycorrhizae in afforestation.' *Jour. Forestry* **34**: 22-9. 1936.
11. Heinicke, A. J.: 'The effect of submerging roots of apple trees at different seasons of the year.' *Proc. Amer. Hort. Sci.* **29**: 205-7. 1932.
12. Jacobs, H. L.: 'Soil improvement for shade trees.' *The Shade Tree* **10**: Nos. 1 and 2. 1937.
13. ———: 'The influence of root environment on shade tree vitality.' *Proc. Fifteenth National Shade Tree Conference*: 51-65. 1939.
14. Kramer, P. J.: *Plant and water relationships.* New York: McGraw-Hill. 1949.
15. Lyon, T. L. and H. O. Buckman: *The nature and properties of soils.* New York: Macmillan. 1922.
16. Mulford, F. L.: *Care of ornamental trees and shrubs.* U.S. Dept. Agr. Farmer's Bul. 1826. 1939.
17. Neal, J. H.: 'Spacing and depth of tile drains.' *Agr. Eng.* 15: 229-32. 1934.
18. Odland, T. E. *et al.*: *The reaction of certain ornamental trees and shrubs to liming.* Rhode Island Agr. Exp. Sta. Bul. 250. 1935.
19. Parker, J.: 'The effects of flooding on the transpiration and survival of some southeastern forest-tree species. *Plant Physiol.* **25**: 453-60. 1950.
20. Russel, E. J. and A. Appleyard: 'The atmosphere of the soil, its composition and the causes of variation.' *Jour. Agr. Sci.* **7**: 1-48. 1915.
21. Salter, R. M.: 'Some soil factors affecting tree growth,' *Science* n.s. **91**: 2365. 1940.
22. Smith, A. L.: *Lichens.* Cambridge: University Press. 1921. 464 p.
23. Waksman, S. A.: *Humus; origin, chemical composition, and importance in nature.* Baltimore: Williams & Wilkins. 1938. 526 p.
24. Yeager, A. F.: 'Root systems of certain trees and shrubs grown on prairie soils.' *Jour. Agr. Res.* **51**: 1085-92. 1935.
25. Yoder, R. E.: 'The significance of soil structure in relation to the tilth problem.' *Proc. Amer. Soc. Soil Sci.* **2**: 21-33. 1937.
26. Young, H. C.: 'The aeration of tree roots.' *Proc. Fifth National Shade Tree Conference*: 28-34. 1929.

Chapter 3 TRANSPLANTING TREES

1. Anderson, P. O.: *Trees and tree planting.* St. Paul: Webb Publishing Co. 1936. 95 p.
2. Anonymous: 'Transplanting of trees and shrubs in the northeastern and north central states.' *Proc. Nineteenth National Shade Tree Conference:* 74-146. 1943. Rev. 1958. Collier Co., Wooster, Ohio.
3. Armstrong, N. C.: 'Trees which ride the rails.' *Proc. Sixth National Shade Tree Conference:* 40-43. 1930.
4. Bailey, L. H.: *The nursery manual.* New York: Macmillan. 1928.
5. ———: 'Transplanting.' *Standard cyclopedia of horticulture.* New York: Macmillan. second ed. 1930.
6. Burkholder, C. L.: *Planting and care of shade trees.* Purdue Univ. Dept. Agr. Ext. Bul. 119. 1932. 7 p.
7. Bushey, D. J.: *The planting and care of shrubs and trees.* Cornell Agr. Exp. Sta. Bul. 185. 1929. 10 p.
8. Calkins, A. B.: 'Tree moving as a business.' *American Nurseryman* **69**: 18-19. Feb. 1, 1939.
9. Chester, K. S., H. J. Harper, R. O. Monosmith, and F. A. Fenton: *A program for shade trees in Oklahoma.* Oklahoma Agr. Exp. Sta. Bul. 234. 1938. 45 p.
10. Coplen, M. G.: 'Mechanics of moving big trees.' *American Nurseryman* **76**: 9-11. Nov. 1, 1942.
11. Cox, L. D.: *A street tree system for New York City.* New York Sta. Col. of Forestry Bul. 8. 1916. 89 p.
12. Davis, J. E.: *Planting and care of shade trees.* Illinois Natural History Survey. Circ. 36. Urbana, Ill. 1940. 23 p.
13. Fernow, B. E.: *The care of trees in lawn, street and park.* New York: Holt. 1910. 392 p.
14. Franke, W. A.: 'Big tree moving.' *Proc. Thirteenth National Shade Tree Conference:* 118-30. 1937.
15. Hottes, A. C.: *The book of trees.* New York: A. T. De La Mare. 1932.
16. Jacobs, H. L.: 'Shifting small trees and shrubs.' Kent, Ohio: *The Arboriculturist* **4**: 5-12. 1938.
17. Mulford, F. L.: *Transplanting trees and shrubs.* U.S. Dept. Agr. Farmer's Bul. 1591. 1929. 34 p.
18. ———: 'Transplanting.' *Proc. Sixth National Shade Tree Conference:* 43-6. 1930.
19. ———: *Planting the roadside.* U.S. Dept. Agr. Farmer's Bul. 1481. 1938. 37 p.
20. Paul, A. F.: *'How to transplant trees.'* Brooklyn Bot. Garden Record. Plants and Gardens **2**: 44-7. Spring 1946.
21. Peirson, H. B. and R. W. Nash: *The planting and care of shade trees.* Maine Forest Ser. Bul. 10. 1936. 28 p.
22. Shield, V. I.: 'Tree moving practices of Shield shade tree specialists.' *Proc. Fourteenth National Shade Tree Conference:* 45-51. 1938.

23. Tilford, P. E.: 'Effect of some synthetic growth substances on root development of transplanted trees.' *Proc. Fourteenth National Shade Tree Conference:* 51-9. 1938.
24. Van Wormer, H. M.: 'Tree moving practices and problems.' *Proc. Thirtieth National Shade Tree Conference:* 76-81. 1954.

Chapter 4 FERTILIZERS AND THEIR USE

1. Beilman, A. P.: 'Experiments on the fertilization of shade trees.' *Proc. Tenth National Shade Tree Conference:* 114-25. 1934.
2. ———: 'Feeding the shade trees.' *Missouri Bot. Gard. Bul.* **24**: No. 4, 95-113. 1936.
3. ———: 'The general care of shade trees.' *Missouri Bot. Gard. Bul.* **21**: No. 7. 1933.
4. Boynton, Damon: 'Spray fertilization of shade trees.' *Proc. Twenty-fifth National Shade Tree Conference:* 6-15. 1949.
5. Bukovac, M. J. and S. H. Wittwer: 'Absorption and mobility of foliar-applied nutrients.' *Plant Physiology* **32**: No. 5, 428-35. Sept. 1957.
6. Cannon, Thomas: 'Foliar analysis as an index of fertilizer requirements of some ornamental trees.' *Proc. Thirty-second National Shade Tree Conference:* 84-93. 1956.
7. Chadwick, L. C.: 'The fertilization of shade trees in the nursery.' *Proc. Amer. Soc. Hort. Sci.* 32 (1934): 357-60. 1935.
8. ———: 'Fertilization of shade trees.' *American Nurseryman* **64**: 9-10. 1936.
9. ———: 'Modern methods of shade tree fertilization.' *Proc. Thirteenth National Shade Tree Conference:* 63-76. 1937.
10. ———: 'Fertilizers' action in the soil.' *American Nurseryman* **82**: 7-8, 29-32. Sept. 15, 1945; 12-16, Oct. 1, 1945.
11. ———: 'Some shade tree fertilization problems.' *Arborist's News* **12**: No. 2, 9-12. Feb. 1947.
12. Chandler, R. F. Jr.: 'The influence of nitrogenous fertilizer applications upon the growth and seed production of beech and sugar maple trees.' *Arborist's News* **4**: 17-19. 1939.
13. Cook, R. L. and C. E. Millar: 'The residual effect of fertilizers.' *Mich. Agr. Exp. Sta. Quart. Bul.* **18**: 227-34. 1936.
14. Deuber, C. G.: 'Soil fertility and root development.' *Proc. Eastern Shade Tree Conference:* 1938. New York. 1939.
15. Gourley, J. H.: 'The advantages of fall application of fertilizers to both peaches and apples.' *Trans. Illinois Sta. Hort. Soc.* **69**: 265-71. 1935.
16. Heinicke, A. J.: 'The physiology of trees with special reference to their food supply.' *Proc. Fifteenth National Shade Tree Conference:* 26-37. 1939.
17. Irish, C. F.: 'The process of aeration and fertilization of tree roots by means of compressed air.' *Proc. Fifth National Shade Tree Conference:* 35-7. 1929.
18. Jacobs, H. L.: *Fertilization of shade trees.* Part 1. 'Fall vs. spring fertilization.' Kent, Ohio: Davey Tree Expert Co. Res. Dept. Bul. 4. 1929. 27 p.
19. ———: *Fertilization of shade trees.* Part 2. 'Chemical fertilizers for conifers.' Kent, Ohio: Davey Tree Expert Co. Res. Dept. Bul. 5. 1930. 15 p.

20. Jacobs, H. L.: 'Fertilizer formulae-tree responses.' *Proc. Thirteenth National Shade Tree Conference:* 51-63. 1937.
21. Kenworthy, A. L.: 'Leaf analysis to determine fertilizer needs of trees and shrubs.' *Trees Magazine:* 8, 18. Mar.-Apr. 1955.
22. Lunt, H. A.: 'The use of fertilizer in the coniferous nursery.' *Conn. Agr. Sta. Bul.* 416: 723-66. 1938.
23. May, Curtis: 'A discussion of tree feeding.' *Proc. Eighth National Shade Tree Conference:* 31-5. 1932.
24. Mitchell, H. L.: 'A method of determining the nutrient needs of the shade trees with special reference to phosphorus.' *The Shade Tree* **8**: Nos. 5, 6, and 7. 1935.
25. Mitchell, H. L. and R. F. Chandler: *The nitrogen nutrition and growth of certain deciduous trees of northeastern United States.* Black Rock Forest Bul. 11. 1939. 94 p.
26. Pirone, P. P.: Foliage application of nutrients. *Proc. Twenty-seventh National Shade Tree Conference:* 23-35. 1951.
27. Teubner, F. G., S. H. Wittwer, W. G. Long, and H. B. Tukey: 'Some factors affecting absorption and transport of foliar-applied nutrients as revealed by radioactive isotopes.' *Quart. Bul. Mich. Agr. Exp. Sta.* **39**: No. 3, 398-415. Feb. 1957.
28. Tukey, H. B.: The uptake of nutrients by leaves and branches of fruit trees. *Rept. Thirteenth Int. Hort. Congress:* 297-306. London. 1952.
29. Wyman, Donald: 'Recent developments in the Cornell shade tree experiments.' *Proc. Tenth National Shade Tree Conference:* 111-14. 1934.

Chapter 5 PRUNING TREES AND TREATING WOUNDS

1. Anonymous: *How to prune.* Oakville, Conn.: Seymour Smith & Son. 1950.
2. Armstrong, Norman: 'Shade tree care.' *Gardener's Chronicle of America* **39**: No. 7. 1935.
3. ———: 'Basic principles of shade tree pruning.' *Proc. Fourteenth National Shade Tree Conference:* 126-40. 1938.
4. ———: 'Pruning and training the young shade tree.' *Arborist's News* **11**: No. 7, 49-51. July 1946.
5. Asplundh Tree Expert Co.: *Line clearance.* Philadelphia: Patterson & White Co. Oct. 1937.
6. Beilmann, A. P.: 'The pruning of ornamental trees and shrubs.' *Missouri Bot. Gard. Bul.* **18**: No. 3. 1930.
7. ———: 'When is the best time to prune a tree? (maples).' *Missouri Bot. Gard. Bul.* **21**: No. 4. 1933.
8. Bromley, S. W.: 'How to prune trees.' *Brooklyn Bot. Garden Record.* Plants and Gardens. 48-9. Spring 1946.
9. Collingwood, G. H.: 'First aids in tree surgery.' *Amer. For. and For. Life* **36**: 105-8. 1930.
10. Collins, J. G.: *Treatment and care of tree wounds.* U.S. Dept. Agr. Farmer's Bul. 1726. 1934. 38 p.

11. Everett, T. H.: Tree pruning pointers. *Garden Jour. New York Bot Garden.* 16-18. Jan.-Feb. 1957.
12. Fenska, R. R.: *The new tree experts manual.* New York: A. T. De La Mare. 1954.
13. Howe, G. H.: 'Effects of various dressings on pruning wounds of fruit trees.' *New York (Geneva) Agr. Exp. Sta. Bul.* **395**: 83-94. 1915.
14. Kemmerer, H. R.: *Pruning trees, shrubs, and roses.* Cir. 779. Ill. Agr. Ext. Service. 1957. 15 p.
15. Kuemmerling, Karl, *et al.*: 'Safe practices for arborists.' *Trees Magazine* **14**: 10-22. Nov.-Dec. 1953.
16. Leiser, A. T.: 'Polyethylene treatment for tree bark wounds.' *American Nurseryman* **107**: No. 9. 13, 49. 1958.
17. Le Seur, A. D. C.: 'Tree wounds.' *Estate Magazine* **28**: 823-6. 1928.
18. ———: 'Care of old trees.' *Jour. Roy. Hort. Soc.* 61: 149-59. 1936.
19. Lesser, M. A.: 'Tree wound dressings.' *Agr. Chem.*: 26-7, 60-61. Aug. 1946.
20. Marshall, R. P.: 'A rubber dressing for tree wounds.' *Phytopathology* **21**: 1091. 1931.
21. ———: 'Water blistering of wound dressings.' *Phytopathology* **21**: 1173-80. 1931.
22. ———: *The relation of season of wounding and shellacking to callus formation in tree wounds.* U.S. Dept. Agr. Tech. Bul. 246. 1931. 28 p.
23. McKenzie, M. A.: *Restoring hurricane-damaged trees.* Mass. State Coll. of Agr. Special Circ. 30. 1938 (mimeographed).
24. McQuilkin, W. E. and J. W. Showalter: 'Lanolin mixtures as dressings for tree wounds.' *Arborist's News* **10**: No. 3, 17-19. March 1945.
25. Parker, W. E.: 'Practical line clearance.' *The Shade Tree* **12**: No. 11. 1939.
26. Roth, E. R.: *The effect of pruning on the heartrot hazard in hardwoods.* U.S. Dept. Agr. Bur. Pl. Ind. Civilian Conservation Corp. June 14, 1939 (mimeographed).
27. Selby, A. D.: 'Dressings for pruning wounds of trees.' *Ohio Agr. Exp. Sta. Circ.* **126**: 163-70. 1912.
28. Strufe, F. W.: 'Safety practices.' *Proc. Fourteenth National Shade Tree Conference*: 41-54. 1940.
29. Tilford, P. E. and H. C. Young: 'Tree wound dressings.' *Gard. Chron. Amer.* **42**: 41-2. 1938.
30. Tilford, Paul: 'Tree wound dressings.' *Proc. Sixteenth National Shade Tree Conference*: 41-54. 1940.
31. Young, H. C. and P. E. Tilford: 'Tree wound dressings.' *Ohio Agr. Exp. Sta. Bimonth. Bul.* **22**: 83-7. 1937.
32. Zimmerman, P. W.: 'Factors affecting healing of wounds with special reference to cuttings and injured bark of trees.' *Proc. Seventeenth National Shade Tree Conference*: 6-8. 1931.

Chapter 6 CAVITY TREATMENTS

1. Armstrong, Norman: 'Importance and place of cavity work in arboriculture.' *Proc. Eleventh National Shade Tree Conference*: 61-8. 1935.

2. Bartlett, F. A.: 'Historical background of cavity work.' *Proc. Eleventh National Shade Tree Conference*: 39-51. 1935.
3. Blair, M. F.: 'The solid type of cavity repair.' *American Nurseryman* **64**: 7-9. 1936.
4. Creager, D. B.: 'Rubber as a cavity filling for trees.' *The Shade Tree* **10**: No. 7. 1937.
5. Freeman, G. K.: 'Cavity treatment—shell and open type.' *American Nurseryman* **63**: 5-6. 1936.
6. Jacobs, H. L.: 'A modern view of twenty-five years of cavity treatment.' *Proc. Eleventh National Shade Tree Conference*: 55-61. 1935.
7. Le Seur, A. D. C.: *The care and repair of ornamental trees*. London: Country Life Ltd. 1934. 257 p.
8. ———: 'The repair of wounds in trees.' *Gard. Chron.* (London) **95**: 76. 1934.
9. Marshall, R. P.: 'Scientific aspects of handling tree cavities.' *Proc. Eleventh National Shade Tree Conference*: 51-5. 1935.
10. Peet, Elbert: *Practical tree repair*. New York: R. M. McBride. 1932.
11. Rankin, W. H.: 'Wound gums and their relation to fungi.' *Proc. Ninth National Shade Tree Conference*: 111-15. 1932.
12. Welch, D. S.: 'The occurrence of active wood-rotting fungi in apparently sound wood of living trees.' *Arborist's News* **14**: No. 2, 18-19. Feb. 1949.
13. ———: 'The cause and treatment of cavities in trees.' *Proc. Twenty-fifth National Shade Tree Conference*: 126-31. 1949.

Chapter 7 BRACING AND CABLING

1. Beilmann, A. P.: *Weak branching in trees*. Missouri Bot. Gard. Bul. 20: No. 6, 1932.
2. Felt, E. P. (editor): 'Cabling and the use of wood screws.' *Bartlett Scientific Tree Topics* **1**: No. 2. Stamford, Conn. 1940.
3. Johnson, L. M.: 'Cabling and bracing structurally weak trees.' *Bartlett Scientific Tree Topics* **1**: No. 3. Stamford, Conn. 1940.
4. Luckner, Wm. Jr.: 'Proper guying practices of transplanted trees.' *Proc. Twenty-ninth National Shade Tree Conference*: 47-50. 1951.
5. Thompson, A. R.: 'Tree cabling materials.' *Proc. Twelfth National Shade Tree Conference*: 30-43. 1936.
6. Wachtel, C. L. 'Proper bracing practices.' *Proc. Twenty-ninth National Shade Tree Conference*: 60-62. 1951.

Chapter 8 TREES SUITABLE FOR VARIOUS LOCATIONS

1. Alden, P. E.: 'Planting.' *The Shade Tree* **12**: No. 3. 1939.
2. Balfour, F. R. S.: 'The planting and after care of roadside trees.' *Quarterly Jour.* **29**: No. 3. 1935.
3. Caddick, J. W.: *Care and selection of trees in Rhode Island*. Rhode Island Agr. Exp. Sta. Misc. Publ. 46. 1955. 10 p.
4. Curtis, R. W. and Donald Wyman: *Ornamental trees for New York State*. Cornell Agr. Exp. Sta. Ext. Bul. 287. 1934. 32 p.

5. Eaton, F. S.: 'Varieties of deciduous trees especially suited for narrow streets.' *The Shade Tree* **12**: No. 5. 1939.
6. Felt, E. P.: *Our shade trees*. New York: O. Judd Publishing Co. 1938. 187 p.
7. Gerling, Jake: 'Better shade trees and more practical use.' *Proc. Twenty-ninth National Shade Tree Conference*: 101-11. 1951.
8. ———: 'Progress towards better trees.' *American Nurseryman* **105**: 97-8. Jan. 1957.
9. Hui-Lin Li: 'The origin and history of the cultivated plane trees.' *Morris Arboretum Bul.* 3-9. March 1957.
10. Mulford, F. L.: *Trees for town and city streets*. U.S. Dept. Agr. Farmer's Bul. 1208. 1927. 30 p.
11. ———: *Trees for roadside planting*. U.S. Dept. Agr. Farmer's Bull. 1482. 1938. 50 p.
12. ———: *Planting the roadside*. U.S. Dept. Agr. Farmer's Bul. 1481. 1938. 37 p.
13. Nelson, P. K. *et al.*: 'The hundred finest trees and shrubs for temperate climates.' *Brooklyn Bot. Garden Record*. Plants and Gardens 3: No. 3. Autumn 1957.
14. Scanlon, E. H.: 'Selection and use of trees to minimize public area conflicts.' *Proc. Thirty-first National Shade Tree Conference*: 183-96. 1955.
15. Solotaroff, W.: *Shade trees in towns and cities*. New York: Wiley. 1911.
16. Stefferud, Alfred: *Trees. Yearbook of agriculture 1949*. Washington: U.S. Dept. of Agr. U. S. Govt. Printing Office. 1949.
17. Stone, G. E.: 'Shade trees, characteristics, adaptation, diseases and care.' *Mass. Agr. Exp. Sta. Bul.* **170**: 123-264. 1916.
18. White, R. P.: 'Trees suitable for street planting.' *Horticultural Topics* **1**: Nos. 9 & 10. New Brunswick, N. J. 1937.
19. Witte, C. P.: 'Trees available for street planting.' *The Shade Tree* **9**: Nos. 11 & 12. 1936.
20. Woodward, K. W.: *Shade trees in New Hampshire*. New Hampshire Col. Agr. Ext. Ser. Circ. 53. 1922. 16 p.
21. Wyman, Donald: *Narrow upright trees in the Arnold Arboretum*. Harvard Univ. Arnold Arboretum Bul. of Popular Information 6: No. 11. 1938.
22. ———: 'Forty-five of the best trees for Massachusetts gardens.' *Arnoldia* **12**: No. 1, 1-20. Mar. 7, 1952.
23. ———: 'Eighty trees for the small place.' *Arnoldia* **16**: No. 3, 9-15. 1956.
24. ———: 'New and rare ornamental and woody plants recently distributed by the Arnold Arboretum.' *Arnoldia* **16**: Nos. 7-9, 33-51. Oct. 1956.

Chapter 9 DIAGNOSING TREE TROUBLES

1. Beilmann, A. P.: 'The use of instruments in tree diagnosis.' *Proc. Eleventh National Shade Tree Conference*: 18-27. 1935.
2. ———: 'Possible use of instruments in diagnosis.' *Trees* **6**: 14, 16. May-June 1941.
3. Chadwick, L. C.: 'Diagnosis of nutritional requirements of ornamental plants.' *Proc. Nineteenth National Shade Tree Conference*: 5-19. 1943.

4. Dodge, A. W.: 'Diagnosing shade tree troubles.' *Proc. Fourteenth National Shade Tree Conference:* 118-26. 1938.
5. Hamilton, C. C.: 'Diagnosis of insect injury.' *Trees* **8**: 18. May-June 1946.
6. Hepting, G. H., E. E. Roth, and Bailey Sleeth: 'Discolorations and decay from increment borings.' *Jour. Forestry* **47**: No. 5, 366-70. 1949.
7. Howe, Hollis: 'Relation of complete facts to diagnosis.' *Trees:* 15, 18. Aug. 1945.
8. Long, W. H.: 'Data to be noted in studying heart rots of living trees.' *Phytopathology* **31**: No. 12. 1199-1200. 1931.
9. Mitchell, J. H.: 'Practical shade tree maintenance.' *Proc. Thirtieth National Shade Tree Conference:* 319-23. 1954.
10. Pirone, P. P.: 'How to get help in diagnosing tree troubles.' *The Shade Tree.* Kearney, N. J.: N. J. Federal Shade Tree Commissions. Jan. 1947.
11. Tilford, P. E.: 'Diagnosing shade tree troubles.' *Proc. Twenty-first National Shade Tree Conference:* 17-25. 1945.
12. Vaughn-Eames, H.: 'The increment borer in diagnosis.' *Proc. Eighth National Shade Tree Conference:* 48-50. 1932.
13. Welch, D. S.: 'The diagnosis of shade tree troubles.' *Proc. Seventeenth National Shade Tree Conference:* 45-50. 1941.
14. White, R. P.: 'Four unusual symptomatic pictures and their diagnosis.' *Proc. Seventh National Shade Tree Conference:* 82-6. 1931.
15. Worley, C. L., H. R. Lesselbaum, and T. M. Mathews: 'Deficiency symptoms for the major elements in seedlings of three broad-leaved trees.' *Jour. Tenn. Acad. Sci.* **6**: 239-47. 1941.

Chapter 10 NONPARASITIC INJURIES

1. Bakke, A. L.: 'The effect of city smoke on vegetation.' *Iowa Exp. Sta. Bul.* **145**: 383-409. 1913.
2. Beilmann, A. P.: 'Lightning and trees.' *Missouri Bot. Gard. Bul.* **28**: 209-16. 1940.
3. Benedict, H. M. and J. P. Nielsen: 'The markings produced by various air pollutants on leaves.' *Proc. Thirty-second National Shade Tree Conference:* 205-15. 1956.
4. Bieberdorf, F. W., C. L. Shrewsbury, H. C. McKee, and L. H. Krough: 'Vegetation as a measure indicator of air pollution.' Pt. 1, *Bul. Torrey Bot. Club* **85**: No. 3, 197-200. May 1958.
5. Bray, D. F.: 'Gas injury to shade trees.' *Bartlett Scientific Tree Topics* **2**: No. 5, 19-22. Stamford, Conn. 1958.
6. Carter, J. C.: 'Non-parasitic tree troubles.' *American Nurseryman* **106**: 110-14. Oct. 15, 1957.
7. Chadwick, L. C.: 'Chlorosis of pin oak.' *Proc. Amer. Soc. Hort. Sci.* 33. 1936.
8. Clopper, H. S.: 'Lightning protection for trees.' *American Forests.* May 1935.
9. Covert, E. N.: *Protection of buildings and farm property against lightning.* U.S. Dept. Agr. Farmer's Bul. 1512. 1926. 32 p.
10. Crocker, William: 'The effect of illuminating gas on trees.' *Proc. Seventh National Shade Tree Conference:* 24-34. 1931.

11. Day, W. R. and T. R. Peace: 'The influence of certain accessory factors on frost injury to forest trees.' IV. Air and soil conditions. *Forestry* (Gr. Brit.) **11**: 92-103. 1937.
12. Deuber, C. G.: 'Observations of shade trees in the vicinity of gas leaks.' *Amer. Gas Assoc. Monthly*. Nov. 1933.
13. ———: 'Effects on trees of an illuminating gas in the soil.' *Plant Physiology* **11**: 401-12. 1936.
14. Dodge, A. W.: 'Lightning damage to trees.' *Proc. Twelfth National Shade Tree Conference*: 43-55. 1936.
15. ———: 'Installation of lightning protection on shade trees.' *Proc. Thirteenth National Shade Tree Conference*: 95-109. 1937.
16. Fowler, M. E. and G. F. Gravatt: *Reducing damage to trees from construction work*. U.S. Dept. Agr. Farmer's Bull. 1967. 1945. 22 p.
17. Guba, E. F.: 'Slime flux.' *Proc. Tenth National Shade Tree Conference*: 56-61. 1934.
18. ———: 'Slime flux of shade trees.' *Trees* **5**: No. 2, 9. March-April 1942.
19. Hartley, C. and T. C. Merril: 'Storm and drought injury to foliage trees.' *Phytopathology* **5**: 20-29. 1915.
20. Hitchcock, A. E., William Crocker, and P. W. Zimmerman: 'Toxicity of illuminating gas in soil.' *Proc. Ninth National Shade Tree Conference*: 34-6. 1932.
21. Jacobs, Homer: '2,4-D, Friend or Foe?' *Proc. Twenty-sixth National Shade Tree Conference*: 23-37. 1950.
22. Jennings, O. E.: 'Smoke injury to shade trees.' *Proc. Tenth National Shade Tree Conference*: 44-9. 1934.
23. Knight, L. J. and William Crocker: 'Toxicity of smoke' *Bot. Gaz.* **55**: 337-71. 1913.
24. Kreag, K. K.: 'Chlorosis studies in Michigan.' *Trees* **2**: No. 4. 1939.
25. ———: 'Nature and control of shade tree chlorosis.' *Proc. Sixteenth National Shade Tree Conference*: 32-40. 1940.
26. Lambert, J.: 'Winter injury to miscellaneous trees and shrubs.' *Arboretum Bul. Assoc. Morris Arb.* **1**: 50-55. 1936.
27. MacDougal, D. T.: 'Effect of lightning on trees.' *Journ. New York Bot. Gard.* **3**: 131-5. 1902.
28. Marsden, D. H.: 'Gas injury to trees.' *Brooklyn Bot. Gard. Record*. Plants & Gardens **7**: No. 2, 156-8. 1951.
29. ———: 'Some common noninfectious diseases of shade and ornamental trees in New England.' *Proc. Twenty-eighth National Shade Tree Conference*: 73-9. 1952.
30. Marshall, R. P.: *Care of damaged shade trees*. U.S. Dept. Agr. Farmer's Bul. 1896. 1942. 34 p.
31. Meserve, A. W.: 'Effect of changing grades around trees and suggested treatment.' *Proc. Thirteenth National Shade Tree Conference*: 36-51. 1937.
32. Middleton, J. T.: 'Response of plants to air pollution.' *Jour. Air Poll. Control Assoc.* 7-11. May 1956.
33. Parker, W. E.: 'Spray damage from lead arsenate.' *The Shade Tree* **11**: No. 6, 1938.

34. Pirone, P. P.: 'Chlorosis of pin oaks and its control.' *The Shade Tree* **13**: No. 12. 1940.
35. ———: '2,4-D injury to shade trees.' *Proc. Twenty-second National Shade Tree Conference*: 107-14. 1946.
36. ———: 'Before you build, safeguard your trees.' *Amer. Home*. 32-3. June 1946.
37. ———: 'Chelates will cure iron deficiency.' *Flower Grower*. 20-21. Nov. 1954.
38. ———: 'Air pollution—an ever-increasing problem.' *Flower Grower*. 28-9. Sept. 1956.
39. Stone, G. E.: *Electrical injuries to trees*. Mass. Agr. Exp. Sta. Bul. 156. 1914. 19 p.
40. ———: 'Frost cracks, winter killing of cork cambium, and sunscald.' *Mass. Agr. Exp. Sta. Bul.* 170: 204-8. 1916.
41. Thomas, H. E. and L. H. MacDaniels: *Freezing injury to the roots and crowns of apple trees*. Cornell Agr. Exp. Sta. Bul. 556. 1933. 23 p.
42. Van Wormer, H. M.: 'Effect and treatment of girdling roots.' *Proc. Thirteenth National Shade Tree Conference*: 30-36. 1937.
43. ———: 'Effects of girdling roots on trees.' *Arborist's News* **5**: 81-4. 1940.
44. Wallace, T.: 'Leaf scorch on fruit trees.' *Jour. Pomol. Hort. Sci.* **7**: 1-31. 1928.
45. Warne, L. G.: 'The distribution of potassium in normal and scorched foliage.' *Ann. Bot.* **48**: 57-67. 1934.
46. Went, F. W.: 'Plant damage due to air pollution and the use of plants as indicators of air pollution.' *Air Poll. Control Assoc. News*. 3-5. Jan. 1958.
47. Whitehead, J. B.: 'Lightning protection for trees.' *Science* **78**: 507-8. 1933.
48. Wilson, J. D. and H. A. Runnels: 'Bordeaux mixture as a factor increasing drought-injury.' *Phytopathology* **21**: 729-38. 1931.
49. Witte, C. P.: 'Frost cracks.' *Proc. Twelfth National Shade Tree Conference*: 181-8. 1936.
50. Young, H. C.: 'Spray injuries.' *Proc. Sixth National Shade Tree Conference*: 83-6. 1930.
51. Zeller, S. M. and O. T. McWhorter: *Physical injuries to trees with special reference to winter injury*. Oreg. Agr. Exp. Sta. Ext. Bul. 485. 1936. 8 p.

Chapter 11 INSECT CONTROL

1. Creighton, J. T.: 'New developments in insecticides and application equipment.' *Proc. Twenty-third National Shade Tree Conference*: 106-21. 1947.
2. Felt, E. P.: 'Fundamentals in the control of insect pests.' *Proc. Tenth National Shade Tree Conference*: 37-44. 1934.
3. Fisher, E. H. (editor): *Entoma*. Twelfth edition. 1957-58. Ent. Soc. America. 1957.
4. Flint, W. P. and M. D. Farrar: 'Dormant spraying materials for shade trees.' *Arborist's News* **5**: No. 1, 1-2. 1940.
5. Frear, D. E.: *Pesticide Handbook 1958*. State College, Pa.: College Science Publishers. 1958.

6. Graham, S. A.: *Principles of forest entomology*. New York: McGraw-Hill. 1929. 339 p.
7. Hamilton, C. C.: 'Experiments in the control of boring insects.' *Proc. Tenth National Shade Tree Conference*: 31-7. 1934.
8. ———: 'Winter spraying—trees and shrubs.' *The Shade Tree* **12**: No. 2. 1939.
9. ———: 'Systemic insecticides; their nature and use in agriculture.' *Proc. Thirty-third National Shade Tree Conference*: 128-45. 1957.
10. ———: 'Insect pest control for ornamental stock.' *American Nurseryman* **107**: 76-87. March 15, 1958.
11. Hartzell, A. and W. J. Youden: 'Efficiency of banding for the control of cankerworms.' *Contr. Boyce Thompson Inst.* **7**: 365-77. 1935.
12. Herrick, G. W.: *Insect enemies of shade trees*. Ithaca, N. Y.: Comstock Publishing Co. 1935. 417 p.
13. Houser, J. S.: 'Borer control experiments.' *Proc. Thirteenth National Shade Tree Conference*: 159-69. 1937.
14. Howard, L. O.: *The insect book*. New York: Doubleday, Page. 1901. 429 p.
15. Martin, Hubert: *Guide to the chemicals used in crop protection*. Third edition. London, Ontario: Sci. Service Lab. 1957 (mimeographed).
16. Metcalf, C. L. and P. W. Flint: *Destructive and useful insects*. New York: McGraw-Hill. Second edition. 1939. 981 p.
17. Middleton, Wm.: 'Some conditions leading to the attack of shade and ornamental trees by borers.' *Proc. Eighth National Shade Tree Conference*: 23-31. 1932.
18. Neiswander, R. B.: 'Insect pests of ornamentals: borers and gall makers.' *Arborist's News* **21**: No. 5, 33-7. 1956.
19. Pirone, P. P.: 'Plan your dormant spray operations now.' *Flower Grower*: 40-43. Feb. 1954.
20. Sanderson, E. D. and L. M. Peairs: *Insect pests of farm, garden, and orchard*. New York: Wiley. 1931. 568 p.
21. Shread, J. C.: *Ants injurious to trees and wood products*. Hillsboro, N. H.: Tree pest leaflet 53. New Eng. Sec. Soc. Am. Foresters. 1950.
22. Snyder, T. E.: *Our enemy the termite*. Ithaca, N. Y.: Comstock Publishing Co. 1935. 196 p.
23. Wilson, J. D.: *Insects and their control*. Plainfield, N. J.: Interstate Printing Corp. 1931. 383 p.
24. Westcott, Cynthia: *The gardener's bug book*. Garden City, N. Y.: American Garden Guild and Doubleday & Co. 1956.
25. Wysong, Noel: 'Termites as tree pests.' *American Nurseryman* **98**: 84-8. July 15, 1953.

Chapter 12 SPRAYING EQUIPMENT AND PRACTICES

1. Anderson, O. G. and F. C. Roth: *Insecticides and fungicides, spraying and dusting equipment*. New York: Wiley. 1923.
2. Anonymous: *How to spray the aircraft way*. U.S. Dept. Agr. Farmer's Bul. 2062. June 1954.

3. Carr, J. P. and J. M. Patterson: 'Application of chemicals with air sprayers.' *Jour. Agr. and Food Chem.* **2**: No. 22. 1102-4. Oct. 1954.
4. Cockerham, K. L.: *A manual for spraying.* New York: Macmillan. 1923.
5. Cleveland, C. R.: 'Spray materials and practices.' *Proc. Fourteenth National Shade Tree Conference:* 64-91. 1938.
6. Hamilton, C. C. (editor): *Entoma.* New Brunswick, N. J.: Eastern Branch of American Assoc. of Economic Entomologists. 1939. 172 p.
7. Harding, G. W.: 'Operator experience with mist blower.' *Proc. Twenty-third National Shade Tree Conference:* 194-9. 1947.
8. Langford, G. S.: 'Tolerance of shade trees to spray materials.' *Proc. Seventeenth National Shade Tree Conference:* 19-30. 1941.
9. Mason, A. F.: *Spraying, dusting and fumigating of plants.* New York: Macmillan. 1936. 539 p.
10. Potts, S. F.: 'Concentrated spray application.' *Proc. Twenty-third National Shade Tree Conference:* 183-94. 1947.
11. ———: *Control of forest and shade-tree insects with concentrated sprays applied by mist blower.* Hillsboro, N. H.: Tree pest leaflet 54. New Eng. Sec. Soc. Am. Foresters. 1950.
12. ———: *Concentrated spray equipment.* Caldwell, N. J.: Dorland Books. 1958. 650 p.
13. Rankin, W. H.: 'Spraying for leaf diseases of shade trees.' *Proc. Eighth National Shade Tree Conference:* 64-70. 1932.
14. Wester, H. V.: *General spraying and other practices.* Tree Preservation Bulletin No. 6. Rev. 1953. Supt. Documents, Washington 25, D.C.

Chapter 13 TREE DISEASES AND THEIR CONTROL

1. Beattie, R. K.: 'Virus diseases of forest and shade trees.' *Proc. Fifteenth National Shade Tree Conference:* 12-22. 1939.
2. Boyce, J. S.: *Forest pathology.* Macmillan. 1938. 600 p.
3. Carter, J. C.: 'Fungicides for use on shade and ornamental trees.' *Proc. Twenty-ninth National Shade Tree Conference:* 118-22. 1953.
4. Felt, E. P. and S. W. Bromley: 'Insecticides and fungicides for ornamentals.' *Proc. Seventh National Shade Tree Conference:* 115-21. 1931.
5. Heald, F. D.: *Manual of plant diseases.* New York: McGraw-Hill. 1926. 891 p.
6. Hartig, Robert: *Diseases of trees.* London: Macmillan. 1894. 331 p.
7. Hubert, E. E.: *An outline of forest pathology.* New York: Wiley. 1931. 543 p.
8. Kienholz, Raymond and C. B. Bidwell: *A survey of diseases and defects in Connecticut forests.* Conn. Agr. Exp. Sta. Bul. 412. 1938. 66 p.
9. Kimmey, J. W.: *Dwarf mistletoes of California and their control.* Berkeley, Cal.: Tech. Paper 19. California Forest & Range Exp. Sta. 1-12. Oct. 1957.
10. Massee, George: *Diseases of cultivated plants and trees.* London: Duckworth & Co. 1915. 605 p.
11. Pope, H.: *Krankheiten und Schadlinge der Zierpflanzen.* Berlin: Paul Parey. 1932.
12. Rankin, W. H.: *Manual of tree diseases.* New York: Macmillan. 1918. 398 p.

13. Smith, E. F.: *An introduction to bacterial diseases of plants.* Philadelphia: Saunders. 1920. 688 p.
14. Stevens, F. L.: *Plant disease fungi.* New York: Macmillan. 1925. 469 p.
15. Stevens, F. L. and J. G. Hall: *Diseases of economic plants.* New York: Macmillan. 1933. 507 p.
16. White, R. P.: 'Shade tree leaf troubles.' *N. J. Agr. Exp. Sta. Nursery Disease Notes* **2**: No. 2. 1929.
17. Young, H. C.: 'Fungicides and their use on ornamental shade trees.' *Arborist's News* **5**: No. 5. 33-6. 1940.

Chapter 14 GENERAL PARASITIC DISEASES

1. Caroselli, N. E.: *Verticillium wilt of maples.* Rhode Island Exp. Sta. Bul. 335. June 1957. 74 p.
2. Day, W. R.: 'Parasitism of Armillaria mellea in relation to conifers.' *Quart. Jour. Forestry* **21**: 9-21. 1927.
3. Dochinger, L. S.: 'Verticillium wilt: Its nature and control.' *Proc. Thirty-third National Shade Tree Conference:* 202-12. 1957.
4. Drain, B. D.: *Fire-blight canker treatments.* Tennessee Agr. Sta. Circ. 62. 1938. 4 p.
5. Dunegan, J. C.: 'Antibiotics for tree disease control.' *Proc. Thirty-third National Shade Tree Conference:* 116-26. 1957.
6. Hildebrand, E. M.: *Fire blight and its control.* Cornell Agr. Exp. Sta. Ext. Bul. 405. 1939.
7. McCown, M. and J. A. McClintock: *Controlling fire-blight by canker treatment and spraying.* Indiana Agr. Exp. Sta. Bul. 208. 1935. 8 p.
8. Reed, G. M.: 'The powdery mildews—Erysiphaceae.' *Amer. Micros. Soc. Trans.* **32**: 219-25. 1913.
9. Salmon, E. S.: 'A monograph of the Erysiphaceae.' *Mem. Torrey Bot. Club.* 9: 1-290. 1900.
10. Strong, F. C.: 'Root and root crown troubles.' *Trees* **4**: No. 5, 10-14. Sept.-Oct. 1941.
11. Thomas, H. E.: 'Studies on *Armillaria mellea,* infection, parasitism, and host resistance.' *Jour. Agr. Res.* **48**: 187-218. 1934.
12. Thomas, H. E. and P. A. Ark: *Fire blight of pears and related plants.* Cal. Agr. Exp. Sta. Bul. 586. 1934.
13. Weiss, Freeman: *Powdery mildew of ornamental plants.* U.S. Dept. Agr. Leaflet 197. 1940. 4 p.
14. White, R. P.: *Fire blight of ornamentals.* N. J. Agr. Exp. Sta. Nursery Disease Notes 5: Nos. 1 and 2. 1932.

Chapter 15 DISEASES AND INSECT PESTS OF LOW-GROWING TREES

1. Arthur, J. C.: *Manual of the rusts in the United States and Canada.* Lafayette, Ind.: Purdue Research Foundation. 1934. 438 p.
2. Beattie, R. K.: Diseases threatening ornamental and forest trees. *Jour. Econ. Ent.* **26**: 621-4. 1933.

3. Carter, J. C.: 'Diseases of oaks and Verticillium wilt of woody plants.' *Proc. Sixteenth National Shade Tree Conference:* 83-91. 1940.
4. Chambers, E. L. and N. F. Thompson: *Some of the more important insects and plant diseases of Wisconsin trees and shrubs.* Wis. Dept. Agr. & Mkts. Bul. 123. 1931. 58 p.
5. ———: *Pests and diseases of trees and shrubs.* Wis. Dept. Agr. and Mkts. Bul. 145. 1933. 87 p.
6. Cook, M. T.: *Common diseases of shade and ornamental trees.* N. J. Agr. Exp. Sta. Circ. 98. 1918. 28 p.
7. Dodge, B. O. and H. W. Rickett: *Diseases and pests of ornamental plants.* New York: Ronald Press. 1948. 638 p.
8. Felt, E. P. and W. H. Rankin: *Insects and diseases of ornamental trees and shrubs.* New York: Macmillan. 1932. 507 p.
9. Gambrell, F. L. and R. M. Gilmer: *Insects and diseases of fruit nursery stocks and their control.* Geneva, N. Y.: N. Y. Agr. Exp. Sta. Bul. 776. 1956.
10. Graves, A. N.: 'Some diseases of trees in greater New York.' *Mycologia* **11**: 111-24. 1919.
11. Herrick, G. W. *Insect enemies of shade trees.* Ithaca, N. Y.: Comstock Publishing Co. 1935. 417 p.
12. Jones, S. G.: 'Life-history and cytology of *Rhytisma acerinum.*' *Ann. Bot.* **39**: 41-75. 1925.
13. Langford, G. S. and E. N. Cory: *Common insects of lawns, ornamental shrubs, and shade trees.* Md. Agr. Exp. Sta. Ext. Bul. 81. 1939. 54 p.
14. Pirone, P. P.: *Diseases of ornamental plants.* N. J. Agr. Exp. Sta. Cir. 385. 1939. 80 p.
15. ———: 'Diseases of shrubs and small trees.' *Proc. Sixteenth National Shade Tree Conference:* 91-112. 1940.
16. Strong, F. C.: 'Some important shade tree diseases of the midwest.' *Proc. Fourteenth National Shade Tree Conference:* 106-16. 1938.
17. Welch, D. S.: 'Crown canker of flowering dogwood.' *Arborist's News* **3**: No. 12. 1938.

Chapter 16 DISEASES AND INSECT PESTS OF TALL-GROWING TREES

1. Ahrens, W. E., *et al.*: *A list of references to literature on the Dutch elm disease.* Morristown, N. J.: U.S. Dept. Agr. Dutch Elm Disease Laboratory. 1935. 29 p. (mimeographed).
2. American Forestry Association: *The American elm—its glorious past, its present dilemma and its hope for protection.* Not dated. Washington, D.C. 31 p.
3. Anderson, H. W. and David Gottlieb: 'Plant disease control with antibiotics.' *Economic Botany* **6**: No. 3, 294-308. 1952.
4. Anderson, P. J. and W. H. Rankin: *Endothia canker of chestnut.* Cornell Univ. Agr. Exp. Sta. Bul. 347. 1914.
5. Anonymous: *Chestnut blight and resistant chestnuts.* U.S. Dept. Agr. Farmer's Bul. 2068. 1954. 21 p.

6. Ashcroft, J. M.: *European canker of black walnut and other trees.* W. Va. Agr. Exp. Sta. Bul. 261. 1934.
7. Atkinson, G. F.: 'Studies on some shade tree and timber destroying fungi.' *Cornell Agr. Exp. Sta. Bul.* 193: 199-235. 1901.
8. Avery, G. S.: 'The dying oaks.' *Scientific American.* 112-22. May 1957.
9. Beattie, R. K. and J. D. Diller: 'Fifty years of chestnut blight in America.' *Jour. Forestry.* **52**: 323-9. 1954.
10. Bier, J. E. *Some common tree diseases of British Columbia.* Bot. and Plant Path. Sci. Service, Dept. Agr. Dominion of Canada. 48 p.
11. Britton, W. E. and R. B. Friend: 'Insect pests of elms in Connecticut.' *Conn. Agr. Exp. Sta. Bul.* **369**: 265-307. 1935.
12. Burgess, A. F.: 'Insect pests of trees.' *Proc. Fifth National Shade Tree Conference:* 63-4. 1929.
13. Clapper, R. B. and G. F. Gravatt: 'The American chestnut: Its past, present, and future.' *Southern Lumberman* **167**: No. 2105, 227-9. Dec. 1943.
14. Clapper, R. B. and R. K. Beattie: 'American and blight-resistant chestnuts.' *Flower Grower* **31**: No. 10, 488-9, 500. 1944.
15. Clinton, G. P. and F. A. McCormick: 'The willow scab fungus.' *Conn. Agr. Expt. Sta. Bul.* **302**: 443-69. 1929.
16. Comstock, J. H. and A. B. Comstock: *A manual for the study of insects.* Ithaca, N. Y.: Comstock Publishing Co. 1906. 701 p.
17. Day, W. R.: 'The watermark disease of the cricket bat willow.' *Oxford Forestry Mem.* **3**: 1-30. 1924.
18. ———: 'The relationship between frost damage and larch canker.' *Forestry* (Gr. Brit.) **5**: 41-56. 1931.
19. Dietrich, Joseph: 'A look at Dutch elm disease control.' *Trees Magazine.* 16-18. Nov.-Dec. 1957.
20. Doane, R. W., E. C. Van Dyke, W. J. Chamberlain, and H. E. Burke: *Forest Insects.* New York: McGraw-Hill. 1936. 463 p.
21. Edgerton, C. W.: 'The physiology and development of some anthracnoses.' *Bot. Gaz.* **45**: 367-408. 1908.
22. Ehrlich, John: 'The beech bark disease.' *Canad. Jour. Res.* **10**. 1934.
23. Felt, E. P.: *Insects affecting park and woodland trees.* Albany: N. Y. Education Dept. Memoir 8. Nos. 1 and 2. 1905.
24. Galloway, B. T.: 'The search in foreign countries for blight resistant chestnuts and related tree crops.' *U.S. Dept. Agr. Circ.* 383. 1-16. 1926.
25. Garren, K. H.: Leader dieback—a 'new' disease of sweet gum. *Plant Disease Reporter.* **33**: No. 9, 351-3. 1949.
26. Grant, T. J. and Perley Spaulding: 'Avenue of entrance for canker forming nectrias of New England hardwoods.' *Phytopathology* **29**: 351-9. 1939.
27. Gravatt, G. F. and L. S. Gill: *Chestnut blight.* U.S. Dept. Agr. Farmer's Bul. 1641. 1930.
28. Hahn, G. G. and T. T. Ayers: 'The European larch canker and its relation to certain other cankers of conifers in the United States.' *Jour. Forestry* **34**: 898-908. 1936.
29. Hartley, C. and G. G. Hahn: 'Notes on some diseases of aspen.' *Phytopathology* **10**: 141-7. 1920.

30. Hedgcock, G. G.: 'Notes on some diseases of trees in our national forests.' *Phytopathology* **4**: 181-8. 1914.
31. Hedgcock, G. G. and N. R. Hunt: '*Dothichiza populea* in the United States.' *Mycologia* **8**: 300-308. 1916.
32. Hedgcock, G. G.: 'The large leaf spot of chestnut and oak associated with *Monochaetia dezmazierii*.' *Mycologia* **21**: 324-5. 1929.
33. Henry, B. W.: '*Chalara quercina*, n. sp., the cause of oak wilt.' *Phytopathology* **34**: 631-5. 1944.
34. Hepting, G. H.: *A vascular wilt of the Mimosa tree*. U.S. Dept. Agr. Circ. 535. 1939.
35. Hiley, W. E.: *The fungal diseases of the common larch*. New York: Oxford University Press. 1919. 204 p.
36. Hutchins, L. M., *et al.*: 'Diseases of shade and forest trees.' *Brooklyn Bot. Garden Record*. Plants and Gardens. **7**: No. 2, 121-58. 1951.
37. Jackson, L. W. R., G. E. Thompson, and H. O. Lund: *Forest diseases and insects of Georgia's trees*. Georgia Forestry Commission. 40 p.
38. Jones, T. W. and T. W. Bretz: *Experimental oak wilt control in Missouri*. Columbia, Mo.: Mo. Agr. Exp. Sta. Research Bul. 657. Feb. 1958.
39. Marshall, R. P.: 'Shade tree diseases.' *Jour. Forestry*. April 1937.
40. Marshall, R. P. and Alma Waterman: *Common diseases of important shade trees*. U.S. Dept. Agr. Farmer's Bul. 1987. 1948. 55 p.
41. McKenzie, M. A.: 'Shade tree diseases of New England.' *Proc. Twelfth National Shade Tree Conference*: 100-103. 1936.
42. Miles, L. E.: 'Leaf spots of elm.' *Bot. Gaz.* **71**: 161-96. 1921.
43. Miller, P. W.: *Walnut blight and its control in the Pacific northwest*. U.S. Dept. Agr. Circ. 331. 1934.
44. Neiswander, R. B.: 'The control of some gall makers on shade trees.' *Trees Magazine* **17**: 20-22. May-June 1957.
45. Nienstaedt, Hans and A. H. Graves: *Blight resistant chestnut culture and care*. Connecticut Agr. Exp. Sta. Circ. 192. 1955.
46. Pirone, P. P.: 'The detrimental effect of walnut to rhododendron and other ornamentals.' *N.J. Agr. Exp. Sta. Nursery Disease Notes* **11**: No. 4. 1938.
47. ———: 'New diseases of Norway maples.' *Trees* **3**: 7, 15, 16, 18. 1940.
48. ———: 'A new disease of sweet gum.' *American Forests* **48**: 130-31. March 1942.
49. ———: 'A new threat to London plane trees.' *Proc. Twenty-fourth National Shade Tree Conference*: 97-101. 1948.
50. ———: 'Fire as predisposing (contributing) factor in canker diseases of street trees.' *Trees* **13**: 9-10, 20. Sept.-Oct. 1953.
51. Rex, E. G. and J. M. Walter: *The canker stain disease of plane trees*. Trenton, N.J.: N.J. Dept. Agr. Circ. 360. 1946. 23 p.
52. Schrenk, H. von, and Perley Spaulding: *Diseases of deciduous forest trees*. U.S. Dept. Agr. Bur. Pl. Ind. Bul. 149. 1929. 85 p.
53. Spaulding, Perley: *Nectria canker of hardwoods in New England*. Mass. Forest & Park Assoc. Tree leaflet No. 10. 1936.
54. Stewart, V. B.: 'The leaf blotch disease of horsechestnut.' *Cornell Agr. Exp. Sta. Bul.* **371**: 411-19. 1916.

55. Strong, F. C.: 'Nature and control of anthracnose of shade trees.' *Arborist's News* **22**: No. 5, 33-9. 1957.
56. Swingle, R. U.: *Phloem necrosis, a virus disease of American elms.* U.S. Dept. Agr. Circ. 640. 1942. 8 p.
57. Walter, J. M.: *Canker stain of plane trees.* U.S. Dept. Agr. Circ. 742. 1946. 12 p.
58. Welch, D. S. and J. G. Matthysse: *Control of Dutch elm disease in New York State.* Cornell Ext. Bul. 932. 1955. 15 p.
59. Welch, D. S.: 'The range and importance of Nectria canker of hardwoods in the northeast.' *Jour. Forestry* 32: 997-1002. 1934.
60. Whitten, R. R. and D. E. Parker: 'Experimental control of shade tree insects with DDT.' *Proc. Twenty-first National Shade Tree Conference*: 13-17. 1945.
61. Zentmeyer, G. A., J. G. Horsfall and P. P. Wallace: *Dutch elm disease and its chemotherapy.* Connecticut Agr. Exp. Sta. Bul. 498. 1946. 70 p.

Chapter 17 DISEASES AND INSECT PESTS OF EVERGREEN TREES

1. Cromwell, I. H.: 'The hosts, life history, and control of the cedar-apple rust fungus.' *Jour. Arnold Arb.* **15**: 163-232. 1934.
2. Davis, W. C., G. Y. Young, D. H. Latham, and Carl Hartley: *Diseases of conifers in forest nurseries.* U.S. Dept. Agr. Div. Forest Pathology. Dec. 1938.
3. Felt, E. P. and W. H. Rankin: *Insects and diseases of ornamental trees and shrubs.* New York: Macmillan. 1932. 507 p.
4. Gambrell, F. L.: 'Studies on some insects of evergreens.' *Proc. Eighth National Shade Tree Conference*: 89-93. 1932.
5. Gambrell, F. L. and F. Z. Hartzell: 'Dormant spray mixtures on conifers.' *Jour. Econ. Ent.* **32**: 206-9. 1939.
6. Gilgut, C. J.: 'Cytospora canker of spruces.' *Proc. Twelfth National Shade Tree Conference*: 113-20. 1936.
7. Haddow, W. R. and F. S. Newman: 'A disease of Scot's pine caused by Diplodia pinea associated with the pine spittle bug.' *Trans. Royal Can. Inst.* **24** (pt. 1): 1-19. 1942.
8. Hahn, G. G. and T. T. Ayers: 'Dasyscyphae on conifers in North America.' *Mycologia* **26**: 73-101, 167-80, 479-501. 1934.
9. Hahn, G. G.: 'Distribution and hosts of cedar blight in the United States.' *Plant Dis. Reptr.* **24**: No. 3. 1940.
10. Hamilton, C. C.: *Insect pests of boxwood.* N.J. Agr. Exp. Sta: Circ. 179. 1926.
11. Hedgcock, G. G.: Notes on the distribution of some fungi associated with diseases of conifers.' *Plant Dis. Reptr.* **16**: No. 4. 28-42. 1932.
12. Hepting, G. H. and R. W. Davidson: 'Some leaf and twig diseases of hemlock in North Carolina.' *Plant Dis. Rept.* **19**: 308-9. 1935.
13. Herrick, G. W.: *Insect enemies of shade trees.* Ithaca, N.Y.: Comstock Publishing Co. 1935. 417 p.

14. Houser, J. S.: 'Some troublesome pests of conifers.' *Proc. Sixteenth National Shade Tree Conference*: 65-82. 1940.
15. Jackson, L. W. R.: 'Winter injury of *Buxus sempervirens*.' *Phytopathology* **28**: 372-4. 1938.
16. Martin, J. F.: 'The application of surgery to blister rust infected trees of ornamental value.' *Proc. Sixteenth National Shade Tree Conference*: 113-20. 1940.
17. McDaniel, E. I.: *Some chewing insects infesting Michigan evergreens*. Mich. Agr. Exp. Sta. Circ. Bul. 141. 1932. 54 p.
18. Palmiter, D. H.: *Rust diseases of apples and their control*. N.Y. Agr. Exp. Sta. Bul. 756. 1952.
19. Schread, J. C.: *Scale insects and their control*. Conn. Agr. Exp. Sta. Bul. 578. 1954. 23 p.
20. Spaulding, Perley: 'The blister rust of white pine.' *U.S. Dept. Agr. Bul.* **975**: 1-100. 1922.
21. Strong, F. C.: 'Spruce branch canker.' *Proc. Twenty-ninth National Shade Tree Conference*: 30-36. 1953.
22. White, R. P.: *Diseases of ornamental plants*. N.J. Agr. Exp. Sta. Circ. 226. 1931. 97 p.
23. White, R. P. and C. C. Hamilton: *Diseases and insect pests of rhododendron and azalea*. N.J. Agr. Exp. Sta. Circ. 350. 1935.
24. White, R. P.: 'Cedar rusts.' *N.J. Agr. Exp. Sta. Nursery Disease Notes* **10**: Nos. 2, 3, 4. 1937.
25. Zentmeyer, G. A.: 'Cytospora canker of Italian cypress.' *Phytopathology* **31**: No. 10. 896-906. 1941.

Index

Abies, 24, 392
Acacia, 24
Acantholyda erythrocephala, 413
Acco-Streptomycin, 239
Acer, 25, 132, 382: *buergerianum*, 118; *campestre*, 118; *cappadocicum*, 118; *cappadocicum* var. *aureum* and *rubrum*, 118; *dasycarpum*, 132; *diabolicum purpurascens*, 118; *ginnala*, 118; *macrophyllum*, 286; *mandshuricum*, 118; *negundo*, 134, 140; *nigrum*, 125, 133; *palmatum* var. *atropurpureum* and *Oshiu-beni*, 118; *pennsylvanicum*, 25; *platanoides*, 132; *platanoides columnare*, 125; *platanoides schwedleri*, 133; *pseudoplatanus*, 133; *pseudoplatanus Leopoldii*, 133; *pseudoplatanus pyramidalis*, 125; *pseudoplatanus Spaethii*, 133; *rubrum*, 25, 133; *rubrum conica* Scanlon, 123; *saccharum*, 133; *spicatum*, 25, 118; *truncatum*, 118
Actias luna, 367
Actidione, 234, 409
Actinothyrium gloeosporioides, 296
Adelopus gaumanni, 393
Aesculus, 129, 133: *carnea*, 129; *carnea briotii*, 129; *glabra*, 24; *hippocastanum*, 24, 129; *hippocastanum Baumanii*, 129; *pavia*, 24
Agrico Water Soluble Plant Food, 44, 58
Agrilus: *anxius*, 262, 366; *bilineatus*, 309, 352; *difficilis*, 344
Agri-mycin 100, 239, 242
Agri-Strep, 239
Agrox, 233
Ailanthus, 301: *glandulosa*, 26
Akebono flowering cherry, 114
Albizzia, 298: *julibrissin*, 298; *julibrissin rosea*, 121
Aldenham crabapple, 115
Alder, 24, 303
Alder Diseases, 303: canker, 304; leaf curl, 303; powdery mildew, 303
Alder Insects, 304: blight aphid, 304; flea beetle, 304; lace bug, 304; psyllid, 304
Aldrin, 203, 234
Aleurodiscus amorphus, 394; *griseo-canus*, 339
Almira maple, 133
Almond, flowering, 24
Alnus, 24, 303
Alsophila pomentaria, 324
Alternaria catalpae, 313
Altica ambiens, 304
Aluminum sulfate, 20
Amelanchier, 26, 297, 403; *canadensis*, 238
American linden, 130
Ammonium sulfate, 49
Amur maple, 118
Anions, 17
Annual rings, 6
Antheraea polyphemus, 279, 367
Antispila nyssaefoliella, 371
Anti-transpirants, 45
Ants in trees, 213; control of, 214
Aonidiella aurantii, 427
Aphis: *fabae*, 275; *medicaginis*, 275
Aphrophora parallela, 415
Apiosporina collinsii, 297
Aplospora nyssae, 370
Apple, 24, 28, 238, 403
Apple bark borer, 298
App-L-Set, 129
Aramite, 203, 234
Arborvitae, 24, 382
Arborvitae Diseases, 382: browning and shedding of leaves, 383; juniper blight, 383; leaf blight, 382; tip blight, 383
Arborvitae Insects, 383: bagworm, 383; cedar tree borer, 385; leaf miner, 384; scale, 384; spider mite, 385
Arceuthobium douglasii, 394
Archips rosaceana, 301
Argyresthia: *cupressella*, 392; *thuiella*, 384
Armillaria mellea, 244, 301, 390
Arnold crabapple, 115

Aronia, 403
Arsenate of lead, 236
Ascochyta: paulowniae, 274; *salicis*, 377
Ash, 23, 24, 28, 114, 126, 253: black, 126, 420; green, 140; red, 126, 420; white, 126
Ash Diseases, 253: canker, 254; leaf spot, 254; rust, 253
Ash Insects, 254: carpenter worm, 255; fall webworm, 255; flower gall, 257; lilac borer, 254; lilac leaf miner, 256; oyster shell scale, 255; sawfly, 254; scurfy scale, 255
Aspidiotus: britannicus, 401; *californicus*, 414; *camelliae*, 418; *hederae*, 418, 427; *juglans-regiae*, 375; *perniciosus*, 265; *uvae*, 360, 397
Asterolecanium: puteanum, 401; *variolosum*, 351
Atlas cedar, 390
Atropellis: pinicola, 411; *tingens*, 411
Atteva aurea, 302
Augustine ascending elm, 125
Automeris io, 297
Autoserica castanea, 263
Azaleas, 22
Azobenzene, 235

Bacteria, 224: dissemination of, 225; overwintering of, 228
Bactericides, 230
Bald Cypress, 24, 385
Bald Cypress Diseases, 385: twig blight, 385; wood decay, 385
Bald Cypress Insects, 385: cypress moth, 385
Basi-cop, 231
Beauty crabapple, 116
Bechtel's crabapple, 116
Beech, 23, 24, 28, 304: American, 304; Darwyck, 306; European, 306; golden-leaved, 306; purple, 306; red fern-leaved, 306; tricolor, 306; weeping, 306
Beech Diseases, 306: bark disease, 307; bleeding canker, 307; leaf mottle, 307
Beech Insects, 308: apple tree tent caterpillar, 309; blight aphid, 308; brown wood borer, 309; datana caterpillar, 309; gypsy moth, 309; scales, 309; two-lined chestnut borer, 309; woolly aphid, 308
Betula, 257: *lenta*, 24; *papyrifera*, 258; *populifolia*, 258
BHC, 234
Bifusella: abietis, 393; *faullii*, 393
Birch, 23, 28, 257: canoe, 258; gray, 258; sweet, 24
Birch Diseases, 258: canker, 258; die-back, 261; leaf blister, 258; leaf spot, 258; leaf rust, 258; wood decay, 260
Birch Insects, 261: bronze birch borer, 262; case bearer, 261; leaf miner, 261; leaf skeletonizer, 261
Blackjack oak, 139
Black locust, 130
Black maple, 133
Black scale, 309, 360, 375
Black walnut, 374
Blight, fire, 238
Blood, dried, 49
Bordeaux mixture, 230, 234, 235: method of preparing, 230; qualities of, 230
Bordo, 231
Borer-Kil, 213
Borers, 211: control of, 212; factors favoring infestations by, 211; preventing attacks by, 212
Borer-Sol, 213
Boron, 47
Bor-tox, 213
Botryosphaeria, 321: *ribis*, 359, 367, 377
Botrytis cinerea, 271, 393
Box, common, 24
Boxelder, 134, 140
Boxwood, 386
Boxwood Diseases, 386: blight, 387; canker, 386; leaf spot, 387; winter injury and sunscald, 387
Boxwood Insects, 388: California red scale, 388; cottony-cushion scale, 388; giant hornet, 388; leaf miner, 388; lesser snow scale, 388; mite, 389; mealy bugs, 388; other pests, 389; oyster shell scale, 388; psyllid, 388; scales, 388
Boxwood nematodes, 389
Brachyrhinus: ovatus, 425; *sulcatus*, 425
Bracing: and cabling, 104; box, 110; cavity, 109; crotch, 107; flexible, 110; hub and spoke, 110; installation procedure in, 111; materials, 111; rigid, 106; rubbing limbs, 109; simple direct, 110; triangular, 110
Bucculatrix canadensiella, 261
Buckeye, 28: Ohio, 24; red, 24
Buisman elm, 128
Butternut, 28, 374
Buttonwood, 137
Buxus sempervirens, 24, 386

Calcium, 17, 18, 47: arsenate, 206; caseinate, 211; chloride injury, 190; nitrate, 49
Calgon, 186
Caliciopsis pinea, 411
Calico scale, 375
California red scale, 375, 388
California tent caterpillar, 295
Caliroa: cerasi, 266; *lineata*, 354
Calligrapha scalaris, 399

Callosamia promethea, 296, 367
Cambium, 6
Camphor scale, 330
Camponotus herculeanus pennsylvanicus, 428
Capitophorus braggii, 296
Captan, 232, 234, 235: 50W, 232
Carbon, 47
Carmine crabapple, 116
Carolina silverbell, 121
Carpinus, 24, 279: *betulus pyramidalis*, 121; *japonica*, 121; *tschonoskii*, 120
Carya, 330
Casein, 211
Castanea, 311: *dentata*, 24; *japonica*, 314; *mollissima*, 314; *pumila*, 24, 312, 346
Catalpa, 24, 310: common, 310; western, 310
Catalpa, 24: *bignonioides*, 310; *speciosa*, 310
Catalpa Diseases, 310: leaf spot, 310; powdery mildew, 310; wood decay, 310
Catalpa Insects, 311: mealy bug, 311; midge, 311; sphinx, 311
Caterpillar, California tent, 295
Cations, 17
Cavities: covered, 94; filled, 95; how started, 86; method of treating, 89; objective of filling, 88; open, 94; rate of development, 88
Cavity: bracing, 92; cleaning out, 89; factors determining whether to fill, 96; filling, 98; filling materials, 100; fills, quality, 100; shaping, 91; size of, 96; treatments, 85; sterilizing and dressing, 93; work, history of, 85
Cecidomyia gleditsiae, 343
Cedar, 390: atlas, 390; deodar, 390; Keteleer red, 403; Lebanon, 390; red, 23
Cedar Diseases, 390: tip blight, 390; root rot, 390
Cedar Insects, 390: black scale, 390; deodar weevil, 390
Cedrus, 390: *atlantica*, 390; *deodara*, 390; *libani*, 390
Celtis, 24, 129, 329: *mississippiensis*, 129, 329; *occidentalis*, 129; *sinensis*, 129
Cenangium abietis, 393, 410
Cephalosporium wilt of elms, 320
Ceratocystis: *fagacearum*, 345; *fimbriata*, 357; *ulmi*, 317, 357
Ceratomia catalpae, 311
Cercis, 26, 294: *canadensis*, 121; *chinensis*, 121
Cercospora, 368: *apiifoliae*, 278; *carii*, 295; *catalpae*, 310; *chionanthi*, 275; *confluens*, 278; *daleae*, 301; *elaeagni*, 295; *glandulosa*, 301; *gymnocladi*, 280; *halesiae*, 299; *ilicis*, 398; *juglandis*, 374; *laburni*, 275; *microsora*, 339; *nyssae*, 370; *oxydendri*, 300; *pulvinula*, 398; *salicina*, 377; *sequoiae*, 417; *thujina*, 383
Cercosporella, 329: *mirabilis*, 278
Ceroplastes floridensis, 268
Chaenomeles japonica, 26, 238
Chain saws, 68: manufactures of, 68
Chalara quercina, 345
Chalepus dorsalis, 342
Chamadaphne calyculata, 421
Chamaecyparis, 24: *thyoides*, 26
Charles F. Irish maple, 133
Chem Bam, 233
Chemical injuries, 188
Chermes: *abietis*, 422; *cooleyi*, 423; *strobilobius*, 338
Cherry, 14, 24, 114: choke, 24; Kwanzan, 114
Cherry (flowering) Diseases, 263: leaf spot, 263; winter injury, 263; yellowing, 263
Cherry (flowering) Insects, 263: Asiatic garden beetle, 263; Japanese beetle, 266; lesser peach borer, 265; oriental fruit worm, 265; peach borer, 265; pear-slug sawfly, 266; San Jose scale, 265; West Indian peach scale, 265
Chestnut, 311: American, 24; Asiatic, 314; Chinese, 420
Chestnut Diseases, 311: blight, 311; twig canker of Asiatic chestnut, 315
Chestnut Insects, 315: brown wood borer, 315; datana caterpillar, 315; Japanese beetle, 315; lecanium scale, 315; maple and oak twig pruner, 315; spring cankerworm, 315; tussock moth, 315; two-lined chestnut borer, 315
Chinese dogwood, 117
Chinese elm, 128
Chinese hackberry, 129
Chinese redbud, 121
Chinquapin, 24, 312, 346
Chionanthus, 275: *virginica*, 24, 120
Chionaspis: *furfura*, 255; *gleditsiae*, 344; *ortholobia*, 344; *salicis-nigrae*, 380
Chipper for disposal of prunings, 76
Chloranil, 234
Chlordane, 203, 234, 235
Chloride, 17
Chlorine, effect on trees, 185
Chlorobenzilate, 203, 235
Chlorophyll, 9
Chloroplasts, 9
Chloroscypha chloromela, 417
Chlorosis, 196; cause of, 196; symptoms of, 197; control with iron chelates, 197; with other materials, 198
Choristoneura fumiferana, 423

Christmas berry, 238
Chrysobothrus femorata, 292
Chrysomphalus obscurus, 351
Chrysomyxa: cassandrae, 421; *ledi*, 421; *ledicola*, 421
Cicadellidae, 290
Cinara sabinae, 405
Cinder injury, 190
Citheronia regalis, 296
Citrus mealy bug, 418
Cladrastis, 19, 139, 380: *lutea*, 139
Cleveland maple, 133
Clitocybe tabescens, 390
Cockspur thorn, 238
Coleophora: caryaefoliella, 332; *laricella*, 337; *limosipennella*, 324; *salmani*, 261
Coleosporium solidaginis, 409
Coliseum maple, 118
Colletotrichum, 387
Collybia velutipes, 335
Colopha ulmicola, 326
Combination sprays, 233
Coniophora puteana, 396
Coniothyrium, 321
Contarinia juniperina, 405
Coposil, 231
Copper, 191: fungicides, fixed, 235; in rainwater, 191; sprays, 230
Copper A, 231
Copper Hydro Bordo, 231
Copper sulfate snow, 235
Cork cambium, 7
Cork trees, 127
Cornus, 268: *florida*, 24, 116; *Kousa*, 117; *Kousa chinensis*, 117
Coromate, 232
Corozate, 233
Coryneum, 347: *cardinale*, 391; *berckmanii*, 383, 391
Corythucha: arcuata, 353; *ciliata*, 359; *mollicula*, 380; *pergandei*, 304; *ulmi*, 328
Cotinus, 300: *coggygria*, 122
Cotoneaster, 238
Cottonseed meal, 49
Cottony-cushion scale, 309, 330, 375, 388
Cottony maple scale, 335, 341, 360
Crabapple, 115, 403: Aldenham, 115; Arnold, 115; Beauty, 116; Bechtel, 116; Carmine, 116; Crimson Brilliant, 116; Dolgo, 115; Dorothea, 116; Frau Louise Dittman, 116; Hyssop, 17; Redfield, 115; Redford, 115; Shakespeare, 116; Van Eseltine, 116; Young America, 115
Crabapple (flowering) Diseases, 266: sprays for disease and insect control, 267
Crape Myrtle, 119, 268
Crape Myrtle Diseases, 268: black spot, 268; powdery mildew, 268; tip blight, 268
Crape Myrtle Insects, 268: aphid, 268; Florida waxscale, 268
Crataegus, 24, 120, 276, 403: *crusgalli*, 238; *lavallei*, 120; *monogyna biflora* or *praecox*, 120; *monogyna stricta*, 125; *oxyacantha*, 238; *oxyacantha paulii*, 120; *phaenopyrum*, 120
Crimean linden, 130
Crimson Brilliant crabapple, 116
Crimson King maple, 133
Cristulariella depraedens, 285
Cronartium ribicola, 408
Crown gall, 240: cause of, 241; control of, 241; symptoms, 241; trees susceptible to, 240
Cryptococcus fagi, 308, 309
Cryptodiaporthe: castanea, 315; *salicina*, 377
Cryptomeria, 390
Cryptomeria Diseases, 390: leaf blight, 390; leaf spot, 391
Cryptorhynchus lapathi, 380
Cryptosporium macrospermum, 394
Cucumber tree, 122
Cupressus, 391: *sempervirens*, 391
Cuprocide, 231
Cup-shakes, 169
Cycloheximide, 409
Cydonia, 403: *vulgaris*, 238
Cylindrosporium, 329, 368: *betulae*, 258; *brevispina*, 278; *crataegi*, 278; *salicinum*, 377
Cynipoidea, 351
Cypress, 14, 24, 391: Italian, 391
Cypress Diseases, 391: blight, 391; canker, 391; crown gall, 391; cytospora canker, 391; twig blight, 391
Cypress Insects, 391: aphid, 391; bark scale, 392; caterpillars, 392; imperial moth, 392; mealy bug, 392; tip moth, 392; Webber moth, 392; white-marked tussock moth, 392
Cyrtepistomus castaneus, 354
Cytospora, 396: *acerina*, 420; *ambiens*, 420; *annularis*, 254, 420; *cenisia* var. *littoralis*, 391; *chrysosperma*, 293, 362, 377, 420; *corni*, 420; *kunzei*, 418, 419; *leucostoma*, 420; *massariana*, 293; *microspora*, 420; *mollisissima*, 420; *pinastri*, 394, 420; *rubescens*, 420

Daikoku flowering cherry, 114
Dasyscypha: ellisiana, 394; *pini*, 411; *pseudotsugae*, 394; *willkommii*, 337
Datana: integerrima, 375; *ministra*, 350
Davidia involucrata, 119
David's striped bark maple, 118
Davis gas tester, 176

Dawn redwood, 122
DCPC, 205
DDT, 203, 234, 235
Dehorning, 72
Delnav, 204
Demeton, 234
Dendrol, 235
Deodar cedar, 390
Dermatea: balsamea, 396; *livida*, 417
Derris extracts, 204
Desmia funeralis, 295
Diagnosing: procedure in, 154; tree troubles, 151
Diagnosis: importance of correct, 152
Diagnostician: qualifications of a tree, 151
Diagnostic tools, 156
Diaporthe, 321: *eres*, 398
Diaspidiotus osborni, 344
Diaspis carueli, 405
Diatrype, 347
Diatrypella oregonensis, 304
Dichlone, 232, 234
Dichloro diphenyl trichloroethane, 203
Dichomeris marginella, 405
Didymascella: thujina, 382; *tsugae*, 396
Dieldrin, 205, 234, 235
Dimite, 205, 234
Dinitro octyl phenyl crotonate, 232
Dinitro: powder, 235; slurry, 235; sprays, 208
Diospyros, 238
Diplodia, 321, 398: *infuscans*, 254; *longispora*, 347; *pinea*, 390, 406; *sophorae*, 280; *theobromae*, 367
Discella carbonacea, 377
Disease control, spraying for, 229
Diseases, general parasitic, 238
Diseases: of evergreen trees, 382; of low-growing trees, 253; of tall-growing trees, 303
Disodium ethylene bisdithiocarbamate, 233
Dithane: *D-14*, 233; *M-22*, 233; *Z-78*, 233
Dithiocarbamates, 232
DMC, 205
DN Dry Mix, 235
Dockmackie, 24
Dogwood, 28, 116, 268, 420: Chinese, 117; flowering, 24; Japanese, 117; native, 116
Dogwood Diseases, 268: crown canker, 268; flower and twig blight, 271; leaf spot, 271; spot anthracnose, 271
Dogwood Insects, 271: borers, 271; club gall, 273; scales, 273; twig girdler, 274; white fly, 273
Dolgo crabapple, 115
Dormant oil, 234
Dormant sprays, 235
Dorothea crabapple, 116
Dothichiza populea, 362
Dothiorella, 347, 359: *fraxinicola*, 254; *ulmi*, 320
Dothistroma pini, 411
Douglas fir, 24, 393
Dove tree, 119
Dreft, 210
Drosophila melanogaster, 344
Du Pont Spreader-Sticker, 211
Dust formulas, 234
Dutch elm disease, 316; trees resistant to, 318

Eacles imperialis, 344, 392
Ecdytolophia insiticiana, 343
Ectostroma, 368
Ehrhornia cupressi, 392
Elaeagnus, 295: *angustifolia*, 121, 139
Elaphidion villosum, 291
Electrical injuries, 174
Elgetol, 235
Elm, 23, 24, 28, 127, 316: American, 420; Augustine ascending, 125; Buisman, 128, 318; Chinese, 128; English, 128; golden, 127; Hansen Manchurian, 128, 319; Siberian, 128
Elm Diseases, 316: canker, 321; dothiorella wilt, 320; Dutch elm, 316; leaf spot, 321; mosaic, 320; other leaf diseases, 322; phloem necrosis, 319; wetwood, 322; wood decay, 323
Elm Insects, 323: borer, 328; carpenter worm, 329; case bearer, 324; cockscomb gall, 326; European bark beetle, 327; European elm scale, 326; fall cankerworm, 324; fall webworm, 329; gypsy moth, 326; lace bug, 328; leaf beetle, 323; leaf miner, 328; leopard moth, 329; mites, 328; scurfy scale, 329; spiny elm caterpillar, 324; spring cankerworm, 324; tussock moth, 324; twig girdler, 328; woolly apple aphid, 326
Elsinoe: corni, 271; *ilicis*, 399
Empoasca pergandei, 344
Empress tree, 274
Empress tree Diseases, 274: leaf spot, 274; wood decay, 274
Enchenopa binotata, 294
Endothia parasitica, 313
Endrin, 234
Englerulaster orbicularis, 398
English elm, 128
English hawthorn, 238
Eotetranychus: carpini borealis, 328; *hicoriae*, 333; *multidigituli*, 343
Epidermis, 8
Epinotia: nanana, 424; *subviridis*, 392

Epizeuxis aemula, 424
EPN, 205
Eriophyes: fraxiniflora, 257; *pyri*, 293
Eriosoma lanigerum, 326
Erwinia: amylovora, 239; *nimipressuralis*, 322; *tumefaciens*, 241, 322
Erysiphe: aggregata, 242, 304; *polygoni*, 298, 369; *lagerstroemiae*, 268
Essential elements, influence on growth, 49
Eucalyptus, 24
Eugene poplar, 138
European fruit lecanium scale, 309, 341
Eurytetranychus buxi, 380
Eutypella parasitica, 287
Exoteleia pinifoliella, 414

Faassen's red-leaf maple, 133
Fabraea: maculata, 297; *theumenii*, 276
Fagus, 24, 304: *grandifolia*, 304; *sylvatica*, 306: *sylvatica* var. *atropunicea*, 306; var. *aurea*, 306; var. *fastigiata*, 306; var. *laciniata*, 306; var. *pendula*, 306; var. *purpureo-pendula*, 306; var. *Rohanni*, 306
Fenusa: pusilla, 261; *ulmi*, 328
Ferbam, 232, 234, 235
Fermate, 232, 235
Ferradow, 232, 236
Ferric dimethyl dithiocarbamate, 232
Fertilizer: formula for home-mixing, 51; formulas, 50; applying with feeding needles, 57; compressed air method of applying, 57; cost of, 60; frequency of applying, 60; how they act, 47; method of applying, 54; punch bar method, 55; rates of application, 52; surface applications of, 55; trenching method of applying, 55; use of, 46
Fertilizing: above-ground parts, 60; broad-leaved evergreens, 53; by spraying leaves, 60; by trunk injections, 64; narrow-leaved evergreens, 52; with Ross root feeder, 59
Fiorinia externa, 397
Fir, 24, 29, 392; balsam, 420
Fir Diseases, 392: canker, 394; leaf cast, 393; needle and twig blight, 392; needle rust, 393; shoestring root rot, 394; sooty mold, 394; witches'-broom, 394; wood decay, 395
Fir Insects, 395: bagworm, 395; balsam fir sawfly, 395; balsam twig aphid, 395; bark beetle, 395; cedar tree borer, 395; cooley gall aphid, 395; hemlock looper moth, 395; pine butterfly, 395; spider mite, 395; spotted tussock moth, 395; Zimmerman pine moth, 395
Fire blight, 238: cause of, 239; control of, 239; symptoms of, 238; use of paints for, 240
Fire damage, 198
Firethorn, 238
Fistulina, 350
Fixed coppers, 234
Flotox wettable sulfur, 232
Flowering ash, 114
Flowering cherry, 263
Flowering crabapple, 266
Fluorine, effect on trees, 183
Flux, slime, 165
Foliar feeding, 60
Foliar sprays, combining pesticides with, 63
Fomes, 350, 385, 395; *annosus*, 337; *applanatus*, 260; *fasciatus*, 281; *fomentarius*, 260; *geotropus*, 281; *igniarius*, 88, 260, 374; *meliae*, 344; *officinalis*, 411; *pini*, 337, 411; *pinicola*, 421; *rimosus*, 341; *roseus*, 337, 411
Fragrant snowbell, 122
Franklinia, 19, 274: *alatamaha*, 24, 120
Franklin tree, 19, 24, 120, 274
Franklin tree Diseases, 274: black mildew, 274; leaf spot, 274
Frau Louise Dittman crabapple, 116
Fraxinus, 24, 126, 253: *americana*, 126; *excelsior pyramidalis*, 125; *Ornus*, 114; *pennsylvanica*, 140; *velutina glabra*, 114
Fringe tree, 24, 28, 120, 275
Fringe tree Diseases, 275: leaf spot, 275; powdery mildew, 275
Frost cracks, development of, 168
Frosts: autumn, 167; spring, 167; susceptibility to spring, 167
Fungi, 224: dissemination of, 225; overwintering of, 228
Fungicides, 230: newer, 232
Fusarium lateritium, 275, 279, 280, 301
Fusarium: oxysporium var. *perniciosum*, 298; *solani* var. *martii*, 399
Fusicladium saliciperdum, 376

Galerucella xanthomelaena, 323
Gall, crown, 240
Gallotox, 233
Galls, oak, 351
Ganoderma lucidum, 288, 396
Ganoderma rot, 329
Gas: constituents of manufactured, 175, 178; constituents of natural, 178; differences between manufactured and natural, 178; effect of manufactured, on tomato, 176; manufactured, 175; natural, 177; symptoms of manufactured, on trees, 176
Gas-injured trees, treatment of, 179
Gas injury, 175: on Carolina poplar, 176; on London planes, 176

Gas leaks, determination of manufactured, 176
Ginkgo, 132, 344: *biloba*, 25, 132; *biloba fastigiata*, 125
Ginkgo, 28: sentry, 125; preventing fruit development in, 132
Girdling roots, 180: development of, 181; how to diagnose for, 181; treatment for, 183
Glastonbury thorn, 120
Gleditsia, 25, 130: *triacanthos*, 130; *triacanthos inermis*, 130; *triacanthos inermis moraine*, 130
Globe locust, 131
Gloeosporium, 349, 368: *ailanthi*, 301; *apocryptum*, 285; *aquifolii*, 398; *aridum*, 254; *betularum*, 258; *crataegi*, 278; *fructigenum*, 296; *inconspicuum*, 321; *liriodendri*, 368; *robergei*, 279; *ulmicolum*, 321
Glomerella cingulata, 314
Glycobius speciosus, 291
Gnomonia: *caryae*, 330; *tiliae*, 338; *ulmea*, 321; *veneta*, 355
Gnomoniella fimbriata, 279
Goldenchain tree, 120, 275
Goldenchain tree Diseases, 275: leaf spot, 275; twig blight, 275
Goldenchain tree Insects, 275: aphids, 275
Golden elm, 128
Goldenrain tree, 28, 120, 275
Goldenrain tree Diseases, 275: canker, 276; leaf spot, 275; wilt, 276
Gordonia Lasianthus, 25
Gossyparia spuria, 326
Gracilaria syringella, 256
Grade changes, damage caused by, 159
Grades: lowering around trees, 164; other methods for raising, 164; preparation for raising, 162
Grape: leaf folder, 295; scale, 360
Graphiolitha molesta, 265
Graphium ulmi, 317
Greedy scale, 375
Green ash, 140
Guard cells, 8
Guignardia aesculi, 334
Guthion, 205
Gymnocladus, 280: *dioicus*, 25, 123
Gymnosporangium: *aurantiacum*, 292; *clavipes*, 404; *globosum*, 277, 292, 404; *juniperinum*, 292; *juniperi-virginianae*, 403; *libocedri*, 293; *nelsoni*, 292; *nootkatense*, 292

Hackberry, 24, 129, 329: Chinese, 129; southern, 129
Hackberry Diseases, 329: Ganoderma rot, 329; leaf spot, 329; witches'-broom, 329
Hackberry Insects, 330: nipple gall, 330; painted hickory borer, 330; scales, 330; spiny elm caterpillar, 330
Hail injury, 191
Halesia, 19, 299: *carolina*, 121; *tetraptera*, 26
Hawthorn, 24, 28, 120, 276: English, 238; Glastonbury, 120; Lavalle, 120; Paul's Scarlet, 120; Washington, 120
Hawthorn Diseases, 276: cedar-hawthorn rust, 277; fire blight, 277; leaf blight, 276; leaf spot, 278; powdery mildew, 278
Hawthorn Insects, 278: gypsy moth, 279; leaf skeletonizer, 278; pear-slug sawfly, 279; western tent caterpillar, 278; western tussock moth, 279
Heartwood, 6
Hedge maple, 118
Heller-Gro, 58
Hemerocampa leucostigma, 324, 344, 392
Hemerophila pariana, 278
Hemlock, 22, 28, 29, 395: Canada, 24; Carolina, 24
Hemlock Diseases, 396: cankers, 396; leaf blight, 396; needle rust, 396; sapwood rot, 396
Hemlock Insects, 396: bagworm, 397; cypress moth, 397; fiorinia hemlock scale, 397; grape scale, 397; looper, 396; spider mites, 397; spotted borer, 397
Hendersonia crataegicola, 278
Heptachlor, 234
Hickory, 23, 28, 330: shagbark, 24
Hickory Diseases, 330: canker, 331; crown gall, 332; leaf spot, 330; witches'-broom, 331
Hickory Insects, 332: bark beetle, 332; cigar case bearer, 332; gall aphis, 332; June bugs, 333; mite, 333; painted hickory borer, 332; twig girdler, 333
Hicoria ovata, 24
Higginsia hiemalis, 263
Holly, 397: American, 24, 29; Chinese, 399
Holly Diseases, 398: canker, 398; die-back, 399; leaf scorch, 399; spine spot, 399; spot anthracnose, 399; tar spot, 398; twig die-back, 399
Holly Insects, 400: bud moth, 401; leaf miner, 400; pit-making scale, 401; scale, 401; southern red mite, 402; white fly, 402
Holy Thorn of Glastonbury, 120
Homadaula albizziae, 299, 343
Honey locust, 130, 341: Moraine, 130; Shademaster, 131; Sunburst, 130; thornless, 130
Hophornbeam, American, 24, 120

Hornbeam, 24, 279: American, 28; Japanese, 121; upright European, 121; Yeddo, 120
Hornbeam Diseases, 279: canker, 279; leaf spot, 279; twig blight, 279
Hornbeam Insects, 279: scale, 279
Horsechestnut, 24, 129, 333: double-flowered, 129; pink-flowered, 129; ruby-red, 129
Horsechestnut Diseases, 333: leaf blotch, 333; leaf scorch, 334; powdery mildew, 335; sapwood rot, 335
Horsechestnut Insects, 335: bagworm, 335; comstock mealy bug, 335; flat-headed borer, 335; Japanese beetle, 335; scales, 335; tussock moth, 335; white scale, 335
Humus, 23
Hyalophora cecropia, 339
Hydrogen, 17, 47
Hydrogen sulfide, effect on trees, 185
Hydroxyl anions, 17
Hylobius pales, 417
Hylurgopinus rufipes, 317
Hymenochaete agglutinans, 304, 396
Hyphantria cunea, 255
Hypodermella: laricis, 336; *mirabilis*, 393; *nervata*, 393
Hyponectria buxi, 387
Hypoxylon pruinatum, 363
Hyssop crabapple, 17

Ichthyura inclusa, 365
Idaho locust, 131
Ilex, 397: *aquifolium*, 398; *cornuta*, 399; *opaca*, 24, 397
Imbibition, 15
Imperial moth, 344
Insecticides, 203
Insect control, 200
Insect Pests: of evergreen trees, 382; of low-growing trees, 253; of tall-growing trees, 303
Insects: biological control of, 214; control methods, 202; life history of, 200; structure of, 202
Io moth, 297
Iron, 47, 50: chelates, 197
Isolan, 210
Isotox 200, 210
Itonida catalpae, 311

Janus integer, 378
Japanese black pine, 139
Japanese dogwood, 117
Japanese hornbeam, 121
Japanese keaki, 319
Japanese maple, 118, 420
Japanese pagoda tree, 123, 279
Japanese Pagoda tree Diseases, 279: canker, 279; twig blight, 280
Japanese snowbell, 122
Japanese walnut, 373, 374
Japellent, 205
Japrocide, 205
Juglans, 26, 238, 371: *ailantifolia*, 373, 374; *cinerea*, 374; *nigra*, 374
Juneberry, 297
Juniper, 24, 402: common, 24; creeping, 24; Hill, 403, mountain, 25; spiny Greek, 403
Juniper Diseases, 402: cedar rusts, 403; twig blight, 402, wood decay, 404
Juniper Insects, 405: bagworm, 406; midge, 405; red cedar aphid, 405; scale, 405; spider mite, 405; taxus mealy bug, 406; webworm, 405
Juniperus, 24, 402: *communis*, 24, 403; *communis saxatilis*, 25; *excelsior* var. *stricta*, 403; *horizontalis*, 24; *sabina*, 26, 403; *scopulorum*, 403; *sibirica*, 403; *virginiana*, 26; *virginiana* var. *Keteleeri*, 403; *virginiana* var. *pyramidiformis hillii*, 403

Kalmia latifolia, 25
Karathane, 232, 234, 236
Karbam: Black, 232; White, 233
Katsura tree, 194
Kelthane, 205, 236
Kentucky Coffee tree, 25, 123, 280
Kentucky Coffee tree Diseases, 280: leaf spot, 280
Kermes pubescens, 351
Kobus magnolia, 117
Koelreuteria, 275: *paniculata*, 120
Kolofog, 232
Krenite, 235
Kromad, 403

Laburnum, 275: *Watereri*, 120
Lagerstroemia, 268: *indica*, 119
Lambdina fiscellaria, 396
Larch, 25, 236
Larch Diseases, 336: European larch canker, 336; leaf cast, 336; needle rust, 336; wood decay, 337
Larch Insects, 337: case bearer, 337; sawfly, 338; woolly larch aphid, 338
Larix, 25, 336
Lavalle hawthorn, 120
Lead arsenate, 205, 234, 236
Leaf absorption of nutrients, 61
Leaf scorch, 194: cause, 194; control, 196; symptoms, 194
Leaves: function of, 9; structure of, 8

Lebanon cedar, 390
Lecanium: caryae, 344; *corni*, 351; *fletcheri*, 384, 427; *nigrofasciatum*, 290, 360; *quercifex*, 351
Ledum glandulosum, 421
Lenzites, 385, 395
Lepidosaphes: beckii, 296, 427; *ulmi*, 255, 360
Leptocorus trivittatus, 292
Lesser peach borer, 298
Lesser snow scale, 388
Lichens, 186: injury by, 186; control of, 188
Lightning: equipment, method of installing, 174; injury, 172; protection, 174; protection equipment companies, 174; repairing, injured trees, 172
Lime, hydrated, 234
Lime requirements of soils, 19
Limestone, 19
Lime sulfur, 206, 231, 234, 236: preparing, 231
Lindane, 206, 234, 236
Linden, 25, 28, 129, 338: American, 130; Crimean, 130; little-leaf European, 130; silver, 129; silver pendent, 129; upright American, 125
Linden Diseases, 338: leaf blotch, 338; leaf spot, 338; nectria canker, 339; powdery mildew, 339
Linden Insects, 339: aphid, 339; borer, 341; cecropia moth, 339; fall cankerworm, 341; gypsy moth, 341; Japanese beetle, 339; leaf beetle, 339; scales, 341; spring cankerworm, 341; tussock moth, 341; twig girdler, 341
Linospora gleditsiae, 342
Liquidambar, 138, 366: *styraciflua*, 26, 138
Liriodendron, 138, 368: *tulipifera*, 26, 138
Lithocolletis hamadryadella, 350
Little-leaf European linden, 130
Live oak, 136
Loblolly bay, 25
Locust, 130, 341: black, 25, 130; globe, 131; honey, 25, 130, 341; Idaho, 131; Moraine, 130; Shademaster, 131; Sunburst, 130; thornless honey, 130
Locust Diseases, 341: leaf spot, 342; witches'-broom, 342; wood decay, 341
Locust Insects, 342: bagworm, 344; borer, 342; carpenter worm, 344; cottony maple scale, 344; fall cankerworm, 344; gall maker, 343; honey locust borer, 344; leaf beetle, 342; leafhopper, 344; mite, 343; pod gall, 343; San Jose scale, 344; scales, 344; spring cankerworm, 344; walnut datana, 344; webworm, 343
Loebner magnolia, 117
Lombardy poplar, 138
London plane, 17, 137
Lophodermium, 410: *autumnale*, 393; *filiforme*, 421; *lacerum*, 393; *laricis*, 336; *piceae*, 421; *pinastri*, 410
Low temperatures, injuries caused by, 167
Luna moth, 367

Macrophoma: candollei, 387; *phacidiella*, 398; *tumefaciens*, 364
Macrosiphum liriodendri, 370
Magnesium, 17, 18, 47: arsenate, 206
Magnetic spray sulfur, 232
Magnolia, 25, 28, 117, 280: Kobus, 117; saucer, 117; southern, 118; star, 117
Magnolia, 25, 280: *acuminata*, 122; *grandiflora*, 114, 118; *Kobus*, 114, 117; *Loebneri*, 117; *soulangeana*, 117; *stellata*, 117; *stellata rubra*, 117
Magnolia Diseases, 280: die-back, 281; leaf blight, 280; leaf spot, 281; nectria canker, 281; wood decay, 281
Magnolia Insects, 281: magnolia scale, 281; tulip tree scale, 281
Maidenhair tree, 25, 132, 344
Maidenhair tree Diseases, 344: leaf spot, 344
Maidenhair tree Insects, 344: grape mealy bug, 344; omnivorous looper, 344; white-marked tussock moth, 344
Malacosoma: americanum, 378; *californicum*, 295; *disstria*, 291; *fragilis*, 214; *pluviale*, 278
Malathion, 206, 234, 236
Malus, 24, 115, 266: *Arnoldiana*, 115; *purpurea aldenhamensis*, 115
Maneb, 233, 234
Manure, 21
Manzate, 233
Maple, 23, 25, 28, 118, 132, 282: Amur, 118; Armstrong, 123; black, 133; coliseum, 118; David's striped bark, 118; fastigiate black, 125; hedge, 118; improved columnar upright, 125; Japanese, 118, 420; mountain, 25, 118; Newton Sentry, 123; Norway, 132; paperbark, 118; red, 25, 133; Scanlon red, 123; Schwedler, 133; silver, 132; sugar, 133; striped, 25; sycamore, 133; trident, 118; wine-leaved sycamore, 133
Maple Diseases, 282: anthracnose, 285; bacterial leaf spot, 286; basal canker, 282; bleeding canker, 284; bull's-eye spot, 285; canker, 287; die-back, 288; ganoderma rot, 133, 288; leaf blister, 287; leaf spot, 284; purple eye spot, 285; tar spot, 285; wilt, 132, 282
Maple Insects, 288: alder blight aphid, 292; bagworm, 292; bladder gall mite, 290;

boxelder bug, 292; carpenter worm, 292; cottony maple scale, 290; fall cankerworm, 292; flat-headed borer, 292; forest tent caterpillar, 291; Japanese beetle, 292; leafhoppers, 290; leopard moth, 291; maple and oak twig pruner, 291; Norway maple aphid, 288; spring cankerworm, 292; sugar maple borer, 291; terrapin scale, 290; tussock moth, 292; white fly, 292; woolly maple scale, 290
Marssonia, 280
Marssonina, 364: *californica*, 374; *juglandis*, 330, 373; *kriegeriana*, 377; *ochraleuca*, 349; *populi*, 364
Matsucoccus resinosae, 416
Mazzard double flowering cherry, 114
Mealy bug: citrus, 418; cypress, 418; yucca, 418
Mechanical injuries, 192
Medullary rays, 6
Megacyllene caryae, 332; *robiniae*, 342
Melampsora: *abietis-canadensis*, 364; *abieticapraearum*, 378; *albertensis*, 394; *bigelowii*, 336, 378; *farlowii*, 396; *medusae*, 336, 364
Melampsoridium betulinum, 258, 336
Melanconis juglandis, 373
Melanconium betulinum, 261
Melanophila fulvoguttata, 397
Meliola cryptocarpa, 274
Mema, 233
Merbam, 403
Mercury compounds, organic, 233
Meria laricis, 336
Mer-Kote, 233
Metasequoia glyptostroboides, 122
Metatetranychus ulmi, 328
Methasan, 233
Methoxychlor, 207, 234, 236
Methyl parathion, 207
MH-30, 132
MH-40, 132
Micronized wettable sulfur, 232
Microsphaeria: *alni*, 242, 304, 339, 359; *alni* var. *vaccinii*, 310
Microstroma juglandis, 331
Microsulfur, 232
Microthyriella cuticulosa, 398
Mike sulfur, 232
Mildew, powdery, 242
Mildex, 232, 234, 236
Milesia, 394
Mimosa, 121, 298
Mindarus abietinus, 395
Mist blowers, 218: manufacturers of, 220
Mistletoe, 394
Mite, spruce, 397
Miticides, 203
Mock-orange, 25
Modesto ash, 114
Molybdenum, 50
Monarthropalpus buxi, 388
Monilinia Johnsonii, 278
Monochaetia desmazierii, 330, 349
Moraine locust, 130
Morus, 25
Mountain Ash, 238, 292: American, 25, 420; European, 25, 420
Mountain Ash Diseases, 292: canker, 293; crown gall, 293; leaf rust, 292
Mountain Ash Insects, 293: pear leaf blister mite, 293; round-headed borer, 293; sawfly, 294; scales, 293; woolly apple aphid, 294
Mountain laurel, 25
Mountain maple, 118
Mulberry, 25
Mycodiplosis alternata, 273
Mycorrhiza, 23: ectotrophic, 23; endotrophic, 23; function of, 23
Mycosphorella: *caroliniana*, 300; *cercidicola*, 294; *nyssaecola*, 370; *platanifolia*, 359; *populorum*, 363; *sequoiae*, 417
Myrioconium comitatum, 377
Myxosporium, 369
Myzocallis kahawaluokalani, 268; *tiliae*, 339

Nabam, 233, 236
Nantucket pine moth, 414
Nectria: *cinnabarina*, 254, 276, 280, 295, 299; *coccinea*, 304, 398; *coccinea* var. *fagi*, 308; *desmazierii*, 386; *galligena*, 259, 279, 296, 331, 372; *magnoliae*, 281
Nemagon, 389
Nematodes on boxwood, 389
Neodiprion: *lecontei*, 413; *sertifer*, 413
Neolecanium cornuparvum, 281
Neutrona, 237
Newton Sentry sugar maple, 123
Niagara Carbamate, 236
Nicotine sulfate, 207, 234, 236
Nitidulid beetles, 345
Nitrate, 17
Nitrogen, 47
NMO, 211
Normal tree, 3
Norway maple, 132: Almira, 133; Charles F. Irish, 133; Cleveland, 133; Crimson King, 133; Faassen's red-leaf, 133
No-Trim Tree Guard, 270
N-trichloromethyl thio tetrahydro phthalimide, 232
Nymphalis antiopa, 324
Nyssa, 370: *sylvatica*, 26

Oak, 23, 28, 135, 345: black, 25; blackjack, 25, 139; chestnut, 25; English, 25; European Turkey, 25; live, 136; overcup, 25; pin, 25, 28, 135; post, 25; red, 25, 135, 139; sand blackjack, 25; scarlet, 25, 135; shingle, 25, 137; shrub, 25; Shumard, 137; southern red, 25; swamp white, 25; upright English, 125; white, 25, 136; yellow, 25; willow, 25, 137
Oak Diseases, 345: anthracnose, 347; basal canker, 348; canker, 348; leaf blister, 348; leaf spot, 349; shoestring root rot, 350; twig blight, 347; wilt, 345; wood decay, 350
Oak Insects, 350: Asiatic oak weevil, 354; carpenter worm, 354; cecropia moth, 354; datana caterpillar, 350; galls, 351; golden oak scale, 351; June bugs, 354; kermes scale, 351; lace bug, 353; lecanium scale, 351; leopard moth, 354; mite, 352; obscure scale, 351; pin oak sawfly, 354; saddle-back caterpillar, 350; twig pruner, 354; two-lined chestnut borer, 352; white blotch leaf miner, 350
Oberea tripunctata, 328
Octachlor, 203
Odonopus calceatus, 297
Oil: emulsified mineral, 211; summer, 234
Oil sprays, 207: dormant, 207; summer, 208; susceptibility to injury by, 207
Oligonychus: *bicolor*, 352; *ilicis*, 402; *ununguis*, 397, 423
Olive parlatoria, 296
Olive, Russian, 121, 139, 295
Omadines, 233
OMPA, 210
Oncideres cingulata, 333
Organic matter, 21
Organic mercury, 234: compounds, 233
Oriental plane, 138
Orthex, 211
Orthocide, 232
Orthocide *406*, 232
Orthol D, 237
Osmosis, 15
Ostrya virginiana, 24, 120
Outer bark, 7
Ovex, 209, 234
Oxydendrum, 19, 300: *arboreum*, 26, 122
Oxygen, 47: in soil, 12
Oyster shell scale, 309, 330, 335, 341, 360, 375, 388

Pachypsylla celtidis-mamma, 330
Pagoda tree, Japanese, 123
Paleacrita vernata, 324
Palisade cells, 9
Paperbark maple, 118
Parandra brunnea, 309
Paratetranychus aceris, 290
Parathion, 209, 234, 237
Parenchyma, 9
Parlatoria oleae, 296
Parzate, 233: Zineb Fungicide, 233
Paulownia (*See* Empress tree)
Paulownia, 274: *tomentosa*, 25
Paul's Scarlet Thorn, 120
Paul's Tree Brace, 41
Peach, 25: lesser, borer, 298
Pear, 25, 238, 403
Peat: moss, 22; sedge, 23
Pecan, 28
Pellicularia koleroga, 280
Periphyllus lyropictus, 288
Persimmon, 238
Pestalotia: *cryptomeriae*, 391; *funerea*, 383, 385, 391, 417, 424
Pezicula carpinea, 279
pH: range for trees, 24; scale, 18
Phacidiopycnis pseudotsugae, 394
Phacidium: *curtisii*, 398; *infestans*, 383
Phaeobulgaria inquinans, 348
Phellodendron, 127: *amurense*, 127; *chinense*, 127; *lavallei*, 127; *sachalinense*, 127
Phenacaspis: *pinifoliae*, 414; *spinicola*, 344
Phenacoccus acericola, 279, 290
Phenyl mercury: solution, 237; triethanol-lactate, 233
Philadelphus, 25
Phix, 233
Phloem, 6
Phleospora, 329: *multimaculans*, 374
Phoma, 321, 347: *conidiogena*, 387
Phomopsis, 281, 390, 411: *crustosa*, 398; *juniperovora*, 383, 391, 402
Phopobota naevana var. *ilicifolia*, 401
Phosphate anions, 17
Phosphorus, 47
Photinia villosa, 238
Photosynthesis, 10
Phygon XL, 232, 237
Phyllactinia: *corylea*, 242, 275, 278, 298, 304, 310, 339, 369, 381; *fagi*, 308
Phyllophaga, 333
Phyllosticta, 329, 349, 364, 368: *ailanthi*, 301; *apicelis*, 377; *argyrea*, 295; *auerswaldii*, 387; *catalpae*, 310; *chionanthi*, 275; *concomitans*, 398; *cookei*, 281; *cytisii*, 275; *glauca*, 281; *ginkgo*, 344; *gordoniae*, 274; *gymnocladi*, 280; *illinoensis*, 296; *lagerstroemia*, 268; *magnoliae*, 281; *minima*, 285; *paulowniae*, 274; *platani*, 359; *terminalis*, 398
Phyllostictina hysterella, 424
Phylloxera caryaecaulis, 332

Phymatotrichum omnivorum, 390
Physalospora: gregaria, 424; *ilicis*, 398; *miyabeana*, 376; *obtusa*, 296, 304
Physokermes piceae, 423
Phytomycin, 239
Phytomyza ilicis, 400
Phytophthora: cactorum, 269, 284, 307; *cambivora*, 283; *cinnamomi*, 14, 283; *ilicis*, 399
Picea, 26, 418: *pungens*, 26
Piggotia fraxini, 254
Pileolaria cotinicoggyriae, 300
Pine, 22, 25, 406: Austrian, 406; Japanese black, 139; mugho, 406; red, 23, 406; Scotch, 406; western yellow, 406; white, 23, 406
Pine Diseases, 406: blight of white pine, 411; cankers, 411; cenangium twig blight, 410; diplodia tip blight, 406; leaf casts, 409; needle blight, 411; needle rust, 409; other rusts, 409; white pine blister rust, 408; wood decay, 411
Pine Insects, 413: bagworm, 417; bark aphid, 413; bark beetles, 414; black pine needle scale, 414; European pine shoot moth, 414; leaf aphid, 413; leaf scale, 414; Nantucket pine moth, 414; needle leaf miner, 414; pales weevil, 417; pine false webworm, 413; red pine scale, 416; sawflies, 413; spittlebug, 415; webworm, 414; white pine weevil, 416
Pineus: pinifoliae, 413; *strobi*, 413
Pinnaspis minor, 299
Pin oak, 135
Pinus, 25, 406: *strobus*, 411; *thunbergii*, 139
Pirostoma nyssae, 370
Pissodes: nemorensis, 390; *strobi*, 416
Pith, 6
Pityokteines sparsus, 395
Plagiodera versicolora, 366
Plagionanthus albatus, 360
Plane, 26, 137, 354: American, 354; London, 137, 354; oriental, 138
Plane Diseases, 354: anthracnose, 354; blight, 354; botryosphaeria canker, 359; cankerstain, 357; other diseases, 359; powdery mildew, 359; scorch, 354
Plane Insects, 359: bagworm, 360; black scale, 360; cottony maple scale, 360; grape scale, 360; lace bug, 359; oyster shell scale, 360; scales, 360; sycamore plant bug, 360; sycamore scale, 360; terrapin scale, 360; white fly, 360
Planococcus citri, 418
Plasmodesma, 10
Platanus, 26, 137, 354: *acerifolia*, 137, 354; *acerifolia* var. *pyramidalis*, 137; *occidentalis*, 354; *orientalis*, 138
Plum: American, 26; beach, 26; common, 26; flowering, 238; purple-leaf, 26
Podosesia syringae fraxini, 254
Podosphaera oxyacanthae, 242, 278, 298
Poliaspis pini, 414
Polyphemus moth, 297, 367
Polyporus, 350, 385, 395: *betulinus*, 260; *catalpae*, 310; *circinatus*, 411; *halesiae*, 299; *hirsutus*, 344; *lacteus*, 344; *robiniophilus*, 342; *schweinitzii*, 337, 411, 421, *spraguei*, 274, 380; *sulphureus*, 374, 421; *tulipiferus*, 344; *versicolor*, 274, 310, 344
Popillia japonica, 339
Poplar, 26, 138, 360: Eugene, 138; Japan, 362; Lombardy, 138; Simon, 362; white, 138
Poplar Diseases, 360: branch gall, 364; cytospora canker, 362; die-back, 365; dothichiza canker, 360; hypoxylon canker, 363; inkspot, 365; leaf blister, 364; leaf rust, 364; leaf spot, 364; septoria canker, 363
Poplar Insects, 365: borer, 365; bronze birch borer, 366; fall webworm, 366; red-humped caterpillar, 365; tentmaker, 365; willow leaf beetle, 366
Populus, 26, 138, 360: *alba*, 138; *balsamifera*, 365; *canadensis eugenei*, 138; *grandidentata*, 365; *maximowiczii*, 362; *nigra* var. *italica*, 138, 360; *simonii*, 362
Poria, 385
Porthetria dispar, 326
Potassium, 17, 18, 47
Powdery mildew, 242: cause of, 242; trees susceptible to, 242
Pratylenchus, 389
Prionoxystus robiniae, 255
Pristiphora: erichsonii, 338; *geniculata*, 294
Prociphilus: imbricator, 308; *tessellatus*, 304
Promethea moth, 367
Prune: how to, 69; when to, 74
Pruners, safety hints to, 78
Pruning: directional, 76; equipment, 66; equipment manufacturers, 68; evergreen trees, 74; for appearance, 66; disease control, 74; health, 65; line clearance, 75; safety, 66; large branches, 70; need for, 65; storm damaged trees, 73; suckers or watersprouts, 73; transplanted trees, 41
Prunings, disposal of, 76
Prunus, 24, 263: *americana*, 126; *avium plena*, 114; *cerasifera* var. *pissardi*, 26; *domestica*, 26; *glandulosa*, 24; *maritima*, 26; *persica*, 25; *Sargentii*, 114; *Sargentii* var. *columnare*, 125; *serrulata*, 114; *triloba* var. *plena*, 238; *virginiana*, 214; *yedoensis*, 114
Pseudaulacaspis pentagona, 265

Pseudococcus: comstocki, 311, 335, 388; *cuspidatae*, 427; *maritimus*, 344, 427; *ryani*, 392, 418
Pseudomonas: aceris, 286; *saliciperda*, 377
Pseudonectria rousselliana, 386
Pseudotsuga, 392: *taxifolia*, 24
Psylla: buxi, 388; *floccosa*, 304
Puccinia: andropogonis var. *onobrychidis*, 300; *caricis-shepherdiae*, 295; *coronata* var. *elaeagni*, 295; *peridermiospora*, 253
Pulvinaria: floccifera, 427; *innumerabilis*, 290, 360
Puratized Agricultural Spray, 233, 237
Putnam scale, 309, 330, 335, 341, 375
Puto yuccae, 418
Pyracantha coccinea, 238
Pyrethrum, 209, 234, 237
Pyrus, 25

Quercus, 345: *alba*, 25, 136; *bicolor*, 25; *borealis maxima*, 25, 135, 139; *brutia*, 346; *catesbaei*, 25; *cerris*, 25; *coccinea*, 25, 135; *falcata*, 25; *ilicifolia*, 25; *imbricaria*, 25, 137; *lusitanica*, 346; *macrocarpa*, 25; *marilandica*, 25, 139; *Muhlenbergii*, 25; *palustris*, 25, 135; *phellos*, 25, 137; *prinus*, 25; *robur*, 25; *robur fastigiata*, 125; *shumardii*, 137; *stellata*, 25; *velutina*, 25; *virginiana*, 136
Quince, 238: flowering, 26, 238
Quinones, 234

Ramularia rosea, 377
Ra-pid-gro, 44, 61
Recurvaria: apicitripunctella, 385; *piceaella*, 424
Redbud, 26, 294: Chinese, 121; eastern, 121
Redbud Diseases, 294: botryosphaeria canker, 294; leaf spot, 294
Redbud Insects, 294: caterpillars, 295; scales, 295; two-marked treehopper, 294
Red cedar, 26
Redfield crabapple, 115
Redford crabapple, 115
Red maple, 133
Red oak, 135, 139
Redwood, 417: dawn, 122
Redwood Diseases, 417: canker, 417; needle blight, 417; shoestring root rot, 417
Redwood Insects, 417: cedar tree borer, 417; citrus mealy bug, 418; cypress mealy bug, 418; greedy scale, 418; mealy bugs, 418; oleander scale, 418; sequoia pitch moth, 418; yucca mealy bug, 418
Reed peat, 23
Rehmiellopsis balsameae, 392
Reticulitermes flavipes, 428
Rex-Keyser watering method, 42
Rhabdocline pseudotsugae, 393
Rhabdogloeum hypophyllum, 393
Rhabdophaga: salicis, 380; *strobiliodes*, 378
Rhizoecus falcifer, 388
Rhizosphaera kalkhoffi, 421
Rhododendron, 22
Rhus typhina, 312
Rhyacionia: buoliana, 414; *frustrana*, 414
Rhynchaenus rufipes, 380
Rhytisma: acerinum, 285; *ilicinicolum*, 398; *salicinum*, 377; *velatum*, 398
Ringwood Repellent, 199
Robinia, 130: *pseudo-acacia*, 25, 130; *pseudo-acacia* × *hispida*, 131; *pseudo-acacia* var. *umbraculifera*, 131
Rodent damage, 199
Root: cap, 11; hairs, 11
Root rot, shoestring, 243
Roots: function of, 11; girdling, 180; structure of, 10
Rosa, 238
Rosellinia caryae, 331
Ross Root Feeder, 59
Rotenone, 204, 209, 234, 237
Round-headed apple tree borer, 298
Russian Olive, 121, 139, 295
Russian Olive Diseases, 295: cankers, 295; leaf spot, 295; rust, 295
Russian Olive Insects, 296: oleander-thistle aphid, 296; scales, 296

Sabulodes caberata, 344
Saissetia oleae, 360, 390
Salebria afflictella, 367
Salix, 26, 375: *fragilis*, 377; *lasiandra*, 377; *lasiolepis*, 377; *nigra*, 13; *repens*, 26
Salt injury, 190
Samia cynthia, 301
San Jose scale, 309, 330, 341
Sanninoidea exitiosa, 265
Saperda: calcarata, 365; *candida*, 293, 298; *tridentata*, 328; *vestita*, 341
Sapwood, 6
Sargents flowering cherry, 114
Sassafras, 296
Sassafras, 28
Sassafras Diseases, 296: canker, 296; leaf spot, 296; yellows, 296
Sassafras Insects, 296: hickory horned devil, 296; promethea moth, 296; weevil, 297
Saucer magnolia, 117
Savin, 26
Saws, chain, 68
Scale: bark, 392; black, 309, 360, 375; black pine needle, 414; calico, 375; Cali-

fornia red, 375, 388, 427; camphor, 330; cotton-cushion, 309, 330, 375, 388; cottony maple, 330, 335, 341, 360; cottony taxus, 427; European fruit lecanium, 309, 341; fiorinia hemlock, 397; fletcher, 427; golden oak, 351; grape, 360, 397; greedy, 375; holly, 401; kermes, 351; lecanium, 351; lesser snow, 388; maple phenacoccus, 335; obscure, 351; oleander, 427; oyster shell, 309, 330, 335, 341, 360, 375, 388; pit-making, 401; purple, 296, 427; Putnam, 309, 330, 335, 341, 375; San Jose, 309, 330, 341; scurfy, 335, 375; sycamore, 360; terrapin, 341, 360; tulip tree, 341; walnut, 335, 341; white peach, 375; willow, 341; willow scurfy, 380
Scalecide, 235
Scanlon red maple, 123
Scaphoideus luteolus, 320
Scarlet oak, 135
Schizura concinna, 365
Schwedler maple, 133
Sclerotinia bifrons, 365
Scoleconectria: balsamea, 394; *scolecospora*, 394
Scolytidae, 414
Scolytus: multistriatus, 317, 327; *quadrispinosus*, 332
Scorch, leaf, 194
Sedge peat, 23
Semanotus ligneus, 385, 417
Sentry ginkgo, 125
Septobasidium curtisii, 279
Septogloeum, 329: *salicinum*, 377
Septoria, 296, 349, 364: *argyrea*, 295; *carpinea*, 279; *caryae*, 330; *chionanthi*, 275; *cornicola*, 271; *crataegi*, 278; *elaeagni*, 295; *eleospora*, 275; *hicoriae*, 330; *ilicifolia*, 398; *platanifolia*, 359
Sequoia, 417
Serviceberry, 26, 238, 297
Serviceberry Diseases, 297: leaf blight, 297; mildews, 298; rust, 297; witches'-broom, 297
Serviceberry Insects, 298: borers, 298; pear leaf blister mite, 298; pear-slug sawfly, 298; willow scurfy scale, 298
Setrete, 233
Sevin, 209
Shadbush, 297
Shademaster honey locust, 131
Shakespeare crabapple, 116
Shingle oak, 137
Shirotae flowering cherry, 114
Shoestring root rot, 243: cause of, 244; control of, 245; symptoms of, 244; trees susceptible to, 243
Shumard oak, 137
Sibine stimulea, 350
Silk tree, 121, 298
Silk tree Diseases, 298: canker, 299; wilt, 298
Silk tree Insects, 299: scale, 299; webworm, 299
Silverbell, 19, 26, 299: Carolina, 121
Silverbell Diseases, 299: leaf spot, 299; wood decay, 299
Silver linden, 129
Silver maple, 132
Simon poplar, 362
Siphonatrophia cupressi, 391
Slime flux, 165: treatment of, 165
Sludge, activated, 49
Smog, 184
Smoke injury, 183: susceptibility of trees to, 186
Smoke tree, 122, 300
Smoke tree Diseases, 300: leaf spot, 301; rust, 300
Smoke tree Insects, 301: oblique-banded leaf roller, 301
Snowbell: fragrant, 122; Japanese, 122
Soaps, 209
Sodium, 17, 18: fluosilicate, 206; nitrate, 49
Soil: adaptations of trees, 24; as an anchor, 20; ball sizes in transplanting, 34; components, 12; improvement, 21; minerals, 17; organic matter, 17; pH, 17; poor aeration of, 14; reaction, 17; relation of, to trees, 12; temperatures, 20; water, 14
Soils, lime requirements of, 19
Solenia: anomala, 304; *ochraceae*, 279
Soot injury, 183
Sophora, 279: *japonica*, 123; *japonica* var. *pyramidalis*, 125
Sorbus, 238, 292, 403: *alnifolia*, 114; *americana*, 25; *Aria*, 114; *aucuparia*, 25; *aucuparia columnare*, 125
Sorrel tree, 122, 300
Sorrell tree Diseases, 300: leaf spot, 300; twig blight, 300
Sourgum, 26, 28, 370
Sourwood, 19, 26, 122
Southern magnolia, 118
Sphaceloma murrayae, 377
Sphaeropsis, 411: *ellisii*, 390, 406; *malorum*, 347; *quercina*, 347
Sphaerotheca: pannosa, 242; *phytoptophyla*, 329
Sphaerulina: polyspora, 300; *taxi*, 425
Sphagnum peat, 22
Spirea, 238
Spirea, 238: *vanhouttei*, 238
Spray compatibility chart, 234
Spray Cop, 231
Spray: formulas, 234; how to, 218; injuries, 191

Spraying: equipment, 215; practices, 215, 220; safety precaution in, 221
Sprayers: hydraulic, 215; manufacturers of, 216
Sprays: combination, 233; compatibility of, 233; dormant, 235
Spreading agents, 210
Spred-O-Stick, 211
Springwood, 6
Spruce, 22, 23, 26, 28, 418: Colorado, 26; Norway, 420
Spruce Diseases, 418: cytospora canker, 418; needle cast, 421; needle rust, 421; wood decay, 421
Spruce Insects, 422: bagworm, 424; bud scale, 423; budworm, 423; cooley gall aphid, 423; epizeuxis, 424; gall aphid, 422; needle miners, 424; spider mite, 423
SS3, 211
Standard trees for streetsides and lawns, 125
Star magnolia, 117
Stem: function of, 7; structure of, 6
Stereum, 350, 395: *gausapatum*, 350; *sanguinolentum*, 411
Stewartia, 26, 122: *koreana*, 122; *malacodendron*, 26
Sticking agents, 210
Stomacoccus platani, 360
Stomata, 8
Storax, 26
Stransvaesia davidiana, 238
Streptomycin, 237
Strumella, 339: *coryneoidea*, 331, 348
Styrax: americana, 26; *japonica*, 122; *obassia*, 122
Sugar maple, 133
Sulfate, 17
Sulfur, 20, 47, 50, 234, 237: dioxide, effect on trees, 183; dusts, 210; lime, 236; sprays, 231; wettable flotation, 232
Sumac, staghorn, 312
Summer wood, 6
Sunburst locust, 130
Sunoco, 235
Sun scald on boxwood, 387
Superior dormant miscible oil, 237
Sweetleaf, horse sugar, 26
Sweetgum, 23, 26, 28, 138, 366
Sweetgum Diseases, 366: bleeding necrosis, 366; blight, 367; leader die-back, 367; other diseases, 367
Sweetgum Insects, 367: bagworm, 367; caterpillars, 367; fall webworm, 367; forest tent caterpillar, 367; leaf tier, 367; luna moth, 367; polyphemus moth, 367; promethea moth, 367
Sycamore, 137
Sycamore maple, 133
Sycamore scale, 360
Symplocos tinctoria, 26
Synanthedon pictipes, 265, 298
Systemic insecticides, 210
Systox, 210, 401
Sytam, 210

Tag 331, 237
Taniva albolineana, 424
Tankage, 49
Taphrina: aceris, 287; *amentorium*, 303; *aurea*, 364; *bacteriosperma*, 258; *carnea*, 258; *carveri*, 287; *coerulescens*, 348; *dearnessii*, 287; *flava*, 258; *lethifer*, 287; *macrophylla*, 303; *occidentalis*, 303; *robinsoniana*, 303; *sacchari*, 287
Taxodium, 385: *distichum*, 24
Taxus, 14, 26, 424: *cuspidata nana*, 427; *wardii*, 427
Tedion, 210
Tennam, 233
Tepp, 237
Termites in trees, 213: control of, 214
Terramycin, 239
Terrapin scale, 341, 360
Tetraleurodes mori, 273
Tetralopha robustella, 414
Tetram, 210
Tetranychus telarius, 405
TDE, 234
Thamnosphecia: pyri, 293; *scitula*, 271
Thimet, 210
Thornless honey locust, 130
Thuja, 24, 382: *orientalis*, 383
Thyridopteryx ephemeraeformis, 383
Tilia, 25, 129, 338: *americana*, 130; *americana fastigiata*, 125; *cordata*, 130; *euchlora*, 130; *petiolaris*, 129; *tomentosa*, 129
Tomostethus multicinctus, 254
Torula ligniperda, 260
Toumeyella liriodendri, 369
Trametes pini, 421
Transplanting: importance of good, 27; preparing trees for, 29; season for, 28; soil ball sizes for, 34; trees, 27
Tree: band injury, 191; bracing, 104; cabling, 104; evaluation, 4; parts of a, 5; normal, 5; supporting the, 39; setting the transplanted, 38; selection, 28; movers, 36; owners, advice to, 102; wrap, 212
Tree Diseases, 222: eradication as a control for, 228; exclusion and immunization as a control for, 228; manipulation of the environment as a control for, 229; nature of parasitic agents, 224; protection as a control for, 228; spraying as a control for, 229; symptoms of, 222

Tree-of-heaven, 26, 301
Tree-of-heaven Diseases, 301: leaf spot, 301; shoestring root rot, 301; twig blight, 301; wilt, 301
Tree-of-heaven Insects, 301: ailanthus webworm, 302; cynthia moth, 301
Tree owners, advice to, 102
Trees: aftercare of street, 144; artificial support for, 106; as windbreaks, 140; can culture of, 35; collected or wildling, 29; container-grown, 35; digging deciduous, 31; digging evergreen, 31; fastigiate, 123; fertilizing transplanted, 44; for California, 146; Northeast, 144; north Pacific area, 147; Plains, 145; Rockies, 146; seashore, 139; Southeast, 145; southern Florida, 145; location of street, 142; low-growing, for streetside and lawns, 114; mulching transplanted, 44; nursery-grown, 29; planting, 37; planting streetside, 140; selection of, for city streets, 141; spacing of street, 142; standard, for streetsides and lawns, 125; suitable for various locations, 112; susceptibility to breakage, 106; transplanting, 27; transporting, 36; types of artificial support for, 106; with upright growth habit, 123; value of, 4; wrapping transplanted, 44
Trident maple, 118
Trithion, 210
Triton: B 1956, 211; X-100, 236
Tsuga, 395: *canadensis*, 24; *caroliniana*, 24
Tulip tree, 26, 28, 138, 368
Tulip tree Diseases, 368: die-back, 369; leaf spot, 368; leaf yellowing, 368; nectria canker, 369; powdery mildew, 369
Tulip tree Insects, 369: aphid, 370; scale, 369
Tupelo, 28, 370
Tupelo Diseases, 370: leaf spot, 370; rust, 370
Tupelo Insects, 371: leaf miner, 371
2, 4-D: injury by, 188; removing from spray tank, 188
2, 3-Dichloro-1, 4, Naphthoquinone, 232
Tympanis pinastri, 411

Uncinula: circinata, 242; *clintonii*, 242, 339; *flexuosa*, 242, 335; *macrospora*, 242; *salicis*, 242
Ulmus, 24, 127, 316: *carpinifolia*, 318; *carpinifolia* Buisman, 128; *carpinifolia wredei*, 128; *parvifolia*, 128; *procera*, 128; *pumila*, 128
Upright American linden, 125
Upright English oak, 125
Uredinopsis, 394

Valsa sordida, 377
Vanadium, 50
Van Eseltine crabapple, 116
Vasates quadripedes, 290
Velsicol 1068, 203
Venturia: chlorospora, 376; *inaequalis*, 278
Vespa crabro var. *germana*, 388
Vespamina sequoiae, 418
Verticillium, 386: *albo-atrum*, 251, 301
Verticillium wilt, 132, 247, 310: cause of, 251; control of, 251; diagnosing for, 250; symptoms of, 248; trees susceptible to, 247
Viburnum acerifolium, 24
Volutella buxi, 386

Walnut, 26, 28, 371: black, 374; Japanese, 373, 374
Walnut Diseases, 371: bacterial blight, 372; canker, 371; die-back, 373; leaf spot, 373; trunk decay, 374; witches'-broom, 374
Walnut Insects, 375: black scale, 375; calico scale, 375; California red scale, 375; cottony-cushion scale, 375; datana, 375; greedy scale, 375; oyster shell scale, 375; Putnam, 375; scale, 335, 375; scurfy scale, 375; white peach scale, 375
Walnut injury to plants, 371
Washington hawthorn, 120
Water: capillary, 14; free, 15; hygroscopic, 14
Waterer laburnum, 120
Watering transplanted trees, 42
Weed killers, 188
White ash, 126
White cedar, 26
White peach scale, 375
White poplar, 138
White oak, 136
Willow, 13, 26, 375: creeping, 26; oak, 137
Willow Diseases, 375: bacterial twig blight, 377; black canker, 376; cytospora canker, 377; gray scab, 377; leaf blight, 375; leaf rust, 378; leaf spot, 377
Willow Insects, 378: apple tree tent caterpillar, 378; bagworm, 380; carpenter worm, 380; cone gall, 378; European willow gall midge, 380; flea weevil, 380; lace bug, 380; leaf beetle, 380; mottled willow borer, 380; red-humped caterpillar, 380; scurfy scale, 380; willow-shoot sawfly, 378
Wilt-Pruf, 45, 388
Wine-leaved sycamore maple, 133
Winter drying of evergreens, 169
Winter-injured plants, care of, 171

Winter injury, 167: prevention of, 170; to boxwood, 387; evergreens, 169
Wound dressings: Arbor Aid 101, 81; asphaltum paints, 80; bordeaux paste, 83; commercial tree paints, 83; creosote paints, 81; grafting waxes, 81; house paints, 82; in aerosol cans, 83; lanolin paints, 82; orange shellac, 80; Tree Healing Dressing, 83; varnish-mercury paint, 82
Wounds, treatment of, 78

Xanthomonas juglandis, 372
Xylem, 6

Yellowwood, 19, 28, 139, 380
Yellowwood Diseases, 380: botryosphaeria canker, 381; powdery mildew, 381; verticillium wilt, 381; wood rot, 380
Yew, 26, 424
Yew Diseases, 424: crown gall, 425; dieback, 424; needle blight, 425; twig blight, 424; twig browning, 425
Yew Insects, 425: ants, 428; black vine weevil, 425; California red scale, 427; carpenter ant, 428; cottony taxus scale, 427; fletcher scale, 427; grape mealy bug, 427; oleander scale, 427; purple scale, 427; strawberry root weevil, 425; taxus mealy bug, 427; termites, 428
Yeddo hornbeam, 120
Yoshino cherry, 114
Young America crabapple, 115
Yucca mealy bug, 418

Zelkova serrata, 319
Zerlate, 233
Zeuzera pyrina, 291
Zinc arsenate, 206
Zincate, 233
Zinc dimethyl dithiocarbamate, 232
Zineb, 233, 234, 237
Ziram, 232, 234, 237